SOCIETY FOR RANGE MANAGEMENT
DENVER, COLORADO

Range Science Series
No. 1 October 1972
Second Edition 1981

RANGELAND HYDROLOGY

Farrel A. Branson
U.S. Geological Survey, Denver, Colorado

Gerald F. Gifford
Utah State University, Logan, Utah

Kenneth G. Renard
Science and Education Administration, USDA, Tucson, Arizona

Richard F. Hadley
U.S. Geological Survey, Denver, Colorado

Edited by **Elbert H. Reid**

KENDALL/HUNT PUBLISHING COMPANY
Dubuque, Iowa, USA • Toronto, Ontario, Canada

FRONT COVER: The towering cumulus collecting energy widely and its release in the form of rainfall of high intensity, short duration, and limited areal extent. This mechanism is common in the southwestern United States during the June to September period.
BACK COVER: A livestock reservoir on grazing lands in eastern Wyoming.
page iii: Moving to higher country on the plains rangeland in the Calgary area.

Library of Congress Catalog Card Number: 81–80610

ISBN 0–8403–2408–1

Printed in the United States of America

B 402408 01

Table of Contents

Preface to Second Edition

Motivation for this second and greatly enlarged edition has been, in part, a response to a greater than expected demand for the first edition and, secondly, a response to those who considered the first edition incomplete. All chapters have been expanded and new chapters added. New chapters are those on Snow Hydrology and Snowpack Management on Rangelands (Chapter 8), Urban Impacts on Rangeland Hydrology (Chapter 9), and Rangeland Hydrologic Models (Chapter 11). Although a major revision objective has been to provide a better text for students, our future range resource managers, it is hoped that the additional material presented will be useful and of interest to a much larger audience.

In this edition information has been drawn more extensively from research on nonrangeland areas where principles derived from these lands seem equally applicable to range. There are interactions between range and non-rangeland, thus the rangeland hydrologist should have some knowledge of both.

Since the first edition, two authors have been added, and one, for the press of other business, dropped himself from the effort. The two new authors, Kenneth G. Renard and Richard F. Hadley, are to be commended for their efforts in making this manuscript into the excellent work that it is. Authors Branson and Gifford have, by their untiring efforts, provided the incentive and inspiration to see this publication through this second, expanded edition. The book will serve members of the Range Management profession for many years to come.

PAUL T. TUELLER
Coordinator, SRM Sciential Committees
University of Nevada
Reno, Nevada

FLOYD E. KINSINGER
Executive Secretary
Society for Range Management
Denver, Colorado

Preface to First Edition

The approximately one-third of the earth's surface which is land (about 34 billion acres) may be classified into five general categories. The three primary categories are nonproductive land, forest land, and rangeland—each with distinctive natural characteristics that have developed over time under the formative influences of climate, geologic materials, and natural organisms. The other two broad categories are cropland and urban-industrial land, both of which have been formed as a result of extensive modification or manipulation of natural land types for specific utilitarian purposes.

It is estimated that about 3 to 4% of all land is given over to urban-industrial development, while approximately 10% of the total land area is under cultivation. Nonproductive lands—so termed because photosynthesis is relatively unimportant—consist of the very high mountain peaks, barren deserts, and those areas covered by glaciers or permanent snow and comprise an estimated 15% of the earth's land area. Forest land extends over another 30% or so.

The remaining 40% or more of all land falls in the broad category of rangeland. The principal characteristics of this kind of land are (1) the potential natural vegetation is predominantly grasses, grasslike plants, forbs (generally low-growing, broad-leaved plants), and shrubs, (2) natural herbivory was an important influence in its precivilization state, and (3) it is more suitable for management by ecological principles than for management by agronomic principles. The more obvious kinds of rangeland are natural grasslands, savannas, shrublands, most deserts, tundra, alpine communities, coastal marshes, and wet meadows. However, the limits of this broad rangeland category certainly are not precise and rangelands often are intermingled with other kinds of land—from sea level to above timberline.

Rangeland, then, is the largest single category of land, not only on a world-wide basis but in many individual countries, and it produces a variety of natural resources beneficial to man, including both tangible products and intangible values: grazeable forage for livestock, wildlife recreational opportunities, natural beauty, minerals, some wood products, and, of course, water—the subject of this book, the first in the Society's Range Science Series.

Range science is a relatively new discipline and, like many of the broad agricultural sciences which have preceded it, it is a multidisciplinary synthesis founded on the basic biological sciences and on such physical sciences as mathematics, physics, and chemistry. Too, other disciplines—hydrology, biometeorology, entomology, genetics, animal science, and others—contribute substantially to this broad field; the individual titles of this series clearly indicate that there is a specificity of application of such knowledge to range ecosystems.

The Range Science Series was conceived in 1968 when the then president of the Society, Dr. E.J. Dyksterhuis of Texas A&M University, created 15 special Sciential Committees (the number was later expanded). As Dr. Dyksterhuis stated at that time, these committees' goals include advancing the recognition of rangeland as a unique and vast natural land resource, and the organization of knowledge concerning this resource. The basic idea here is that rangelands are distinctive kinds of ecosystems for which it is more important to develop specific understanding regarding structure, function, management, and use. Accordingly, each of the committees was charged with preparing a summary review of the principles and problems, together with a selected bibliography, of its particular aspect of rangeland management.

The people who have served and are serving on these Sciential Committees are acknowledged authorities in their respective specializations of range science and, hence, this series of booklets will serve to provide a comprehensive view of the breadth of both rangeland and its management. These publications are not intended to be exhaustive studies but, as indicated previously, a brief description of the applicable principles and the problems that may be encountered when undertaking an ecological approach to the rational uses of a distinctive kind of land. As such, the Range Science Series will be most useful to many—not only teachers and students but natural resource managers, ranchers, or, for that matter, anyone who is interested in the betterment of the total productiveness of the world's largest category of land.

We believe Dr. Branson, Dr. Gifford, and Mr. Owen have opened this series in a most commendable fashion; we look forward to wide acceptance and use of Rangeland Hydrology, and to equally valuable contributions from the other committees.

PAUL T. TUELLER
Coordinator, SRM Sciential Committees
University of Nevada
Reno, Nevada

FRANCIS T. COLBERT
Executive Secretary
Society for Range Management
Denver, Colorado

October 1972

Chapter 1
Introduction

Rangeland hydrology, or rangeland watershed management, is the study of hydrologic principles as applied to range ecosystems. Although rangelands encompass many different climatic zones, this book deals chiefly with rangeland hydrology in arid and semiarid regions. Approximately 40% of the world's land surface that is classified as rangeland, of which more than 80% is within arid and semiarid zones. Generally, these lands are characterized by extremes in the various components of the hydrologic cycle, examples being low rainfall, high evapotranspiration potential, and low water yield. Intermittent streams common on arid and semiarid rangelands have different characteristics than do permanent streams of more humid areas. Long time may be required to obtain statistically reliable hydrologic results from studies on rangeland because of the low frequency of runoff events, relatively small runoff quantities, and highly variable precipitation.

When rangelands do occur in the more humid regions, such as subtropical, tropical, alpine, and coastal marsh areas, the hydrologic characteristics are different from those outlined above; but the vast extent and the unique characteristics of arid and semiarid lands make it appropriate that this book emphasize hydrology of these lands.

Ever-increasing demands on renewable natural resources necessitate a more comprehensive understanding of the factors affecting them. Since the need for water for domestic, agricultural, and industrial use is critical and since rangelands comprise such a vast watershed area, yields of water and sediment from rangelands are receiving increasing attention. Various aspects of rangeland hydrology have been reviewed by Chapline (1929), Forsling (1932), Bailey and Connoughton (1936), U.S. Forest Service (1940), Allred (1940), Lawhon (1947), Munns (1947), Lassen, Lull, and Frank (1952), Harper (1953), McArdle (1960), Love (1958), Storey, Hobba, and Rosa (1964), Dunford (1967), Gifford (1968), Wolff (1970), American Society of Engineers (1975), and Heady, Falkenberg, and Riley (eds.) (1976).

The major concerns of rangeland hydrology may be divided into four categories: (1) runoff or water yields from watersheds, (2) erosion and sedimentation, (3) quality of water, and (4) occurrence of groundwater. We will deal chiefly with the first two items. A comprehensive treatment of water quality principles and problems is found in a recent publication by Hem (1970). Examples of references on groundwater include Meinzer (1923), Todd (1959), Davis and DeWiest (1966), and Walton (1970).

The amount of information available on the subject of rangeland hydrology is impressive. However, much remains to be learned and only partial solutions are available for many problems. Some of the results of investigations reported in this book may appear to be in conflict with each other. This emphasizes the need for further investigations to resolve these conflicts. The reader is encouraged to consult original references for additional detail regarding specific studies of interest.

One interesting problem, for which after many decades of debate no satisfactory explanation has been offered, is the spectacular gully erosion in the West. Was the cause of trenching of formerly productive valleys due to a change in climate, catastrophic storms, or increased runoff and erosion following depletion of vegetation cover by excessive grazing use? Convincing arguments have been presented for each of these as the cause of the valley trenching and related types of erosion. However, without concurrent measurements of precipitation characteristics, grazing use, runoff, and erosion over a period of time the answer is a matter of conjecture.

Very little information is available on the effects of upland treatment practices on water yields, sediment yields, and dissolved solids in runoff. Upland treatments referred to here include conversions of woody vegetation, chiefly pinyon-juniper, sagebrush, and chaparral, to herbaceous vegetation; mechanical land treatments such as furrowing, ripping, and pitting; and various grazing treatments. Both watershed and plot studies are required to provide this information.

When choosing among watershed management practices, information is needed on a range of treatment alternatives as applied to most watershed conditions. A second need is for data on multiple use effects of water-yield improvement treatments. For example, consideration must be given to the relative benefits of off-site versus on-site uses of the water resource from rangelands. Research to date has indicated that numerous rangeland treatments result in lower water yield. However, the reduced runoff generally has been accompanied by lower sediment yields and increased forage production for both domestic livestock and wildlife. Brown (1971) states:

> Only when responses of all the land products are viewed together in each production situation can the manager have an adequate basis for choice of a management practice. Depending on his objectives, he then can choose a practice that will optimize the overall multiple use value, or he can knowledgeably choose a practice that will optimize the particular product(s) that are most important to his particular situation. In either event he would be making his decision with a full understanding of the changes and trade-offs that can be expected for all products.

Following this line of reasoning, one must consider the economics of various watershed management practices. For example, near Cisco, Utah, in the Upper Colorado River drainage basin, it was estimated that treating 1,280,000 acres of frail lands with gully plugs and contour furrows would result in a benefit-cost ratio of 0.17 to 1 even if the maximum potential benefits were received in each category considered, if the minimum treatment costs were not exceeded, and if the treatments had an infinite life expectancy (Workman and Keith 1971). With this in mind Green (1971) states:

> The most difficult job is to guard against negating the benefits of spending by accommodating public pressures for certain types of land use that can either act counter to the goals of

watershed management or increase the management cost of rehabilitation activity. The public land manager is no less responsible for fiscal integrity than he is for biological, ecological, and physical concern for the resource.

The idea of using a systems approach as an aid in managing watersheds will increase greatly in the future. Eisel (1972) has recently developed a systems approach to wildland management using a hypothetical example. He notes that the justification for watershed management practices has traditionally been based on the preservation of the soil resource for sustained yield production of forest and grassland products, the prevention of reservoir sedimentation, the production of high quality water from wildland watersheds, and so forth. Results of the study suggested, however, that management emphasis in the hypothetical system should be transferred from traditional objectives to those of preserving the quality of natural environments for outdoor recreation. The revised objective would be that of maintaining and restoring environmental quality in wildland areas rather than with major concern for soil erosion. A more general discussion of the problem has been given by Cooper (1969).

Because of the strong interaction, nutrient cycling as related to hydrological cycle is gaining increased attention. Borman and Likens (1969) and Likens et al. (1970) have discussed the general watershed-nutrient cycling aspect as applied primarily to hardwood forest ecosystems in the northeastern United States. These types of studies should also be expanded to include various rangeland ecosystems to better understand biogeochemical interactions and implications resulting from large-scale changes in habitat or vegetation.

After nearly 30 years of research, much is known about physical responses of weather modification efforts, but little is know about ecological responses caused by weather modification trials. Hydrologic, economic, social, and legal ramifications of such environmental manipulation are poorly understood. The use of evapotranspiration suppressants to increase runoff, treatment of soils with hydrophobic chemicals, and the hydrologic effects of recreational use of rangelands need much additional study. Also, with more people viewing our once remote rangelands each year, we need new criteria for judging the aesthetic effects of management and treatment programs applied to rangelands.

Only some of the more obvious deficiencies in our knowledge of rangeland hydrology can be mentioned in this brief section. In almost all phases of the subject, additional information is needed to permit accurate predictions of effects of land treatments on rangeland ecosystems. It is hoped that the rather extensive bibliography will aid those who desire additional information on the various facets of rangeland hydrology, and also that information presented in this book will help expand public awareness of the importance of our rangeland resource.

The reader is encouraged to keep in mind these deficiencies in our knowledge as the various chapters are read or studied. A number of these problem areas are treated far more thoroughly in this second than in the first edition.

Literature Cited

Allred, B.W. 1940. Range conservation practices for the Great Plains. U.S. Dep. Agr. Misc. Pub. 410. 21 p.

American Society of Civil Engineers. 1975. Watershed management. Proc. Symposium at Logan, Utah, 781 p.

Bailey, R.W., and C.A. Connoughton. 1936. The western range, its social and economic function in watershed protection. p. 303-339. *In:* Senate Doc. 199. 74th Congress, 2nd Session.

Borman, F.H., and G.E. Likens. 1969. The watershed-ecosystem concept and studies of nutrient cycles, p. 309-324. *In:* The Ecosystem Concept in Natural Resource Management, G.M. Van Dyne, [ed.], Academic Press, New York.

Brown, H.E. 1971. Evaluating watershed treatment alternatives. Amer. Soc. Civil Eng. Proc., J. Irrigation and Drainage Div. (Pap. 7952) 97: 93-108.

Chapline, W.R. 1929. Erosion on rangelands. Amer. Soc. Agr. J. 21: 423-429.

Cooper, C.F. 1969. Ecosystem models in watershed management, p. 309-324. *In:* The Ecosystem Concept in Natural Resource Management, G.M. Van Dyne, [ed.], Academic Press, New York.

Davis, S.N., and R.J.M. DeWiest. 1966. Hydrogeology. John Wiley & Sons, Inc., New York.

Dunford, E.G. 1967. Techniques in grassland watershed research. XIV Cong. Intern. Union Forest Res. (Munich), Proc. p. 444-462.

Eisel, L.M. 1972. Watershed management: A systems approach. Water Resources Res. 8: 326-338.

Forsling, C.L. 1932. Erosion on uncultivated lands in the Intermountain region. Sci. Monogr. 34: 311-321.

Gifford, G.F. 1968. Rangeland watershed management—a review. Univ. Nevada Agr. Exp. Sta. Pap. R52. 50 p.

Green, A.W. 1971. Some economic considerations of watershed stabilization on national forests. U.S. Forest Service Res. Pap. INT-92, 10 p.

Harper, V.L. 1953. Watershed management—forest and range aspects in the United States. Unasylva 3: 105-114.

Heady, H.F., D.F. Falkenborg, and J.R. Riley (eds.) 1976. Watershed management on range and forest lands. Proc. 5th Workshop of U.S./Australia Rangelands Panel. Utah Water Res. Lab., Logan, Utah. 222 p.

Hem, J.D. 1970. Study and interpretation of the chemical characteristics of natural water. U.S. Geol. Surv. Water-Supply Pap. 1473. 363 p.

Lassen, L., H.W. Lull, and B. Frank. 1952. Some plant-soil water relations in watershed management. U.S. Dep. Agr. Circ. 910. 64 p.

Lawhon, L.F. 1947. Soil moisture relationships on rangeland. J. Soil Water Conserv. 2: 149-152.

Likens, G.E., F.H. Borman, N.M. Johnson, D.W. Fisher, and R.S. Pierce. 1970. Effects of forest cutting and herbicide treatment on nutrient budgets in the Hubbard Brook watershed-ecosystem. Ecol. Monogr. 40: 23-47.

Love, L.D. 1958. Rangeland watershed management Proc. Annual Meeting Soc. Amer. Forest (Salt Lake City, Utah) p. 198-200.

McArdle, R.E. 1960. Watershed management on wildlands. J. Forest. 58: 259-265.

Meinzer, O.E. 1923. The occurrence of ground water in the United States with a discussion of principles. U.S. Geol. Surv. Water-Supply Pap. 4-9. 329 p.

Munns, E.N. 1947. Hydrology of western ranges. J. Soil Water Conserv. 2: 139-144.

Storey, H.C., R.L. Hobba, and J.M. Rosa. 1964. Hydrology of forest lands and rangelands, p. 22-1 to 22-52. *In:* Ven T. Chow, Handbook Appl. Hydrol. McGraw-Hill, New York.

Todd, D.K.T. 1959. Ground water hydrology. John Wiley & Sons, Inc., New York 336 p.

U.S. Forest Service. 1940. Influence of vegetation and watershed treatments on runoff, silting and streamflow. U.S. Dep. Agr. Misc. Pub. 397. 80 p.

Walton, W.C. 1970. Groundwater resource evaluation. McGraw-Hill, New York. 664 p.

Wolff, D.N. 1970. Grassland infiltration phenomena. Grassland Biome Tech. Rep. No. 54. 125 p.

Workman, J.P., and J.E. Keith. 1971. Economic considerations of the Cisco soil treatment study. Unpublished manuscript, Utah State Univ., Logan. 11 p.

Chapter 2
Precipitation

The rangeland hydrologist is concerned chiefly with precipitation as it reaches the land surface. However, some understanding of the basic processes involved in delivery of water to the earth's surface is a necessary requisite to data interpretation and adequate measurement of precipitation input.

The total amount of water vapor in the atmosphere is very large. In fact, is has been estimated that the amount carried across the land by air currents is more than six times the amount of water carried by all our rivers (Fig. 2-1).

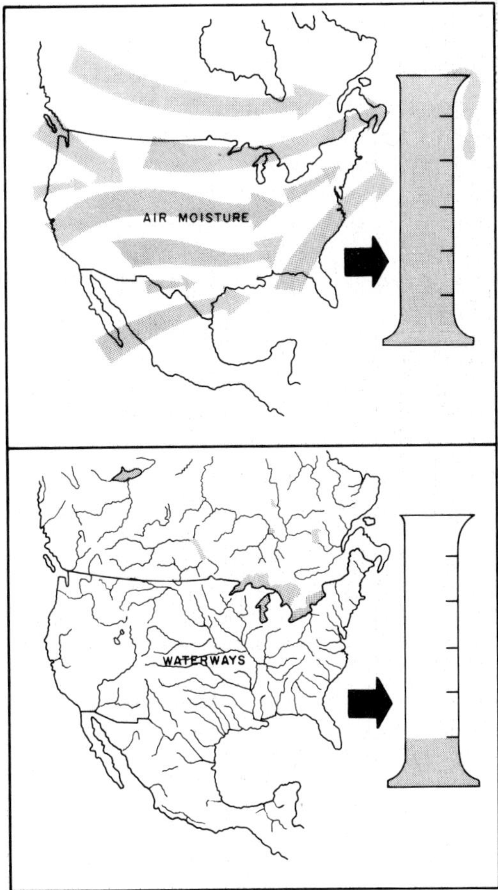

Fig. 2-1. The total amount of water vapor that flows across the land on air currents originating over water is estimated to be more than six times the water carried by all our rivers (from Schroeder and Buck 1970).

The amount of water vapor which the air can carry without loss by condensation depends on the air temperature. The higher the temperature, the more vapor the air can carry. When moist air cools sufficiently, there is too much water for the air to hold as vapor. Clouds form and precipitation develops when the atmosphere becomes saturated with moisture. At saturation the atmospheric vapor pressure is equal to the saturation vapor pressure at the existing temperature and pressure.

What causes the atmosphere to cool so that vapor condenses as rain or snow? The principal cause is the lifting of warm air to higher and cooler altitudes, the resultant cooling causing the atmospheric vapor pressure and saturation vapor pressure to attain the same value, saturation (100% relative humidity). Addition of moisture to the air can also produce the same results.

Sources of Moisture

Water evaporates from the ground surface, from all open bodies of water (lakes, rivers), and, or course, from the ocean (Fig. 2-2). Transpiration (moisture loss through parts of plants, primarily the leaves) is an important source of water vapor and may often produce more vapor than evaporation from land surfaces, lakes, and streams. However, the most important source of moisture in the air is evaporation from the oceans, and particularly the warmest parts of the ocean which are generally nearest the equator.

Fig. 2-2. Although the oceans are the principal source of atmospheric moisture, transpiration from plants is also important. But in arid areas, transpiration adds little moisture to the atmosphere (from Schroeder and Buck 1970).

Vast amounts of solar energy (heat) are required to change water from liquid to vapor in the familiar process known as evaporation. The air carries away the heat with the vapor, and the heat is given up when the vapor condenses to form clouds. Thus, the earth's atmosphere is actually a vast heat engine powered by the sun. The nature of the hydrologic cycle is therefore evident: Through the energy provided by the sun (Fig. 2-3), water evaporates

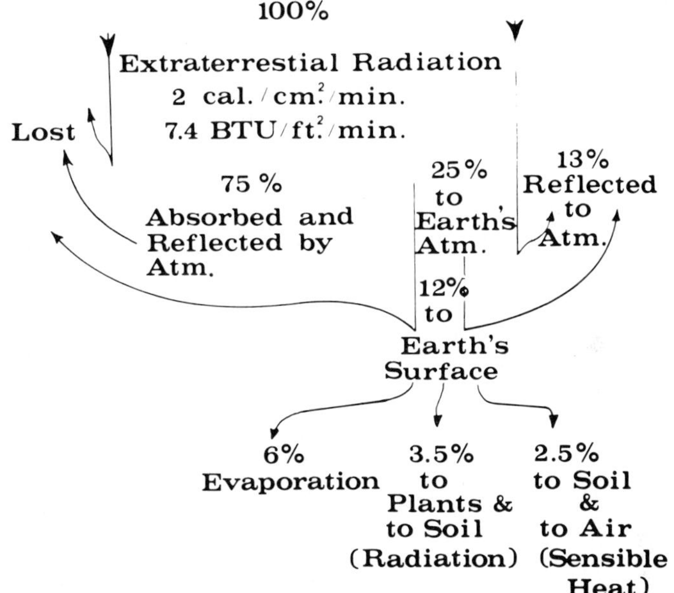

Fig. 2-3. Dissipation of solar energy on an average day in the northern hemisphere (after Gates 1962).

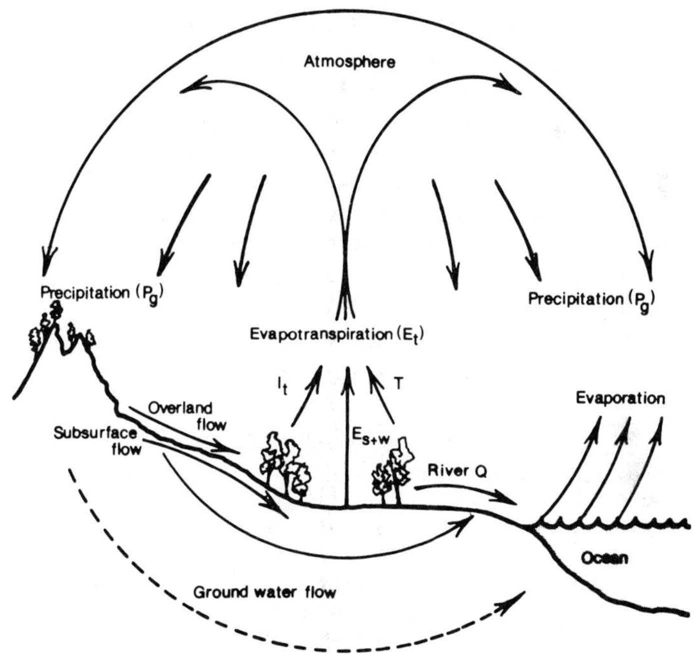

Fig. 2-4. The hydrologic cycle, where Q = streamflow; I_t = interception loss; E_{s+w} = evaporation from soil and water surfaces; T = transpiration loss.

 from the land and ocean, is carried as vapor in the air, somewhere to fall as rain or snow, returning to the ocean or the land to again go through the same process, (Fig. 2-4).

Lifting and cooling of the Air

The lifting and the resultant adiabatic expansion of air, is the most important method of cooling that results in precipitation. Expansion cools the air by allowing its molecules to spread further apart, thus reducing the frequency of their collision. If cooling is sufficient, vapor condenses as droplets of water and these droplets form rain. The condensation process is helped by the presence of small particles of dust or of salt that are always present in the air. The lifting is accomplished by thermal, orographic, or frontal action (Fig. 2-5).

Localized heating results in thermal lifting (Fig. 2-6). Heated surface air becomes buoyant and is forced aloft where cooling takes place. It is most pronounced in the warm seasons and reaches a maximum expression in the southwestern United States during July and August of each year. Rainfall associated with thermal lifting is likely to be scattered in geographic extent, with greatest convective activity over the hottest surfaces.

Orographic lifting (Fig. 2-7) occurs when air is forced up the windward side of slopes, hills, and mountain ranges. It is an important process in production of clouds and precipitation, and, as in thermal lifting, the air is cooled by the adiabatic process.

Frontal lifting (Fig. 2-8) results when air is forced up the slope of warm and

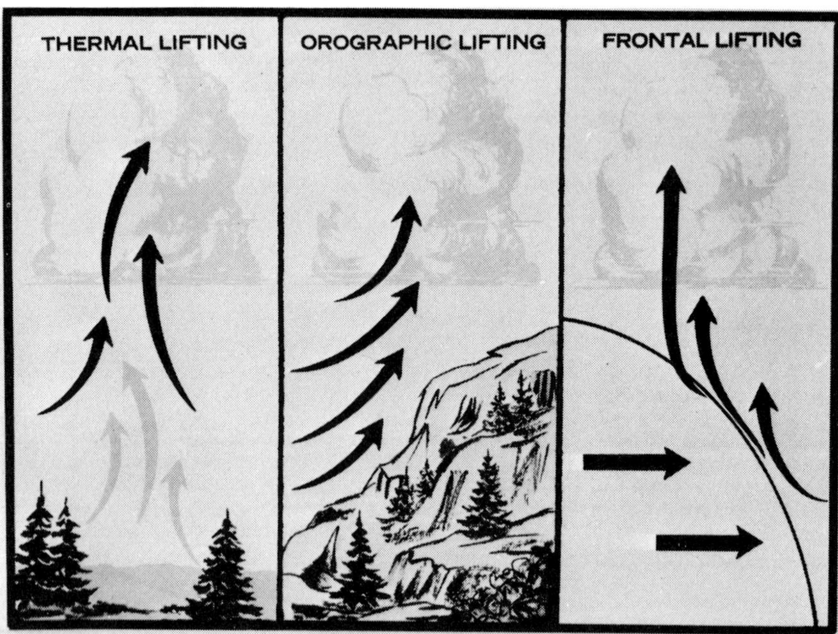

Fig. 2.-5. The most important method of cooling air to saturation is adiabatic cooling because of lifting. Lifting may be thermal, orographic, or frontal (from Schroeder and Buc, 1970).

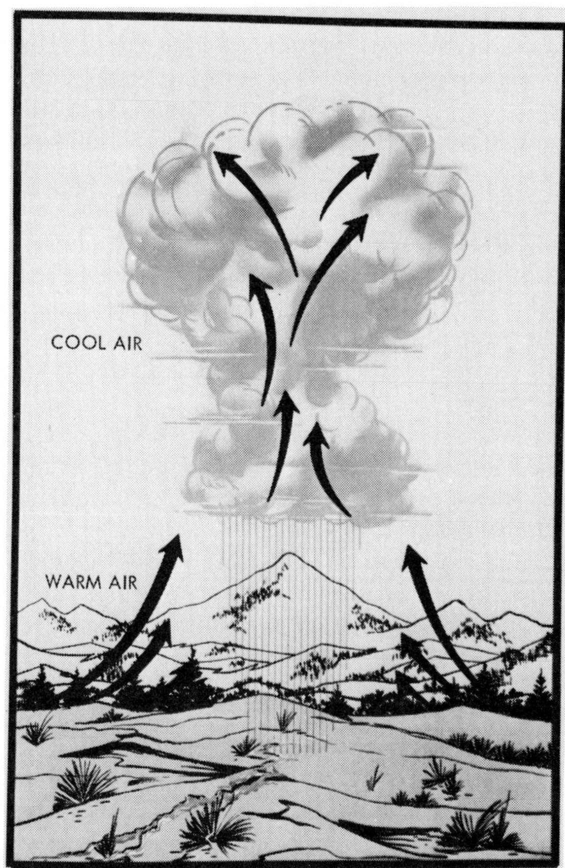

Fig. 2-6. Thermal lifting usually produces cumulus clouds. Continued heating in most air will result in showers and possibly thunderstorms (from Schoreder and Buck 1970).

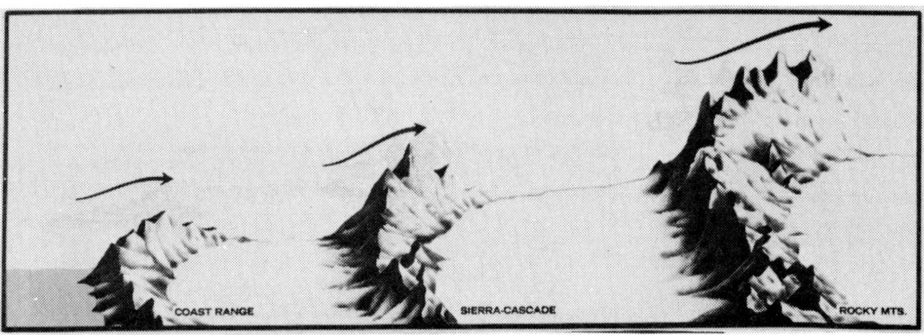

Fig. 2-7. Lifting of moist air over mountain ranges is an important process in producing clouds and precipitation. In the western U.S. the winter precipitation is heaviest on the western slopes of the Coast Ranges, Sierra-Cascades, and Rocky Mountains. Lowlands to the east of the ranges are comparatively dry (from Schroeder and Buck 1970).

 cold fronts. Warm fronts, because of the gradual slope associated with these frontal surfaces, produce steady rains over extensive areas. Cold fronts, which have generally steeper and faster moving leading surfaces, frequently produce intense rainfall along the front or along a squall line ahead of the front. The rainfall, however, is usually more scattered and of shorter duration than that produced by a warm front.

Precipitation Characteristics on Rangelands

In general, where long-term weather records are available, average annual precipitation is usually considered a reliable characteristic for the area; however, total precipitation data are often not sufficient to form generalized hydrologic conclusions since the water yield to be expected for individual storm events in a watershed is extremely variable and is affected by many factors.

Type of precipitation is of great importance. Precipitation may occur in the form of snow, sleet, hail or rain. Precipitation falling as rain may produce high runoff, but the same quantity falling as snow on rangelands may produce little or no runoff. A rainstorm of high *intensity* (quantity per unit of time) may produce high runoff, but low intensity storms that did not exceed soil infiltration capacities may result in no runoff. *Storm durations* usually have to be such that watershed storage capacities are exceeded before runoff

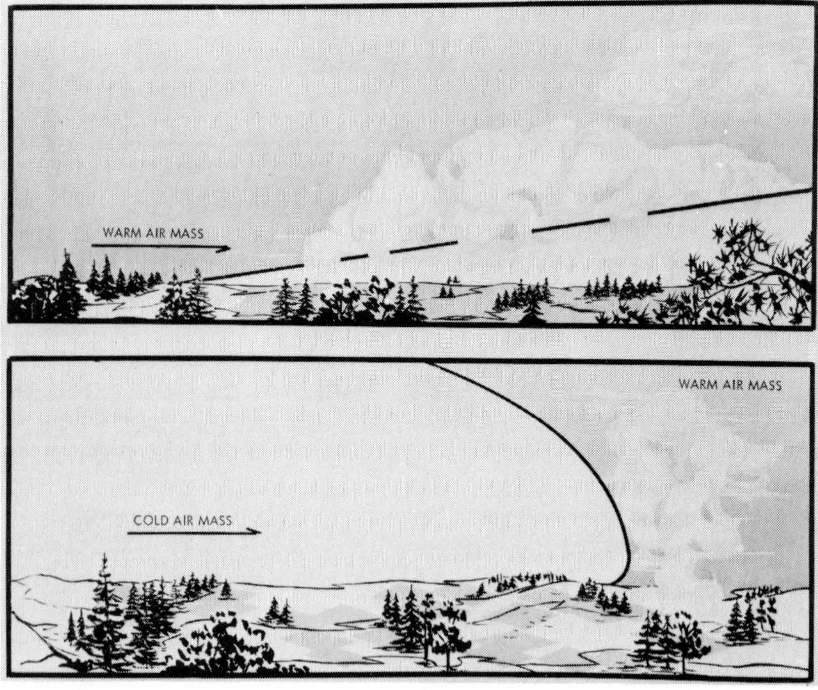

Fig. 2-8. (Top) Lifting of warm, moist air, as it is forced up over cooler air, produces widespread cloudiness and precipitation.
(Bottom) The steepness and speed of cold fronts result in a narrow band of cloudiness and precipitation as warm, moist air ahead of the front is lifted (from Schroeder and Buck 1970).

occurs. *Rainfall distribution* patterns are also important; storms that are distributed uniformly over a watershed generally cause less runoff per unit area of landscape than intense storms that cover only a small portion of a watershed. If *storm direction* is toward the head of a drainage, lower peak flows are produced than for storms moving toward the mouth of a drainage. *Antecedent moisture* (that stored in the soil by previous storms) causes runoff to be greater than when precipitation falls on dry soil. And factors that affect evapotranspiration (e.g., wind velocities and temperatures) also affect runoff.

In view of the multiplicity of factors affecting runoff, it is not surprising to learn that the quantity of precipitation that falls during a season or a year, which is a direct cause of runoff, is sometimes poorly correlated with runoff.

Figure 2-9 illustrates that an inverse relationship often exists between runoff and seasonal precipitation. Some factors contributing to these unexpected results are (1) antecedent soil moisture at the time of runoff events, (2) whether or not the ground is frozen at the time of snowmelt and (3) the storm intensity pattern.

Although runoff from the individual watersheds is often poorly correlated with specific rainfall events and seasonal precipitation, there is a general relationship between annual precipitation and annual runoff for large geographic areas (Fig. 2-10). In certain areas (as shown in the data for Coshocton, Ohio) some upland infiltration returns to downstream channels causing an

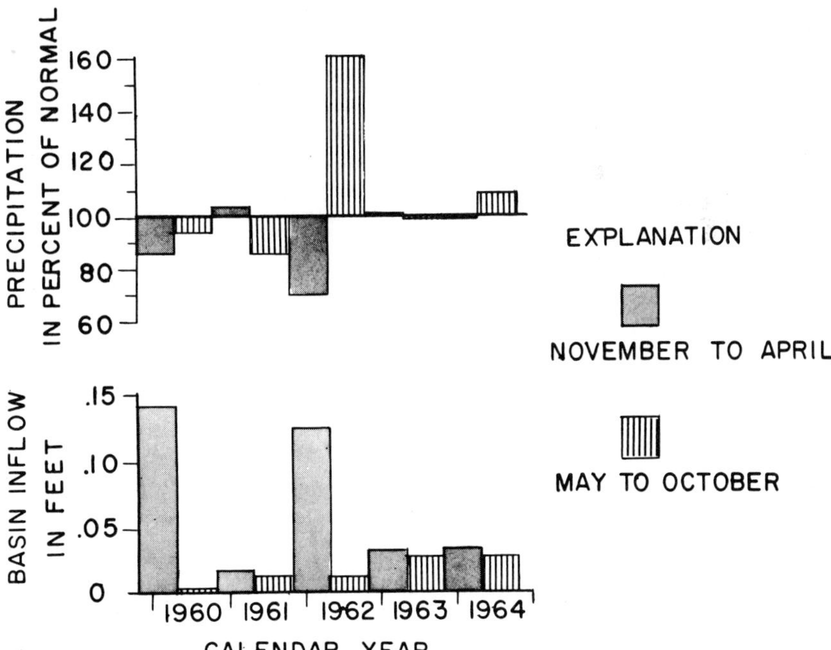

Fig. 2-9. Basin inflow as related to seasonal precipitation for a small watershed in North Dakota (after Eisenlohr and Sloan 1968). Neither seasonal nor annual precipitation were directly related to inflow; an inverse relationship was present during some of the years (multiply ft by 30.48 to obtain cm).

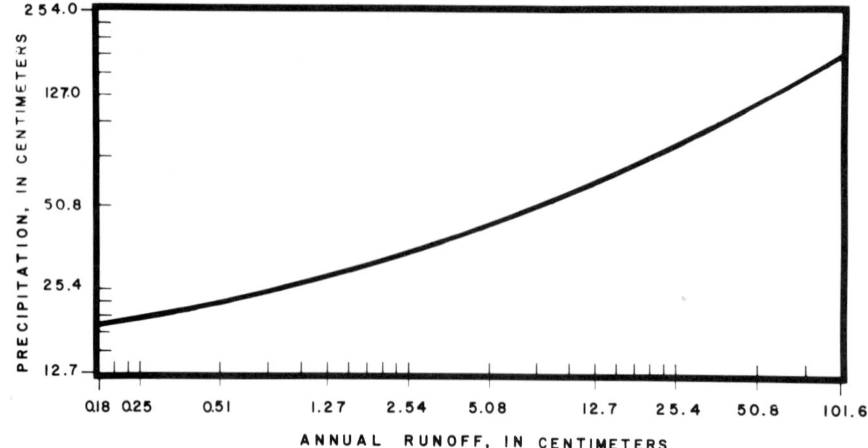

Fig. 2-10. Relationship between runoff and annual precipitation. Runoff data used were adjusted to that which would occur at a mean annual temperature of 10° C (50° F) (after Langbein and Schumm 1958). (Divide cm by 2.54 to obtain inches).

increase in streamflow per unit of drainage area as the as increases (Fig. 2-11). On the other hand, in ephemeral streams where the channels have a potential to absorb large quantities of runoff (transmission losses), runoff per unit of area decreases with increasing size of the contributing watershed. Thus, all combinations of these conditions may be encountered with a greater preval-

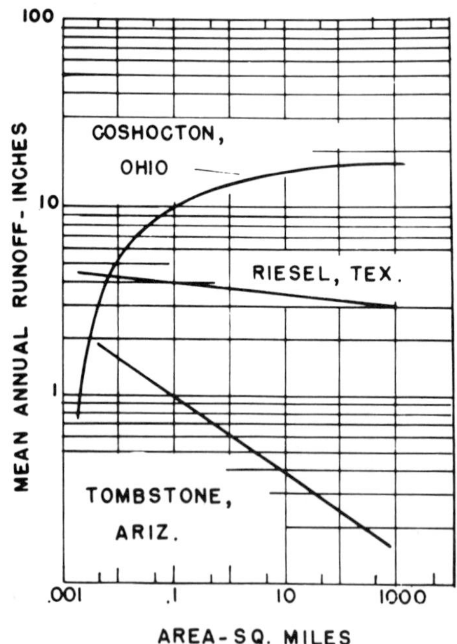

Fig. 2-11. Relationship between mean annual runoff and area (from Osborn and Renard 1970). (Multiply inches by 2.54 to obtain cm and square miles by 2.59 to obtain square km.)

ence of the latter expected in arid and semiarid rangeland.

Annual rainfall variation on rangelands, expressed as coefficients of variation of annual precipitation, varies from 15 to 20% in the northwestern United States to 55 to 60% in the extreme southwest (Hershfield 1962). Also, individual storms show large variations in total amount, duration, distribution, and intensity. Information on how storm behavior varies with season, time of day, and topography on rangelands is increasing, but as yet is still limited.

That the precipitation pattern differs from north to south and from east to west in the rangeland areas of the western United States can be seen in Figure

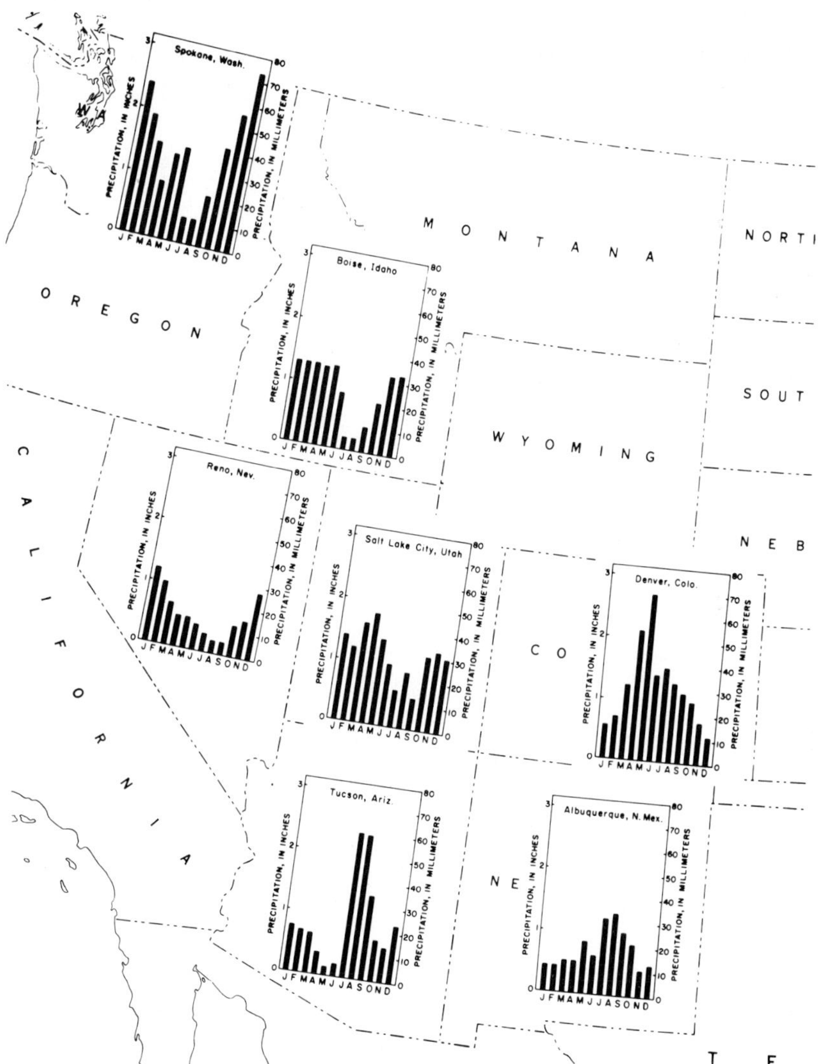

Fig. 2-12. Precipitation variation by month for selected stations in the rangeland areas of the western United States.

 2-12. Not only the amounts differ but the occurrence of the rainfall within the year is different. For example, the Arizona, New Mexico, and Colorado gages show a greater proportion of the annual precipitation in the summer (July-August-September), whereas the northern locations, Boise and Spokane, show their greatest precipitation in the winter.

The local type thunderstorms often lead to statements from disappointed ranchers that it seems to rain everywhere but on their ranches. Exclamations such as "We were completely surrounded by storms but somehow they all veered around us and we didn't get a drop all afternoon" are quite common. A graphical model to describe this thunderstorm illusion was developed by McDonald (1959) and modified by Renard and Brakensiek (1976) as follows:

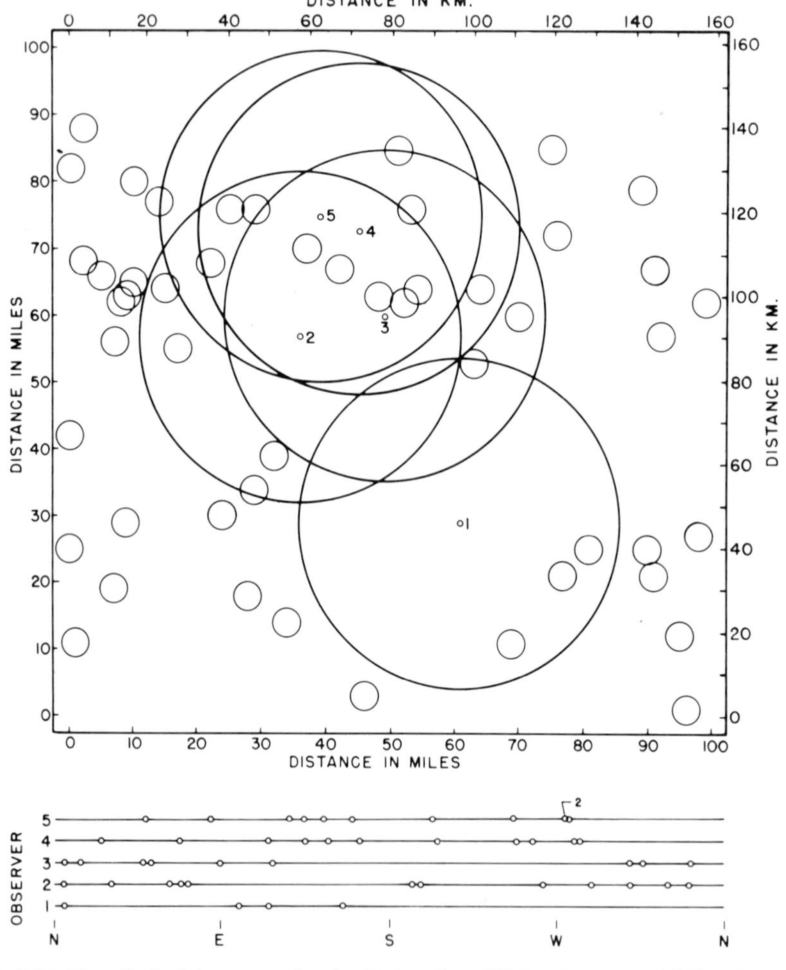

Fig. 2-13. Hypothetical storm map showing the location of 50 thunderstorms with five observers randomly located. Each observer circle was 25 miles (40 km) radius, while each storm had a 1-mile (1.6 km) radius. The lower portion shows the location of the storms in relation to each observer (Renard and Brakensiek 1976 modified from McDonald 1959).

Fifty storms were located in a 10,000 square mile (25,900-km^2) area using a table of random numbers (true air-mass storms occurring in level country are apparently randomly located). Five observers were also randomly located with the constraint that each observer was 40 km (25 miles) inside the boundary to insure that the observer's circle of shower detection was within the model area (Fig. 2-13).

Inspection of Figure 2-13 shows that each observer except number 1, had storms fairly well distributed around the horizon. To illustrate, the storm positions were measured relative to the observer and are presented in the lower portion of Figure 2-13. Observer No. 1, in addition to seeing fewer storms, had two quadrants without any storms. McDonald showed that only 8% of the observer-quadrants will be storm free. The expected value of the number of storms seen per observer is 9.82 whereas the mean for the five observers was 9.4. It can also show that 5.9% of the entire model area will be receiving rain. Hence the probability is only 0.059 that rain will actually fall on any observer.

Studies in the southwestern United States have provided detailed information concerning individual storm behavior: McDonald 1956; Fletcher 1961; Keppel 1963; Osborn and Reynolds 1963; Osborn 1964; Kincaid, Osborn, and Gardner 1966; Osborn and Hickok 1968; Drissel and Osborn 1968; Osborn and Renard 1969; Osborn and Lane 1969; Renard 1970; Osborn and Renard 1970; Osborn 1971; Battan and Green 1971; Osborn, Lane, and Hundley 1972; Osborn and Lane 1972; Mills and Osborn 1973; Smith and Schreiber 1973, 1974. The following important facts (some of which are applicable on a broad scale) have emerged from these studies:

1) Most convective storms in the region are multicellular, random, of short duration (one hour or less), of low volume (generally less than 1.7 cm), and of limited areal extent (Column 2, Table 2-1.) Two examples of storm cellular development on the Walnut Gulch watershed (Arizona) are shown in Figures 2-14 and 2-15. Figure 2-16 is an example which shows a single storm cell's size, speed, and magnitude as determined from a dense raingage network in southern Illinois. At Walnut Gulch, thunderstorm cells are associated with

Table 2-1. 1963 storms over Walnut Gulch Watershed (Arizona) ranked by volume of runoff produced.

Date	Area of watershed covered by 1.5 cm or more km^2 (mi^2)	Ranking by area covered
8/19	108.8 (42)	1
8/24	49.2 (19)	6
8/31	64.8 (25)	2
8/10	28.5 (11)	7
8/22	67.3 (26)	2
9/7	62.2 (23)	2
7/31	59.6 (23)	2
8/12	13.0 (5)	8

Data from Osborn (1964).

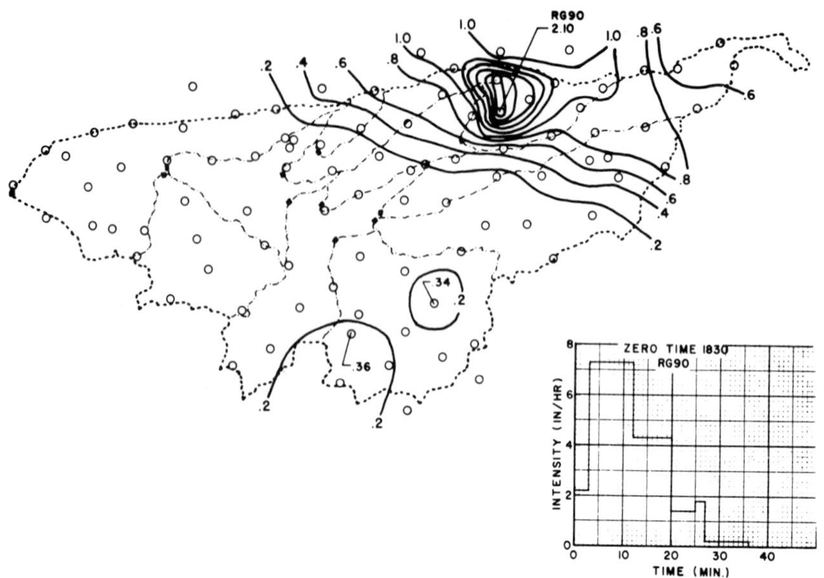

Fig. 2-14. Isohyetal map of total storm depth from a single cell event on August 5, 1968, on the Walnut Gulch watershed (from Renard 1970). Multiply inches by 25.4 to obtain mm).

thermal convection within air masses while in Illinois they are most frequently associated with frontal-activated storms.

2) Although peak rainfall usually occurs during the first 10 minutes of the rainfall duration, and rainfall may occur in 2 to 4 bursts, as defined primarily by studies in Illinois, in Arizona and New Mexico, there does not appear to be a dominant position in the storm for the highest intensity.

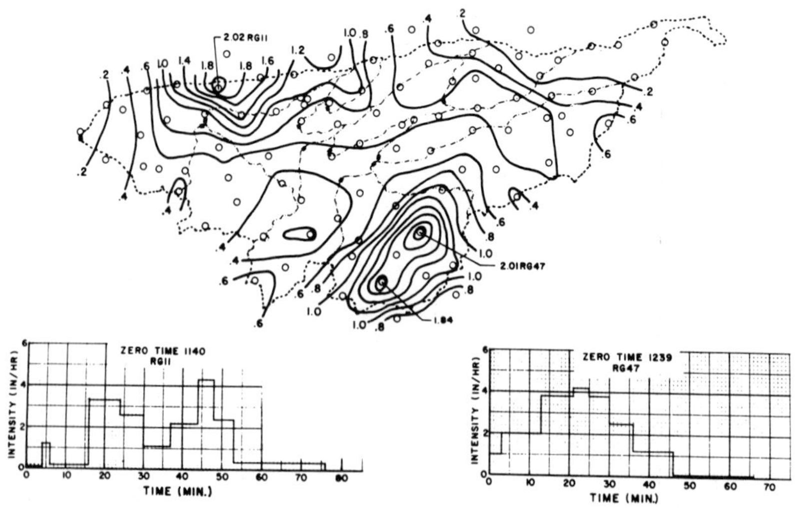

Fig. 2-15. Isohyetal map of total storm depth from a multicellular event on August 31, 1968, on the Walnut Gulch watershed (from Renard 1970). (Multiply inches by 25.4 to obtain mm).

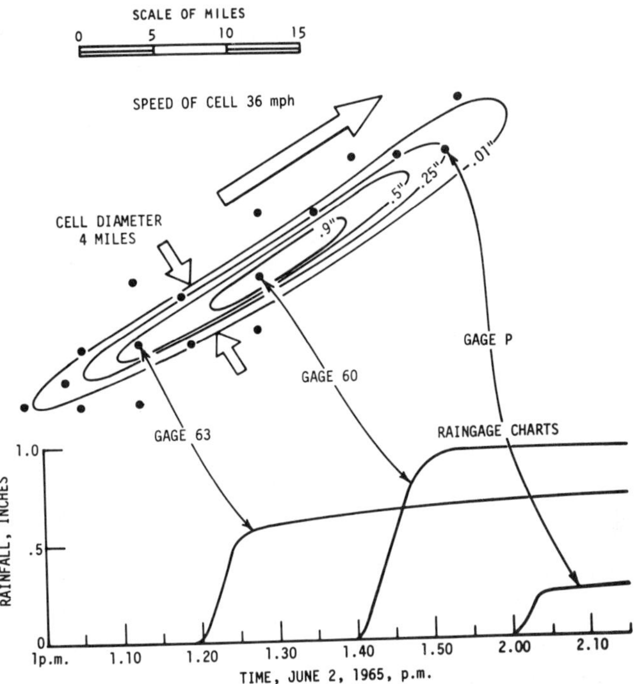

Fig. 2-16. A single storm cell's size, speed and magnitude, for the storm of June 2, 1965, on the little Egypt dense raingage network in southern Illinois (from Stall and Huff 1971). (Multiply in by 25.4 to obtain mm. Multiply mi by 1.61 to obtain km.)

3) Comparatively few storms produce channel flow.

4) Runoff-producing storms, defined as those that result in channel flow, are of relatively high intensity; such storms cause most of the floodwater damage, surface erosion, arroyo formation, and sediment deposition. Storm intensity values from some selected experimental watersheds are presented in Table 2-2.

Table 2-2. Number of times intensities were exceeded for given periods for storms greater than 1.9 cm (0.75 in) on four Safford, Arizona, watersheds.

Maximum intensity	Time interval (minutes)					
cm/hr (in/hr)	5	10	15	20	25	30
20.3 (8.0)	1					
17.8 (7.0)	4	1				
15.2 (6.0)	6	3	1	1		
12.7 (5.0)	15	6	1	1		
10.2 (4.0)	35	19	9	4	1	
7.6 (3.0)	70	46	23	15	8	1
5.1 (2.0)	95	90	77	61	25	3
2.5 (1.0)	105	105	105	103	102	44

Twenty-two years of record; total of nine 12-hour recording rain gages on four watersheds. Data from Osborn and Reynolds (1963).

5) Individual exceptional storms produce as much as surface runoff as several years of normal runoff; an example is the 1967 storm on the Walnut Gulch watershed (Arizona), which produced 3.35 in (85 mm) of rainfall in 45 minutes at one point on the watershed (Osborn and Renard 1969).

6) Approximately 65 to 70% of the annual precipitation, and all of that producing runoff, falls during the summer. This generalization must be modified for rangelands found elsewhere in the United States.

7) The orographic effect is minimal in summer.

8) Records from a few scattered gages are inadequate for the prediction—with the accuracy necessary for hydrologic planning on small watersheds—of areal distributions, volumes, and intensities of convective rainfall. This introduces the problem faced at the beginning of any hydrologic study, that of collection of accurate data which will adequately represent the variations in the applicable atmospheric parameters. This requires the design of suitable data networks and appropriate instrumentation. Without getting into the problem of raingage network design, it is reasonable to assume that the strength of the relationship with which rainfall may be estimated at a given

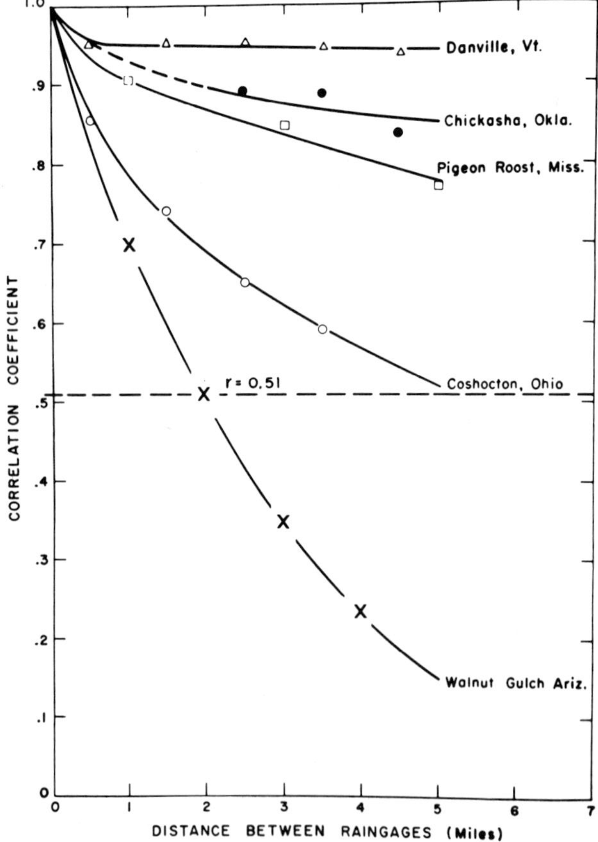

Fig. 2-17. Generalized relationship between correlation coefficient and distance between raingages for five watersheds (from Hershfield 1967). (Multiply miles by 1.61 to obtain km).

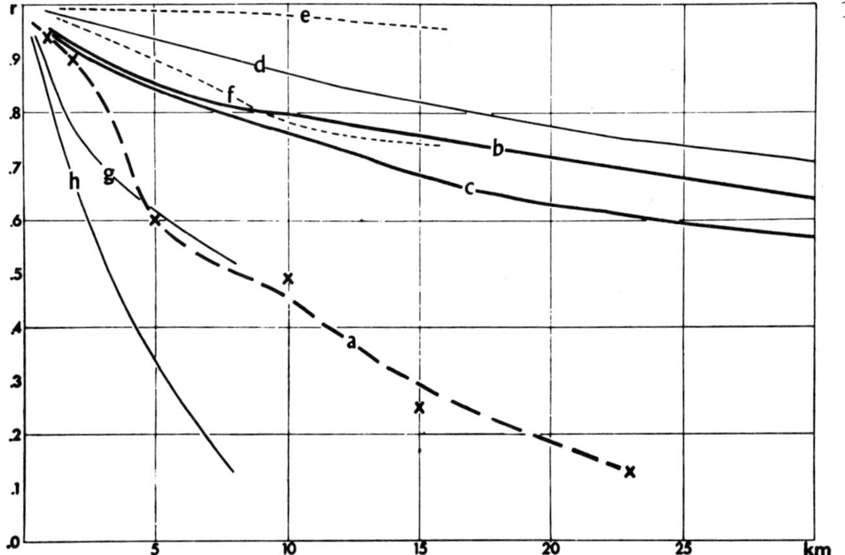

Fig. 2-18. Correlation functions of daily rainfall totals in various networks: (a) Southern Arava, ungrouped data, n = 26-96; (b) Lower Jordan Valley, ungrouped data, n = 260; (c) Lower Jordan Valley, days with less than 25 mm, n = 250; (d) East Central Illinois, summer rain, n = 296 (Huff and Shipp 1969); (e) East Central Illinois, summer, low center rainfall, n = 28; (f) East Central Illinois, summer, air mass storms, n = 73; (g) Coshocton, Ohio, largest summer storms only, n = 15 (Hershfield, 1967); (h) Walnut Gulch, Ariz., largest summer storms only, n = 15 (Hershfield 1968) (after Sharon 1972).

point from known observations at neighboring points must depend upon their intercorrelations. As an example of how correlation coefficients derived from individual storm data vary as a function of distance between raingages in various parts of the country, Figure 2-17 has been extracted from a study by Hershfield (1967). If r = 0.9 is considered an acceptable standard for the degree of linear association between storm amounts at two gages, then the average distance separating gages should be less than 1.6 km (1 mile) at Walnut Gulch, and greater than 6.4 km (4 miles) at Danville, Vermont. In fact, Osborn, Lane, and Hundley (1972) reported that for satisfactory representation of thunderstorm rainfall at Walnut Gulch (without regard to runoff), the optimum raingage density, to maintain a correlation of 0.9 between adjacent gages, would require raingages to be located at 304 m (1,000 ft) intervals. Figure 2-18 shows the same types of relationships, but includes data from Illinois and also Israel for comparative purposes.

9) Suitable predictive values for convective rainfall for specific localities may be developed from records from dense networks of recording rain gages operated over a relatively few years.

10) During the summer, rainfall is most likely to occur from 5 pm to 9 pm local time.

A study on the Reynolds Creek Watershed, in the Owyhee Mountains of

 Table 2-3. Monthly distribution of high intensity storms in southwestern Idaho, 1961-1964.

Month	Percentage of all high intensity events	
	Over 1.2 cm/hr (0.5 in/hr)	Over 5.0 cm/hr (2.0 in/hr)
January	1.5	0.0
February	1.5	3.5
March	0.0	0.0
April	4.5	3.5
May	22.0	24.0
June	33.0	38.0
July	3.0	7.0
August	18.0	24.0
September	4.5	0.0
October	3.0	0.0
November	7.5	0.0
December	1.5	0.0
Total	100.0	100.0

Southwestern Idaho, reinforced observations from the Southwest that one or two rain gages per 260 km^2 (100 square miles) are not adequate to determine the probability of infrequent high intensity rains within a reasonable time (Cooper 1967). This study also indicated that:

1) practically all high-intensity rain during the 4-year study period occurred from April through August (Table 2-3).

2) high-intensity rainfall resulted chiefly from spring and summer storms with a strong convective component; almost all intense thunderstorms during the study period began between 4 pm and 8 pm local time.

3) there was no apparent relationship between elevation and peak storm intensity, or between elevation and total amount of precipitation falling at any given intensity; this suggests that data from accessible valley stations could be used to estimate the relative occurrence of high-intensity rains throughout an area of appreciable elevation range. Kidd (1964) has provided some data on probable return periods of rainstorms in central Idaho.

Price and Evans (1937), Lull and Ellison (1950), Croft and Marston (1950), and Farmer and Fletcher (1971, 1972) all have studied rainfall characteristics along the Wasatch Mountain front and elsewhere in Utah. These studies, in contrast to those of Cooper (1967) in southwestern Idaho, indicate the zone between 2,000 m (6,.500 ft) and 2,462 m (8,000 ft) elevation is expected to receive the highest rainfall intensities. There is a trend toward reduced intensities and greater rainfall amounts with increasing elevation, but the trend is not uniform at all locations. Also, average storm length varies inversely with elevation. Farmer and Fletcher further indicate that most storms in the Great Basin area occur between 10 am and 6 pm hours, peaking at about 1 pm hours (Fig. 2-19). In contrast, patterns are more nearly uniform in northern Utah (Davis County). There is a slight but steady increase in storm occurrence until 8 pm hours, followed by a decrease until midnight. Few storms exceed 6 hours in length (Fig. 2-20).

The question of areal applicability of intensity data has been explored

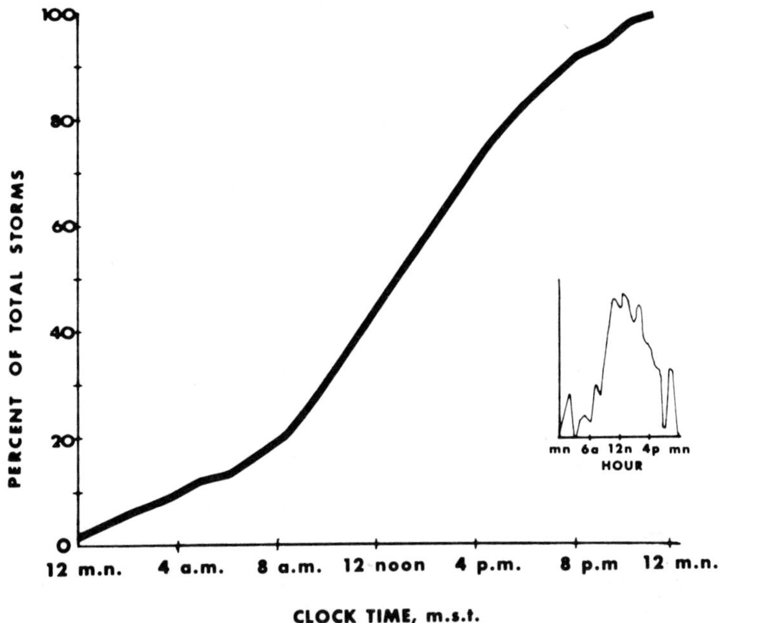

Fig. 2-19. Mass curve of accumulated percent of total storms by hour of storm beginning, for precipitation Zone 3 of the Great Basin Experimental Area. The insert is a schematic showing the distribution of storms by hour (from Farmer and Fletcher 1971).

briefly by Farmer and Fletcher (1971). They combined Utah data with data available for 41 additional intensity gages located at various elevations in New Mexico, Arizona, Utah, and Idaho. The additional data included maximum 2-minute intensities and intensity-frequency data for 60-minute durations. The trend, as shown in Table 2-4, is toward greater intensities as one moves to the more southerly stations. The maximum 2-minute intensity data provided some evidence that rainfall intensities in all areas may be reasonably comparable for durations shorter than 30 minutes. Two-minute intensities for selected stations include: 43.4 cm/hr (17.1 inches/hr), Reynolds Creek, Idaho; 36.6 cm/hr (14.4 inches/hr), Oaks Climatic Station, Utah; 46.3 cm/hr (18.2 inches/hr), Tombstone, Arizona; and 108.2 cm/hr (42.6 inches/hr), Alamogordo Creek, New Mexico. The maximum differences in total rainfall

Table 2-4. Ranges of 60-minute rainfall intensities, cm/hr (in/hr) for 10- and 50-year recurrence intervals by regions.

Region	10 years	50 years
Central and southern Idaho	0.6-1.0 (0.22-0.45)	1.0-2.0 (0.38-0.77)
Northern Utah	0.9-2.2 (0.37-0.86)	2.4-5.6 (0.96-2.20)
Central Utah	1.6-2.2 (0.62-0.86)	2.5-3.7 (0.98-1.20)
Central Utah and northern Arizona	2.8-3.8 (1.11-1.50)	4.3-5.1 (1.69-2.00)
Southern Arizona	4.1-5.1 (1.61-1.99)	5.6-6.6 (2.22-2.60)

Data from Farmer and Fletcher (1971).

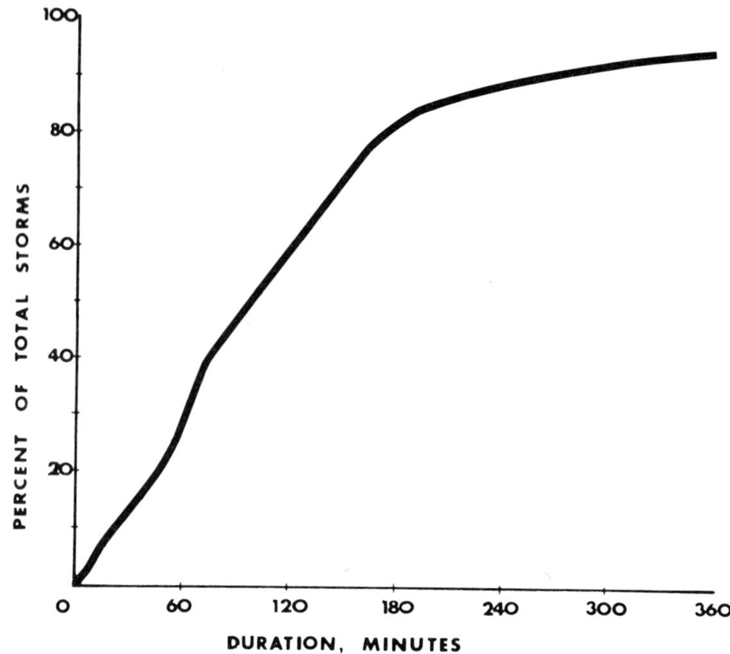

Fig. 2-20. Mass curve of accumulated percent of total storms by storm duration for precipitation Zone 4 of the Davis County Experimental Watershed (from Farmer and Fletcher 1971).

amount resulting from these intensities was 2.4 cm (0.94 inch) in 2 minutes. It seems likely that when pure air mass thunderstorms are being considered, the depth-area-duration conditions may be somewhat similar. On the other hand, the southernmost locations have a greater frequency of storm occurrence.

Osborn (1964) has emphasized that for runoff estimates for larger watersheds two probability estimates may be needed: (1) the probability of storms of certain intensities and areal coverage falling on small tributary watersheds, and (2) the probability of storms developing over a multitributary system in such patterns as to produce important volumes and peaks of runoff.

Caution should be used in extrapolating Weather Bureau rainfall data to rangeland watersheds. An example is given by Renard, Drissel, and Osborn (1970) after analyzing several exceptional runoff-producing storms from the Alamogordo Creek Experimental Watershed in northeastern New Mexico:

> Point rainfall estimates from the US Weather Bureau can be misleading if applied to predicting runoff peaks from small watersheds in northeastern New Mexico. Maximum point rainfall values for durations of less than 6 hours are quite low according to the data presented (in TP-40). The 100-year, 1-hour rainfall of 7.1 cm (2.8 inches) was exceeded twice in 7 years at the same gage on the watershed, and the 50-year, 6-hour maximum point rainfall was exceeded in just 2 hours in the same storm. In another storm, Rain Gage 61 received 12.8 cm (5.02 inches) in 6 hours, which would be classified as the 1,000-year, 6-hour event for the region. These records seem to indicate that the current US Weather Bureau maximum point rainfall values for 6-hour periods are too low and probably represent no more than the 2-hour values.

Storm Rainfall Analysis

Stall and Huff (1971) have suggested that there are 6 critical parameters and 4 additional minor ones to consider in any storm rainfall analysis. These parameters are given below:

Critical Parameters	*Minor Parameters*
Depth	Antecedent conditions
Area	Shape of storm area
Duration	Orientation of storm
Frequency	Direction of storm movement
Sequence	
Place	
Intensity (= Depth ÷ Duration)	

Over the years, hydrologists have attacked this matrix by developing understanding of 2 or 3 of these parameters at a time (also see Smith and Schreiber 1973; Smith and Schreiber 1974; Smith 1974), as illustrated in the following.

Depth-frequency

One of the most common storm rainfall curves is a depth-frequency relation such as illustrated in Figure 2-21. This type of curve relates point rainfall (in centimeters or inches) to the recurrence interval in years. The recurrence interval indicates only the *average interval* between events equal to or greater than a certain size. When two dimensions of the rainfall matrix are treated in this fashion, the other four parameters are treated as follows: areas—specified as at a point; duration—specified, as in this case as 2 hours; place—specified in case as north-central Illinois; sequence—usually specified or assumed as rainfall at a uniform rate.

Another method for estimating storm-depth frequency was presented by Smith and Schreiber (1974) when they showed that the daily rainfall depth for

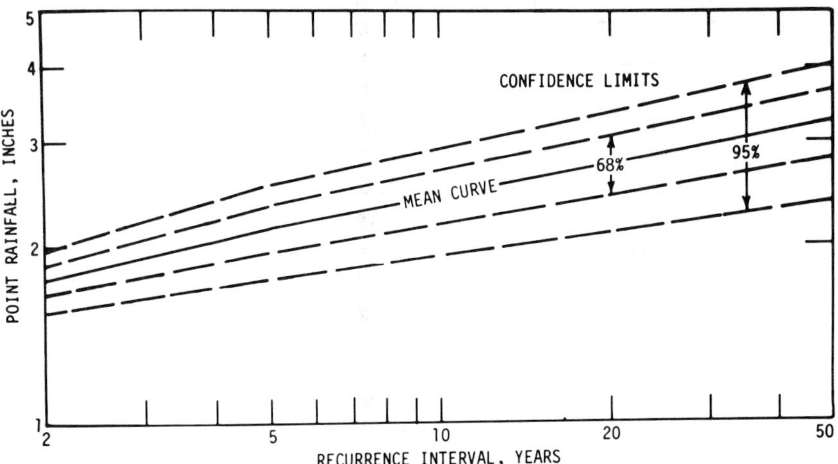

Fig. 2-21. Depth-frequency curve for 2-hour storm rainfall in north-central Illinois (from Stall and Huff 1971). (Multiply inches by 2.54 to obtain cm.)

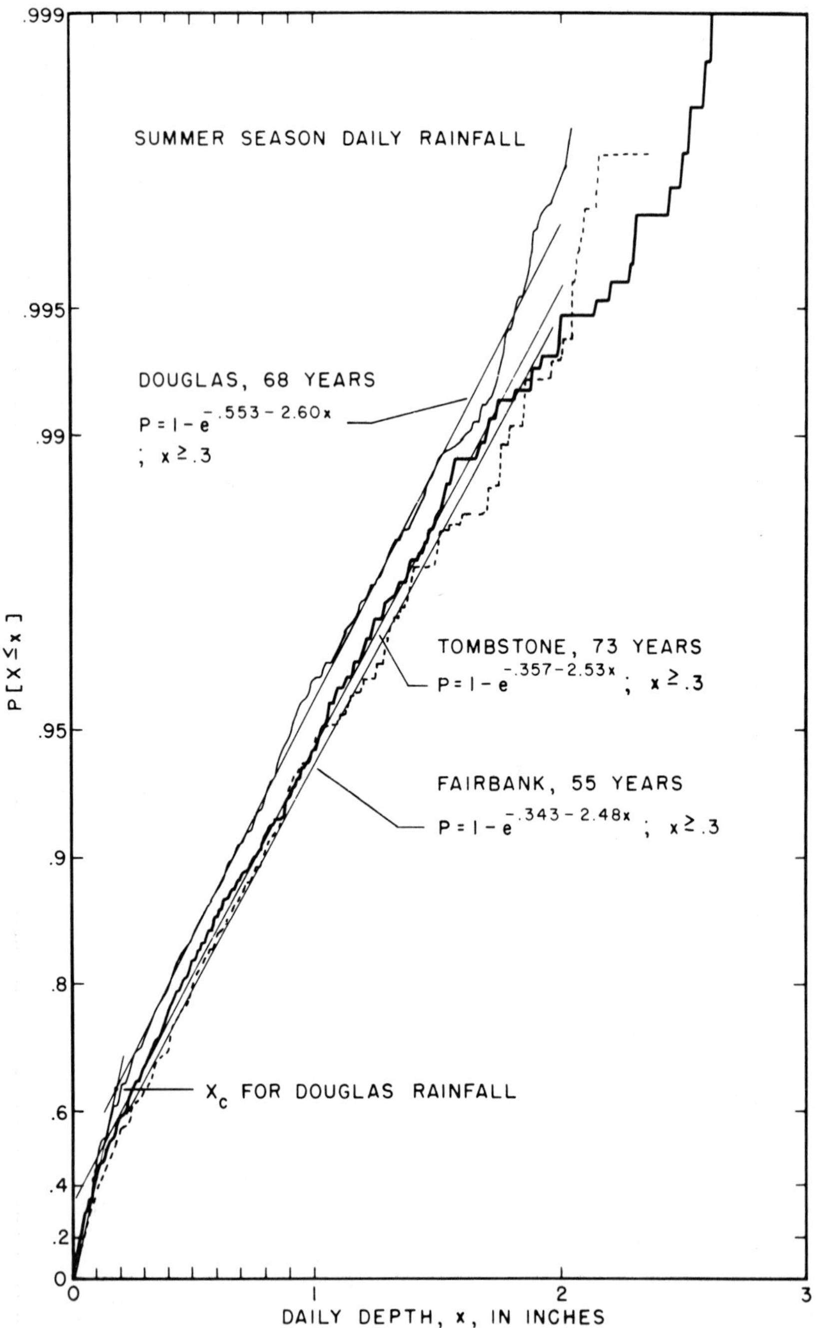

Fig. 2-22. Distributions of summer seasonal daily rainfall depths for three sampling stations in southeastern Arizona. P[X ≤ x] is the probability that daily depth ≤ the x value. (Smith and Schreiber 1974) (inches x 2.54 = cm).

three locations in southeastern Arizona was describable with a compound or mixed exponential of the form shown in Figure 2-22. The curves for the three locations may be approximated by two segments divided at an inflection point, X_c, or by using a mixed exponential of the form

$$P(X \geq x) = 1 - \alpha e^{-\lambda_{1X}} - (1-\alpha) e^{-\lambda_{2X}} \qquad \textbf{(2-1)}$$

Where α is a weighting factor ($0 \leq \alpha \leq 1$) and λ_1 and λ_2 are different parameters, e is the base of the natural logrithom, X is the depth of daily rainfall, and x is a random variable. The skewed shape of the density function increases the uncertainty of the sample probabilities with increasing rainfall depth. Such a relationship has also been used successfully for precipitation stations in Colorado. Smith and Schreiber (1973) also showed that there was an average of 31 stormy days per summer thunderstorm season at the Douglas gage. Thus, from Figure 2-22, a 1-inch storm depth has a 0.955 probability, which divided by 31 gives an annual probability of 0.031. Thus, a 1-inch storm is expected once every 1.05 years.

Many other methods have been specified and used in simulation schemes to estimate daily or shorter precipitation depths. Most of these efforts involve using a statistical distribution of some form. A partial listing of some such generating schemes is given in Table 2-5. There are obviously many other such generation techniques being published regularly. It is important to note, however, that the variability of precipitation patterns and the objective for which the synthetic data are to be used justify the large number of differing schemes; i.e., there is not presently any one distribution determined to be superior to the others.

Depth-place

Place variability of rainfall depth is generally given in the form of maps. Figure 2-23 shows 2-hour storm rainfall in inches expected at a 10-year recurrence interval (US Weather Bureau 1961). In showing depth and place relations (as in Figure 2-21), the other 4 critical parameters are treated as follows: area—specified as at a point; duration—specified as being two hours, frequency—specified as being 10-year recurrence interval; sequence—usually specified as rain falling at a uniform rate.

Depth-area

Figure 2-24 represents generalized storm depth-area curves for various parts of the country. The curves (from Hershfield 1967) show the percent to which point rainfall values must be diminished to represent rainfall over an area. The curve labeled *I* may be considered applicable for major storms in a subhumid region such as exists in northeastern United States, whereas curve *III* might be representative of many of the major storms of the arid Southwest. Curve *II* might apply to some intermediate rainfall regime. The remaining 4 critical rainfall parameters are treated as follows: duration—specified for any particular curve; frequency—usually not specified, the relation being

Table 2-5. Daily or shorter precipitation generation models[1].

Author	Date	Storm depth simulation
Brakensiek	1958	Log-probability and Gumbel extreme value distributions.
Fogel, Duckstein, Sanders	1971	Geometric probability for point depth with a Poisson distribution for the number of rainfall events per season.
Franz	1971	Multivariate normal distribution with persistence provision using a Markov sequence.
Grace & Eagleson	1966	Multiple-stage model with storm length selected from a distribution which in turn was related to storm depth.
Khanal & Hamrick	1971	Markov Chain (first-order).
Kotz & Neumann	1963	Gamma distribution.
Nicks	1971	Maximum daily rainfall on a network was a skewed normal distribution with a Markov Chain for occurrence.
Osborn, Lane & Kagan	1971	Complex process using number of cells (Poisson distribution with a three cell minimum) with the center depth approximated by a negative exponential separation empirically determined from sample data. Point storm depths are the sum of the depth for all cells.
Pattison	1965	Sixth-order Markov Chain using hourly rainfall states (each state implies a depth range).
Sariahmed	1969	Weibull distribution for storm duration (depth a linear function of duration).
Skees & Shenton	1971	Censored Gamma distribution and numerous transformations to approximate a normal or gamma shape.
Smith & Schreiber	1973	Mixed exponential distribution for storm depth with Markov Chain for occurrence.
Todorovic & Woolhiser	1971	Exponential for daily rainfall with the total amount of precipitation in an N-day period from a Markov Chain or a binomial distribution.
Wiser	1971	Mixed distribution generated from: 1) a storminess parameter (2nd-order autoregressive scheme) and 2) local process deterministic component (e.g., orographic effects with a random selection that determines the simulated precipitation.

[1]From Renard & Brakensiek (1976).

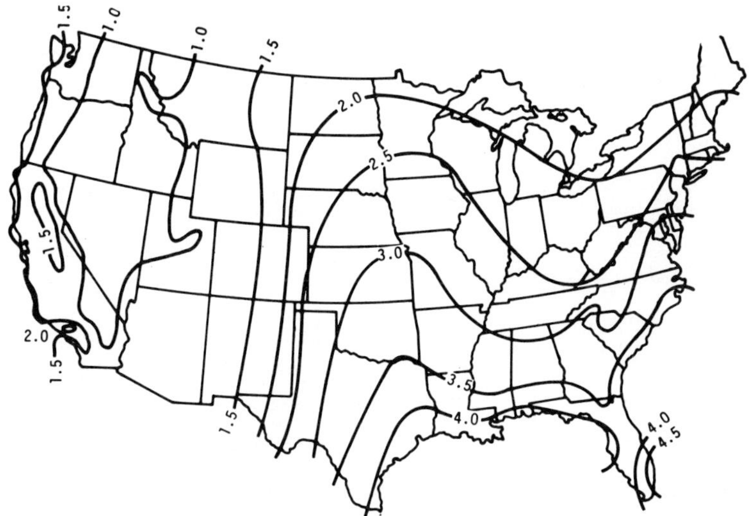

Fig. 2-23. Depth-place map of rainfall, showing 2-hour storm rainfall, in inches (x 2.54 for cm) expected at a 10-year recurrence interval for the United States (U.S. Weather Bureau of Tech. Paper 40, 1961).

assumed universal; place—usually not specified, though it may be in certain instances; sequence—usually specified as rain falling at a uniform rate.

An early and noteworthy treatment of a thunderstorm's areal pattern was the paper by Court (1961) comparing formulas set forth by earlier investigators (Figure 2-25). The relationships reported by Court were empirical and related the average rainfall inside an isohyet to various powers of the area, or to its logarithm or exponential. Court then compared his proposed Gaussian (normal) model with that of the other investigators assuming a storm center depth of 2 inches (5.08 cm) as shown in Figure 2-25. Interestingly, the models reflect the differing storm patterns studied from the limited extent air-mass

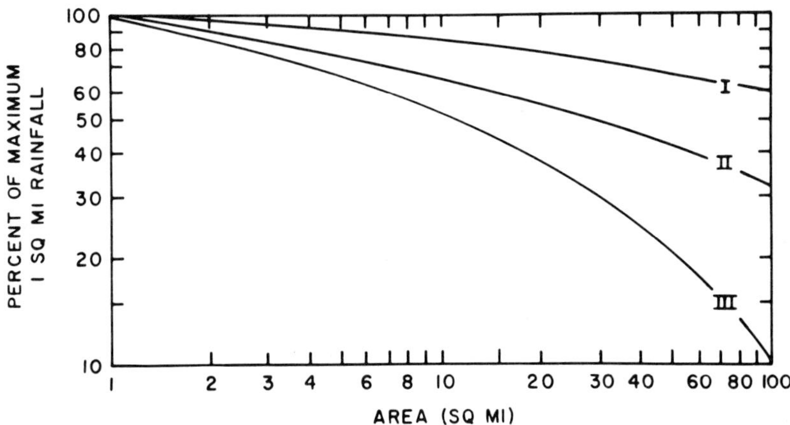

Fig. 2-24. Generalized storm depth-area curves (from Hershfield 1967) (sq. miles × 2.59 = km²).

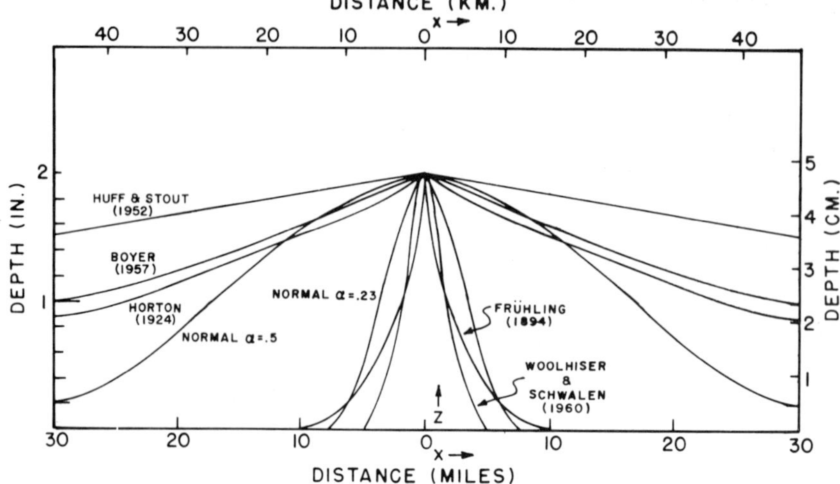

Fig. 2-25. Variation of isohyetal value z (inches) with distance × (miles) from storm center, assuming central precipitation to be m = inches, according to various formulas. (Multiply in by 25.4 to obtain mm. Multiply mi by 1.61 to obtain km) (Court 1961).

storms of Woolhiser and Schwalen (1959) in Arizona to the broad frontal storms of Huff and Stout (1952) experienced in Illinois.

Recently, Smith (1974) investigated the areal properties of thunderstorms in Arizona and described the storm pattern with a monotonic dimensionless depth-area relationship. He also expressed the depth-area relations for air-mass thunderstorms proposed by three other investigators in dimensionless form as shown in Figure 2-26. Assuming storms are occurring randomly, uniformly distributed in space, the rainfall population at any point may be considered to be composed of samples taken with equal likelihood from any point within the associated storm. With probability statistics Smith developed from this assumption, a general relation was presented between normalized storm isohyetal pattern, center depth probability, and point rainfall probability. This general relationship and the dimensionless depth-area relationships were used to obtain a record of the point rainfall depths which compared quite favorably with the historical record at the Tombstone rain gage (Figure 2-27).

Depth-duration

Two techniques for plotting depth-duration information are given in Figure 2-28. The vertical scale in Figure 2-28 is rainfall intensity in inches per hour, which is merely depth divided by duration; consequently, it remains primarily a measure of depth. The upper curve relates rainfall intensity to the duration of rainfall in hours for a specific site and for a 5-year recurrence interval. The lower curve is a curve of storm pattern and is used to provide a series of average intensities for various durations. These intensities are then combined with judgment to provide a storm pattern, or time distribution, for the occurrence of a particular rainfall depth. The remaining 4 critical parame-

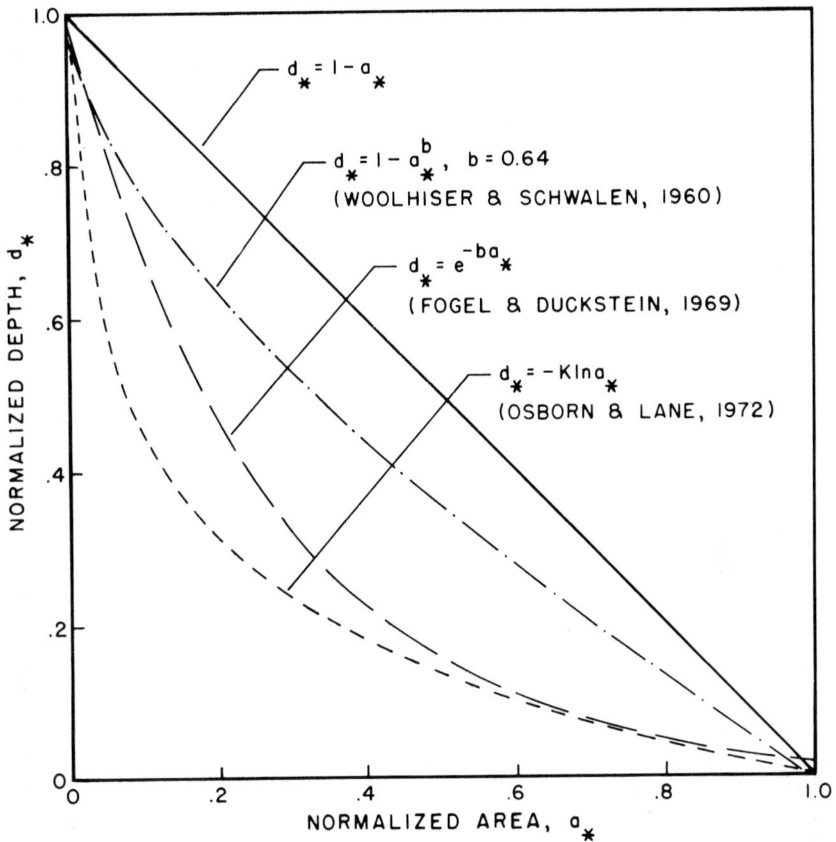

Fig. 2-26. Normalized depth-area relations for air-mass thunderstorms proposed by three investigators (Smith 1974).

ters are treated as follows: area—specified as being at a point; frequency—specified, in this case for a 5-year event; place; sequence—usually is specified as rain falling at a uniform rate during each of the durations plotted or used.

Depth-sequence

The sequence in which a series of particular depths of rainfall occur is an important aspect of rainfall occurrence. To estimate this sequence it is necessary to predict a mass curve of rainfall during a particular duration. For example, Figure 2-29 gives mass curves of rainfall for 261 storms in Illinois. The curves represent the median time distributions for the occurrence of rainfall during each of the four groups. For instance, the left curve is for storms during which the heaviest rainfall occurred in the first quarter of 25 percent of the storm time. The other 4 important parameters of storm rainfall are treated as follows; area—specified; duration—specified for all storms studied; frequency—specified as deciles of all occurrences; place—specified.

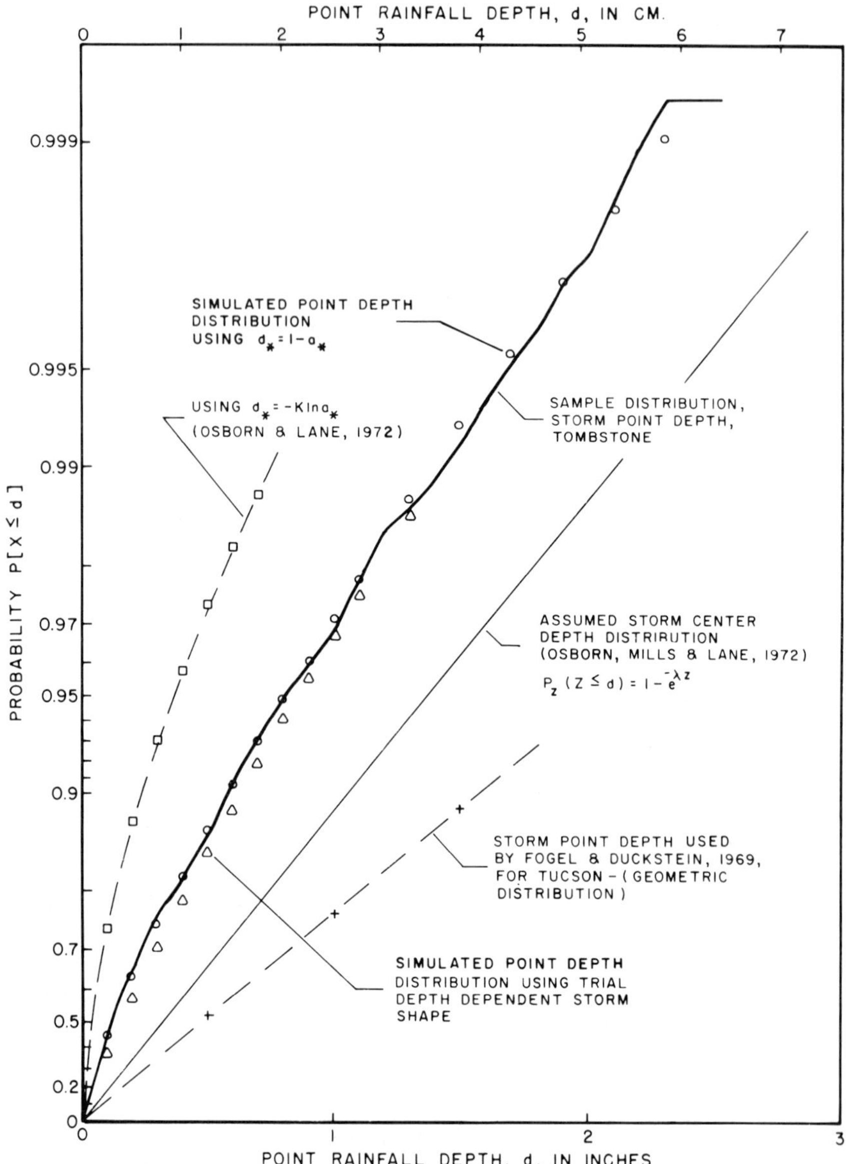

Fig. 2-27. Storm center-depth distribution and several point depth distributions, including measured data for Tombstone, Arizona, and simulated distributions. (Smith 1974) (inches × 2.54 = cm).

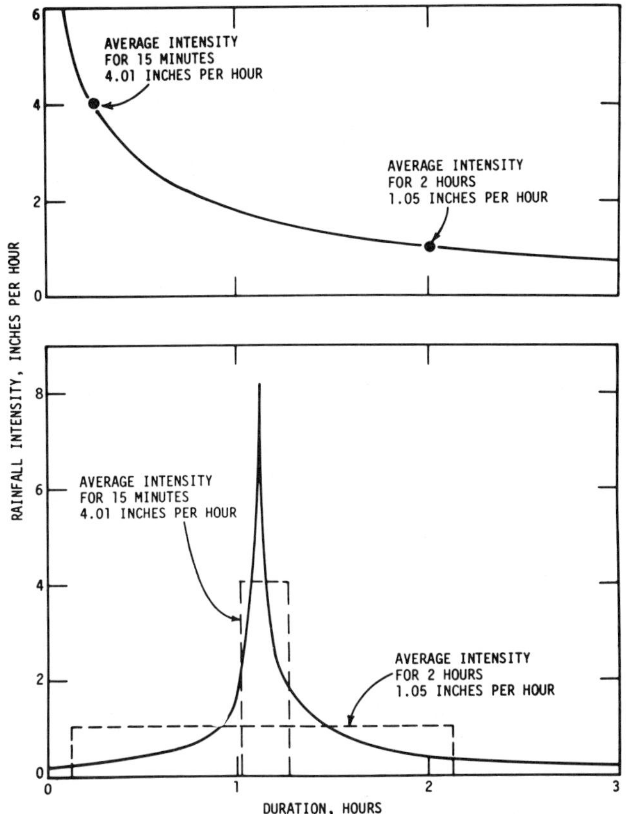

Fig. 2-28. Depth-duration curves for 5-year recurrence interval. Upper gives intensity-duration, lower gives storm pattern or average intensities for various durations (from Stall and Huff 1971). (Multiply inches by 2.54 to obtain cm).

Figure 2-30 shows the hyetographs and the dimensionless distribution graphs for four precipitation events on the Walnut Gulch Experimental Watershed which produced unusually large runoff. Interestingly, the dimensionless distribution graphs exhibit the same general shapes as the ones shown in Figure 2-29. Commonly, all combinations of the dimensionless graphs are encountered on Walnut Gulch. Differences in these distribution patterns can be very important to the resulting hydrograph as will be shown in a subsequent section.

Measurement and Estimation of Precipitation

It is not within the scope of this text to discuss types of rain gages, accuracy of point rainfall measurements, selection of rain gage sites, precipitation variability and its effects on rain gage network designs, and extrapolation of point precipitation values to areal estimates. The reader is simply referred to the extensive review by Corbett (1967) or to other references such as Gregory and Walling (1973) which contains many excellent illustrations. References given below in the next section (modeling) may also be of value. For a

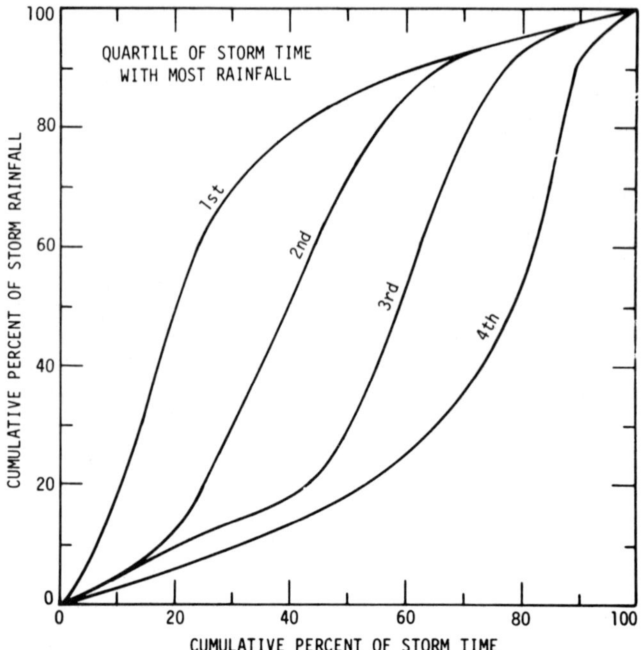

Fig. 2-29. Depth-sequence curves showing the median time distribution of storm rainfall in Illinois for storms with heaviest rain in 1st, 2nd, 3rd, and 4th quartile of storm duration (from Stall and Huff 1971).

discussion on snow, see Chapter 8.

Modeling Thunderstorm Runoff

Considerable effort is currently being expended in modeling thunderstorm behavior (rainfall delivery rates, relation to rain-gage network design, etc.). Because it is beyond the scope of this text to explore the modeling aspects in detail, the reader is referred to such references as Hershfield (1967); Eagleson (1967); Fogel and Duckstein (1969); Morgan (1970); Fogel, Duckstein, and Kisiel (1971); Osborn, Mills, and Lane (1972); Phanartzis and Kisiel (1972); Grayman and Eagleson (1972); Huff and Schickedanz (1972); and Rodriguez-Iturbe and Mejia (1974).

Additional Data

Precipitation data for the United States, as well as evaporation and wind movement information, are compiled by the National Oceanic and Atmospheric Administration (formerly Weather Bureau), and are available by subscription through the Superintendent of Documents, Government Printing Office, Washington, D.C. Rainfall analysis data also are available from the same source. A useful publication is Technical Paper No. 40, (US Weather Bureau [1961]) which contains information for the conterminous 48 states on rainfall frequencies for storm durations of 30 minutes to 24 hours and for return periods of from 1 to 100 years. However, results presented in Technical

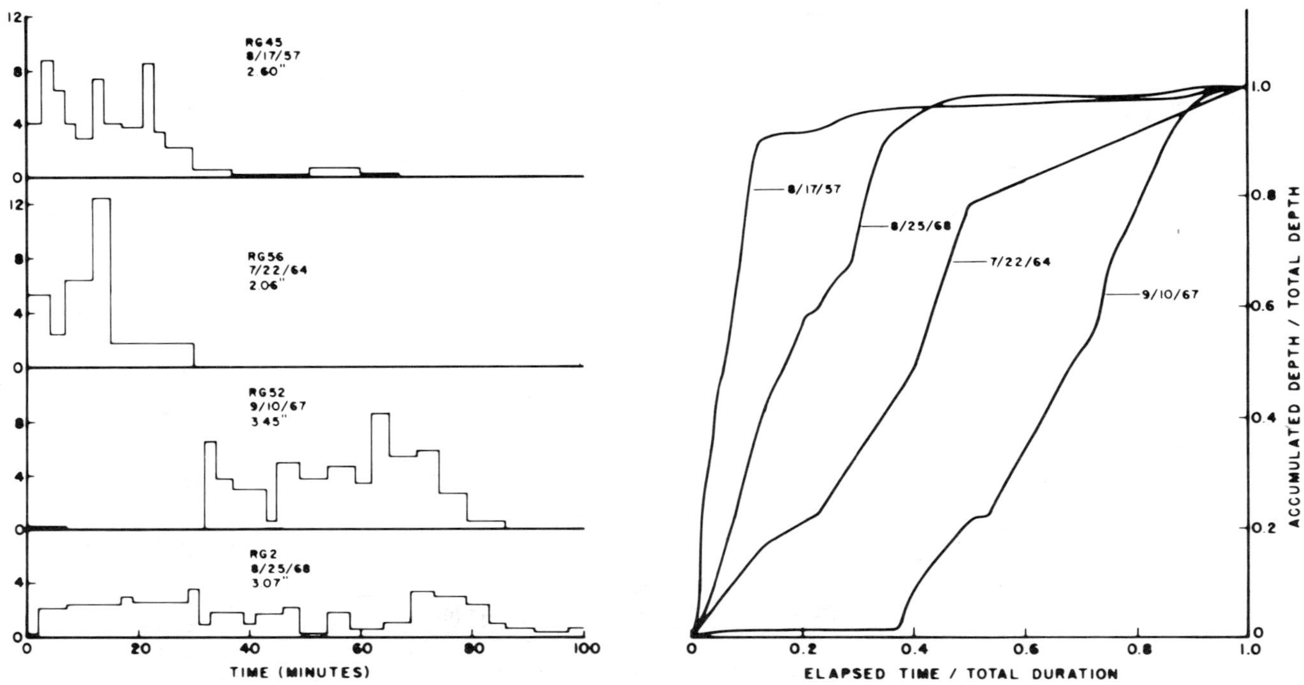

Fig. 2-30. Precipitation hyetograph and distribution graphs for selected storms on the Walnut Gulch watershed (Renard 1970).

 Paper No. 40 are most reliable in relatively flat plains. A relatively new publication, National Oceanic and Atmospheric Administration Atlas 2, US Dept. of Commerce (1973) entitled "Precipitation-Frequency Atlas of the Western United States," contains precipitation-frequency maps for 6- and 24-hour durations for return periods from 2 to 100 years for each of the 11 western states (west of about 103° W.). This new series of maps differs from previous publications through greater attention to the relation between topography and precipitation-frequency values. Technical Paper No. 49, (US Weather Bureau [1964]), contains rainfall frequency data for storm durations of 2 to 10 days. A paper by Gifford, Ashcroft, and Magnuson (1967) gives probabilities of selected precipitation amounts in the western region of the United States.

Literature Cited

Battan, L.J., and C.R. Green. 1971. Summer rainfall on the Santa Catalina Mountains. Univ. of Ariz., Inst. Atmos. Physics Tech. Rep. No. 22: 11 p.

Boyer, M.C. 1957. A correlation of the characteristics of great storms. Trans. Amer. Geophys. Union 38: 233-238.

Brakensiek, D.L. 1958. Fitting a generalized log normal distribution to hydrologic data. Trans. Amer. Geophys. Union 39: 469-473.

Cooper, C.F. 1967. Rainfall intensity and elevation in southwestern Idaho. Water Resour. Res. 3: 131-137.

Corbett, E.S. 1967. Measurement and estimation of precipitation on experimental watersheds. p. 107-129. *In:* Proc. Intern. Symp. on Forest Hydrology, Penn. St. Univ. Aug. 29-Sept. 10, 1965.

Court, Arnold. 1961. Area depth rainfall formulas. J. Geophys. Res. 66. 1823-1831.

Croft, A.F., and R.B. Marston. 1950. Summer rainfall characteristics in northern Utah. Trans. Amer. Geophys. Union, 31:83-95.

Drissel, J.C., and H.B. Osborn. 1968. Variability in rainfall producing runoff from a semiarid rangeland watershed. Alamogordo Creek, New Mexico. J. Hydrol. 6: 194-201.

Eagleson, P.S. 1967. Optimum density of rainfall networks. Water Resour. Res. 3: 1021-1033.

Eisenlohr, W.S., Jr., and C.E. Sloan. 1968. Generalized hydrology of prairie potholes on the Coteau de Missouri, North Dakota. U.S. Geol. Surv. Circ. 558. 12 p.

Farmer, E.E., and J.E. Fletcher. 1971. Precipitation characteristics of summer storms at high-elevation stations in Utah. USDA, Forest Serv. Pap. Int-110. 24 p.

Farmer, E.E., and J.E. FLetcher. 1972. Rainfall intensity-duration-frequency relations for the Wasatch Mountains of northern Utah. Water Resour. Res. 8: 266-271.

Fletcher, J.E. 1961. Some characteristics of precipitation associted with runoff from Walnut Gulch Watershed, Arizona. Trans. Amer. Geophys. Union, 42nd Annual Meeting (Washington, D.C.), April.

Fogel, M.M., and L. Duckstein. 1969. Point rainfall frequencies in convective storms. Water Resources Res. 5: 1229-1237.

Fogel, M.M., L. Duckstein, and C.C. Kisiel. 1971. Space-time validation of a thunderstorm rainfall mode. Water Resources Bull. 7: 309-316.

Fogel, M.M., L. Duckstein, and J.L. Sanders. 1971. An even-based strochastic model of areal rainfall and runoff. p. 247-261. *In:* USDA Misc. Pub. 1275. (Issued 1974).

Franz, D.D. 1971. Hourly rainfall generation for a network. Proceedings Symp. on Statistical Hydrology. USDA Misc. Pub. 1275: 147-153. (Issued 1974.)

Fruhling, A. 1894. Ueber Regan- uad Abflussmengen fur stadische Entwisserungakanale, Der Civilingenieur (Leipzig), ser. 2, 40, 541-558, 623-643, (ref. on p. 558).

Gates, D.M. 1962. Energy exchange in the biosphere. Harper & Row Biol. Mong., New York: 151 p.

Gifford, R.O., G.L. Ashcroft, and M.D. Magnuson. 1967. Probability of selected precipitation amounts in the western region of the United States. Univ. of Nev. Western Regional Res. Publ. T-8. Unpaged.

Grace, R.A., and P.S. Eagleson. 1966. The synthesis of short-time-increment rainfall sequences. Mass. Inst. Technol., Hydrodynamics Lab. Rep. 91.

Grayman, W.M., and P.S. Eagleson. 1972. The design of optimal precipitation measuring networks for flood forecasting. p. 379-394 *In:* Proc. Int. Symp. on Uncertainties in Hydrologic and Water Resource Systems, Dec. 11-14, Univ. of Ariz., Tucson.

Gregory, K.J., and D.E. Walling. 1973. Drainage basin form and process. John Wiley & Sons, New York. 456 p.

Hershfield, D.M. 1962. A note on the variability of annual precipitation. J. Appl. Meteorology 1: 575-578.

Hershfield, D.M. 1968. Rainfall input for hydrological models. Symp. p. 177-188. *In:* Geochem. Precip. Evapor, Soil Moisture Hydrometry, Bern, 1967. Int. Ass. Sci. Hydrology Pub. 78.

Horton, R.E. 1924. Discussion of the distribution of intense rainfall and some other factors in the design of storm-water drains, by Frank A. Marston. Proc. Amer. Soc. Civil Engineers, 50: 660-667.

Huff, F.A., and W.L. Shipp. 1969. Spatial correlations of storm, monthly and seasonal precipitation. J. Appl. Meterol. 8: 542-550.

Huff, F.A., and G.E. Stout. 1952. Area depth studies for thunderstorm rainfall in Illinois. Trans. Amer. Geophys. Union 33: 496-498.

Keppel, R.V. 1963. A record storm event on the Alamogordo Creek watershed in eastern New Mexico. J. Geophys. Res. 68: 4, 877-4, 880.

Khanal, N.N., and R.L. Hamrick. 1971. A stochastic model for daily rainfall data synthesis. p. 197-210. *In:* Proc. Symp. on Statistical Hydrology, (1971) U.S.Dep. Agr. Misc. Pub. 1275.

Kidd, W.J., Jr. 1964. Probable return periods of rainstorms in central Idaho. USDA, Forest Serv. Res. Note INT-28. 8 p.

Kincaid, D.R., H.B. Osborn, and J.L. Gardner. 1966. Use of unit-source watersheds for hydrologic investigations in the semi-arid Southwest. Water Resour. Res. 2: 381-392.

Kotz, S., and J. Neumann. 1963. On the distribution of precipitation amounts for periods of increasing length. J. Geophys. Res. 68: 3635-3640.

Langbein, W.B., and S.A. Schumm. 1958. Yield of sediment in relation to mean annual precipitation. Trans. Amer. Geophys. Union. 39: 1076-1084.

Lull, H.W., and L. Ellison. 1950. Precipitation in relation to altitude in central Utah. Ecology 31: 479-484.

McDonald, J.E. 1956. Variability of precipitation in an arid region: a survey of characteristics for Arizona. Univ. Arizona Inst. Atoms. Phys. Tech. Rep. 1. 88 p.

McDonald, J.E. 1959. It rained everywhere but here—the thunderstorm encirclement illusion. Weatherwide V 12: 158-174.

Mills, W.C., and H.G. Osborn. 1973. Stationarity in thunderstorm rainfall in the southwest, p. 26-31. *In:* Hydrology and Water Resour. in Arizona and the Southwest Vol. 3.

Morgan, D.L. 1970. A cumulus convection model applied to thunderstorm rainfall in arid regions. U.S. Army Corps of Engineers, Hydrologic Eng. Center Res. Note No. 1 (609 Second St., Davis, Calif. 24 p.

Nicks, A.D. 1971. Stochastic generation of the occurrence pattern and location of maximum amount of daily rainfall p. 154-171. *In:* Proc. Symp. on Statistical Hydrology, U.S. Dep. Agr. Misc. Pub. 1275. (Issued 1974.)

Osborn, H.B. 1964. Effect of storm duration on runoff from rangeland watersheds in the semiarid southwestern United States. Intern. Ass. Sci. Hydrol. 9: 40-47.

Osborn, H.B. 1971. Some regional differences in runoff producing thunderstorm rainfall in the Southwest, p. 13-27. *In:* Hydrology and Water Resources in Arizona and the Southwest.

Osborn, H.B., and R.B. Hickok, 1968. Variability of rainfall affecting runoff from a semiarid rangeland watershed. Water Resources Res. 4: 199-203.

Osborn, H.B., and L.J. Lane. 1969. Precipitation-runoff relationships for very small semiarid rangeland watersheds. Water Resour. Res. 5: 419-425.

Osborn, H.B., and L.J. Lane. 1972. Depth-area relationship for thunderstorm rainfall in southeastern Arizona. Trans. Amer. Soc. Agr. Eng. 15: 670-673.

Osborn, H.B., L.J. Lane, and J.F. Hundley. 1972. Optimum gaging of thunderstorm rainfall in southeastern Arizona. Water Resources Res. 8: 259-265.

Osborn, H.B., L.J. Lane, and R.S. Kagan. 1971. Stochastic models of spatial and temporal distribution of thunderstorm rainfall, p. 211-231. *In:* Proc. Symp. on Statistical Hydrology. U.S. Dep. Agr. Misc. Pub. 1275. (Issued 1974.)

Osborn, H.B., W.C. Mills, and L.J. Lane. 1972. Uncertainties in estimating a runoff-producing rainfall for thunderstorm rainfall runoff models, p. 189-202. *In:* Proc. Intern. Symp. on Uncertainties in Hydrologic and Water Resource Systems, Dec. 11-14, Univ. of Ariz., Tucson, Vol. I.

Osborn, H.B. and K.G. Renard. 1969. Analysis of two runoff-producing southwest thunderstorms. J. Hydrol. 8: 282-302.

Osborn, H.B., and K.G. Renard. 1970. Thunderstorm runoff on the Walnut Gulch experimental watershed, Arizona, U.S.A., p. 455-464. *In:* Proc. IASH Symp. on Results of Research on Representative and Experimental Basins, Wellington, New Zealand, Dec. 1-8.

Osborn, H.B., and W.N. Reynolds. 1963. Convective storm patterns in the southwestern United States. Intern. Ass. Sci. Hydrol. Bull. 8: 71-83.

Pattison, A. 1965. Synthesis of hourly rainfall data. Water Resources Res. 1: 489-498.

Phanzrtzis, C.A., and C.C. Kisiel. 1972. Uncertainties in point and areal sampling of precipitation and their importance in hydrologic computations, p. 341-357. *In:* Proc. Intern. Symp. on Uncertainties in Hydrologic and Water Resource Systems, Dec. 11-14, Univ. of Ariz. Tucson, Vol. I.

Price, R., and R.B. Evans. 1937. Climate of the west front of the Wasatch Plateau in central Utah. Monthly Weather Rev. 65: 291-301.

Renard, K.G. 1970. The hydrology of semiarid rangeland watersheds. U.S. Dep. Agr., Res. Serv. 41-162. 26 p.

Renard, K.G., and D.L. Brakensiek. 1976. Precipitation on Intermountain rangeland in the western United States. Fifth US/Australia Range Science Workshop, Boise, Ida., June, 1975.

Renard, K.B., J.C. Drissel, and H.B. Osborn. 1970. Flood peaks from small southwest range watershed. Amer. Soc. Civil Eng. Proc., J. Hydraulics Div., (Pap. 7161) HY3: 773-785.

Rodriguez-Iturbe, I., and J.M. Mejia. 1974. The design of rainfall networks in time and space. Water Resour. Res. 10: 713-728.

Sariahmed, A. 1969. Synthesis of sequences of summer thunderstorm volumes for the Atterbury Watershed in the Tucson Area. M.S. Thesis, Committee of Hydrology and Water Resources, Univ. of Ariz.

Schroeder, M.J., and C.C. Buck. 1970. Fire weather. USDA Handbook 360: 229 p.

Sharon, D. 1972. The spottiness of rainfall in a desert area. J. Hydrol. 17: 161-175.

Skees, P.M., and L.R. Shenton. 1971. Comments on the statistical distribution of rainfall per period under various transformations. p. 172-196 *In:* Proc. Symp. on Statistical Hydrology. USDA Misc. Pub. 1275. (Issued 1974.)

Smith, R.E. 1974. Point process of seasonal thunderstorm rainfall. 3. Relation of point rainfall to storm and areal properties. Water Resour. Res. 10(3): 424-426.

Smith, R.E., and H.A. Schreiber. 1973. Point process of seasonal thunderstorm rainfall. 1. Distribution of rainfall events. Water Resour. Res. 9: 871-884.

Smith, R.E., and H.A. Schreiber. 1974. Point process of seasonal thunderstorm rainfall. 2. Rainfall depth probabilities. Water Resour. Res. 10: 418-423.

Stall, J.B., and F.A. Huff. 1971. The structure of the thunderstorm rainfall. Preprint 1330, Paper Presented at ASCE National Water Resources Engineering meeting, Phoenix, Ariz., Jan. 11-15: 30 p.

Todorovic, P., and D.A. Woolhiser. 1971. Stochastic model of daily rainfall. p. 232-246 *In:* Proc. Symp. on Statistical Hydrology. USDA Misc. Pub. 1275. (Issued 1974).

U.S. Weather Bureau. 1961. Rainfall frequency atlas of the United States U.S. Dep. Commerce, Weather Bur. Tech. Pap. 40, Washington, D.C. 61 p.

U.S. Weather Bureau. 1964. Two- to ten-day precipitation for return periods of 2 to 100 years in the contiguous United States. U.S. Dep. Commerce, Weather Bur. Tech. Pap. 49; Washington, D.C. 29 p.

U.S. Weather Bureau 1973. Precipitation-frequency atlas of the western United States: Vol. 1, Mont.; Vol. II, Wyo.; Vol. III, Nev.; Vol. VIII, Ariz.; Vol. IX, Wash.; Vol. X, Ore.; Vol. XI, Calif. Nat. Oceanic and Atmos. Admin. Atlas 2. U.S. Department of Commerce, Weather Bureau.

Wiser, E.H. 1971. A precipitation data simulator using a second-order autoregressive scheme. p. 120-134. *In:* Proc. Symp. on Statistical Hydrology. (1971) U.S. Dep. Agr. Misc. Pub. 1275.

Woolhiser, D.A., and H.A. Schwalen. 1960. Area-depth frequency relations for thunderstorm rainfall in southern Arizona. Univ. Arizona Agr. Exp. Sta. Tech. Pap. 527. 7 p.

Chapter 3

Interception, Stemflow, and Throughfall

Four things may happen to the raindrop that falls on a watershed: (1) it may be intercepted by leaves, twigs, or stems and be returned to the atmosphere as water vapor, (2) it may follow the stem to the soil surface as stemflow or drip from the foliage, (3) it may proceed through the foliage to the soil surface as throughfall, or (4) in areas of sparse vegetation, the raindrop may directly intersect the land surface (bare soil, erosion pavement, and litter).

Definitions of Terms

In most instances, intercepting vegetation can be divided into canopy and litter layers. The canopy includes leaves, twigs, branches, and stems of both overstory and understory. Litter generally consists of recently fallen leaves, twigs, and fruits. Total interception loss (i.e., rainfall that never reaches mineral soil) includes water retained on both canopy and litter. Interception is therefore a two-part process in which (1) gross rainfall is caught by the canopy and redistributed as throughfall, stemflow and evaporation from the standing vegetation, and (2) the remainder is caught on the litter layer and evaporated without adding to moisture in the mineral soil. Though usually applied with respect to vegetation, the concept of interception (Figure 3-1) may be applied equally well to tin roofs, asphalt parking lots, or artificial turf, although the magnitude of the interception will be smaller.

Interception loss terminology as given by Helvey and Patric (1965) is given below:

Gross rainfall is rainfall per storm measured in the open or above the vegetative canopy.

Total interception loss is rainfall per storm retained by the canopy and litter and evaporated without adding to moisture in the mineral soil.

Net rainfall is water that enters the mineral soil after penetrating the forest canopy and litter, sometimes called "effective rainfall."

Throughfall is that part of the gross rainfall which directly reaches the litter through spaces in the vegetative canopy and as drip from leaves, twigs, and stems.

Stemflow is that part of the gross rainfall which is caught on the canopy and reaches the litter or mineral soil by running down the stems.

Canopy interception loss is rainfall retained on standing vegetation and evaporated without dripping off or running down the stems.

Litter interception loss is rainfall retained in the litter layer and evaporated without adding to moisture in the underlying mineral soil.

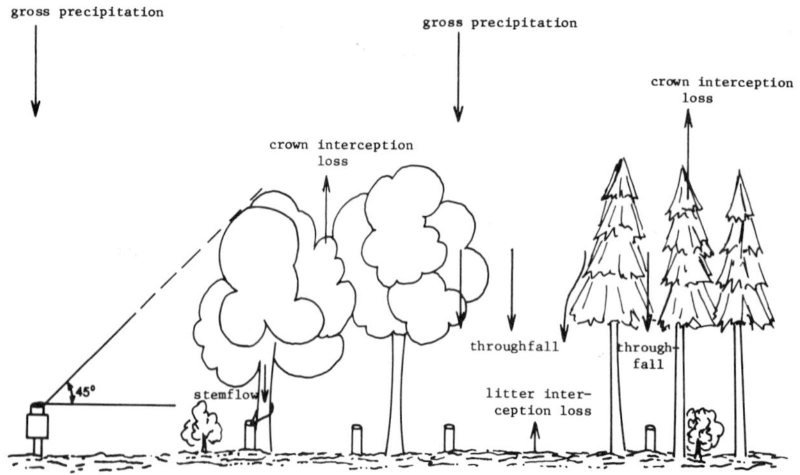

Fig. 3-1. Conceptual picture of the process of interception. Crown interception loss is usually considered as being the difference between gross precipitation and throughfall plus stemflow. Litter interception loss must be added to crown interception loss to obtain total interception (after Hewlett and Nutter 1969).

Factors Affecting Interception

Interception rate is variable and is dependent primarily upon the *interception capacity* (ability of the vegetation surfaces to collect and retain precipitation) of the vegetal cover. *Rainfall frequency* can be an important influencing factor in that short showers separated by periods of clear weather result in maximum interception capacity at the beginning of each shower. If rainfall is prolonged the plants become wetted and interception rate finally equals the evaporation rate from the wetted surfaces; thus, *rainfall duration* becomes important by determining the balance between reduced storage of water on the vegetation surfaces and increased evaporative losses. *Rainfall amount* is closely related to rainfall duration, and, provided that initial interception capacity is filled and the rainfall rate equals or exceeds the evaporation rate, it has no influence upon the amount of interception losses. Other factors affecting interception loss include *type of precipitation* e.g., rain or snow, and *type and morphology of the vegetation cover,* which are discussed more fully in the following sections.

Reported interception values are usually restricted to specific periods such as annual losses or losses for specific storm events.

Interception by Woody Species

A tabulation of results of some interception studies in forests and woodlands is presented in Table 3-1. In general, interception storage for trees varies from 0.05 to 0.9 cm (0.02 to 0.36 in), depending on the factors previously mentioned. It is generally agreed that, within forested ecosystems, canopy interception loss is greatest in the spruce-fir-hemlock type, intermediate in

Table 3-1. Average annual interception, stemflow, and throughfall for forest and woodland vegetation types.[1]

Vegetation type	Interception (%)	Stemflow (%)	Throughfall (%)	Reference
Second growth ponderosa pine	12	3	84	Rowe and Hendrix 1961
Oak-pine forest	13	—	—	Wood 1937
Softwood	37	—	—	Beall 1934
Hardwood	21	—	—	Beall 1934
Pine	17-18	1	—	Kittredge for example 1941
Ponderosa pine	32	—	—	Connoughton, 1934
Lodgepole pine	23	1	—	Dunford and Niederhof 1944
Aspen	16	1	—	Dunford and Niederhof 1944
Lodgepole pine	24	—	—	Miner and Trappe 1957
Utah juniper	17	—	83	Skau 1964
Chaparral	8	—	—	Rowe and Colman 1951

[1]See Helvey and Patric (1965), Zinke (1967), and Helvey (1971) for additional data, including linear regression equations for calculating interception rates.

pine, and least in broadleaved deciduous forests (Table 3-2) (Helvey 1971). However, as can be seen from Table 3-2, the differences are rather minimal, the error of measurement perhaps being larger than the differences shown. For any given storm, plant biomass is undoubtedly a very important factor.

Woody plant interception values are obtained by subtracting throughfall and stemflow from gross precipitation. Depending on the factor to be emphasized, the relation between interception (I), gross precipitation (P),

Table 3-2. Computed throughfall and stemflow (T + S) at 2.54 cm (1.0 in) of gross rainfall by species based on various throughfall plus stemflow equations.[1]

Species	T + S inches	T + S cm
Red pine	2.2	0.85
Loblolly pine	2.2	0.85
Shortleaf pine	2.2	0.86
Eastern white pine	2.2	0.86
Ponderosa pine	2.2	0.87
Average (pines)	2.2	0.86
Spruce-fir-hemlock	1.9	0.74
Mature mixed hardwoods		
Growing season	2.3	0.90
Dormant season	2.4	0.95

[1]Data from Helvey (1971).

 stemflow (S), and throughfall (T) may be expressed as

$$I = P\text{-}S\text{-}T \quad \ldots\ldots\ldots\ldots\ldots\ldots\ldots\ldots \quad (3\text{-}1)$$

$$\text{or, } T = P\text{-}S\text{-}I \quad \ldots\ldots\ldots\ldots\ldots\ldots\ldots\ldots \quad (3\text{-}2)$$

Throughfall and gross precipitation are usually measured by placing rain gages under trees and in adjacent open areas (Wilm 1943). Stemflow is measured by attaching lead collars to stem bases and directing flow into containers for volumetric or weight measurements.

Studies by Rowe (1948) and Hamilton and Rowe (1949) in the chaparral type of central and southern California found that annual interception losses ranged from 4.3 to 9.9 cm (1.7 to 3.9 in) (the overall average amounted to about 8% of the overall annual rainfall), depending on the total annual rainfall and the character of storms producing it. Throughfall, stemflow, and interception loss generally were directly proportional to storm size. For small storms, the interception loss was as much as 50% to 75%, and for large storms as little as 3% to 6% of the gross rainfall. Another study (Rowe and Colman 1951) provided similar results.

In the pinyon-juniper type in Arizona, Collings (1966) found that (1) throughfall for Utah juniper *(Juniperus osteosperma)* was the same as for pinyon *(Pinus edulis);* (2) direction of gage from the tree bole affects rainfall catch (the south side had less throughfall than the north and east sides); (3) throughfall was not affected by the size of trees tested (the three classes of tree bole diameters were 2.5–7.6, 7.6–25.4, and more than 25.4 cm); (4) throughfall was greater near the edge of the tree crown than near the bole; (5) broad-leaved oaks *(Quercus arizonica)* has less throughfall than pinyon and juniper for storms greater than 0.2 cm (0.08 inch); and (6) interception of storms increased to a maximum of 1.2 cm (0.50 inch) of precipitation and then became constant. Skau (1964), also in Arizona, found that throughfall could be predicted with reasonable accuracy for either Utah juniper or alligator juniper *(Juniperus deppeana)* by a single equation. Stemflow accounted for only 1 to 2% of the total precipitation. Average interception in the Utah juniper type during a single year of measurement was about 17.2% of the average annual precipitation for that year.

Townsend (1966), in a study on the west side of the Sierra Nevada Mountains, found that retention-storage of chapparral leaves varied from 0.03 to 0.18 cm (.01 to .07 in) on a per storm basis. Grah and Wilson (1944) found that interception storage capacity for kidneywort *(Baccharis pilularis),* an evergreen brush species, varied between 0.05 to 0.15 cm (0.02 and 0.06 inch), also on a per storm basis.

Hicks (1943) used artificial sprinkling in studying interception by the shrubs chamise *(Adenostoma fasciculatum)*, curlleaf mountain mahogany *(Cercocarpus ledifolius),* and Christmasberry *(Photinia arbutifolia).* Storage capacity of the foliage per storm, was 0.13 to 0.15 cm (0.05 to 0.06 inch), 0.1 cm (0.04 inch), and 0.03 to 0.18 cm (0.01 to 0.07 inch), respectively.

Hull and Klomp (1974) and West and Gifford (1976) found that spring-summer-fall rainfall interception by big sagebrush *(Artemisia tridentata)* ranged from 4 to 31% depending on density and size of plants. Potential interception per shrub per rainfall event ranged from 0.11 to 0.15 cm (0.04 to 0.06 inch). Snow interception averaged 37% on plots studied by Hull and Klomp.

West and Gifford (1976 found that shadscale *(Atriplex confertifolia)* intercepted about 4% of the average April 1 to November 31 rainfall in northern Utah. Potential interception per shrub per rainfall event was about 0.15 cm (0.06 inch).

Recently, the validity of the interception that all intercepted water represents water loss has been questioned (Goodell 1963). This may be particularly true in the case of snow interception. Logic and experimental results (Burgy and Pomeroy 1958; McMillan and Burgy 1960) indicate that wet leaves result in lower transpiration losses than dry leaves; thus, intercepted moisture may not result in total loss from watersheds (Goodell 1963). In other words,

$$I_n = I_g - E_D = I_{SF} + E_R \quad \textbf{(3-3)}$$

where I_n is net loss of water to the soil due to vegetation interception; I_g is gross interception loss and refers to the sum of interception storage and evaporative loss from the canopy during rainfall; E_D is the hypothetical evaporation or transpiration from the surface, had it been dry; I_{SF} is the part of the interception loss which remains on the canopy at the end of the storm during the rainstorm. Couturier and Ripley (1973) have used the above approach and they found net interception losses of 14 and 22% during two growing seasons within an *Agropyron-Koeleria* grass prairie in southwestern Saskatchewan. Gross interception ranged from 21 to 32%.

Recently, the validity of the interpretation that all intercepted water represents water loss has been questioned (Goodell 1963). This may be particularly evaporates at a greater rate than transpiration from the same type of vegetation in the same environment. They state the following.

> Our analysis. . .allows us to identify the energy source for the more rapid evaporation of the intercepted water as being an increase in net radiation through a decreased long-wave reradiation and a decreased or even negative sensible heat flux. Furthermore, the model demonstrates that the enhanced evaporation of the intercepted water can occur for forests of large areal extent, where horizontal advection may be negligible. Thus it reaffirms the conclusions from empirical experiments that interception of precipitation does represent a loss of water to the soil and to the streamflow under field conditions.

Factors in the environmental complex that need further study are (1) plants subject to high total soil moisture stress may absorb moisture through their stomates, (2) energy available for evapotranspiration from wet leaves may be greater than for dry leaves, (3) interception may increase the availability of water for evaporation, and (4) interception storage on plant surfaces that do not transpire may result in evaporation loss.

Interception by Herbaceous Vegetation

Measurement of interception by herbaceous vegetation is difficult, but several imaginative attempts have been made (note Couturier and Ripley

 [1973] reference previously cited). Crouse, Corbett, and Seegrist (1966) have indicated that water storage capacity of grasses is proportional to the product of average height and percent of ground cover. Total interception loss for a storm is a function of the storage capacity and the number of showers per storm. Actual interception losses for storms with total rainfall greater than the storage capacity of plants involved varied from .03 to .9 cm (.01 to .35 inch).

Amisial, et al. (1969) developed a model of the surface runoff process for use in sparsely vegetated watersheds subjected primarily to thunderstorm runoff. The analytic model contained a retention rate which defined losses due to the combined effects of depression storage and vegetation interception. The retention storage was give as

$$R_{cr}\, t = R_r\,(R_{cs} - R_s\, t) \qquad \textbf{(3-4)}$$

in which R_r is a constant less than or equal to one and proportional to the soil and vegetation characteristics, R_{cs} is the retention storage capacity of vegetation and land surface, R_{cr} is the retention capacity rate, and R_s is the amount of rainfall in retention storage and is obtained from

$$R_s\, t = \int_o^t R_t\, dt \qquad \textbf{(3-5)}$$

where R_t is retention quantity at time t, $\int_o^t$ is the intigration from time 0 and it is time increment.

The actual retention rate, R_r, is given by the following equations

$$R_r = O \text{ if } P_r = O \qquad \textbf{(3-6)}$$

$$R_r = P_r \text{ if } O < P_r < R_{cr} \qquad \textbf{(3-7)}$$

$$R_r = R_{cr} \text{ if } P_r \geq R_{cr} \qquad \textbf{(3-8)}$$

where P_r is the precipitation rate.

Calibration of this model with data in southeastern Arizona was obtained using a value of 0.15 inch (0.4 cm) for a mixed grama grassland area. Merriam (1961) tabulated surface storage potentials for several species of grass, and his tabulation (along with other study results) appears in Table 3-3.

Interception Loss From Litter

Little work has been done regarding interception loss from litter. Corbett and Crouse (1968) have stated that the amount of water evaporated from litter is governed primarily by the moisture-holding capacity of the litter and the evaporation potential during and after the storm. In studying interception loss from various amounts of annual grass litter, interception losses from small storms were high while those for larger storms ranged from 2% to 5%. Regression equations were developed for predicting litter interception, and these equations were applied in deriving a distribution of annual interception losses for grass, brush and grass litter during 32 years of accumulated rainfall data (Table 3-4). The time of the storm occurrence and the relative amount of precipitation in large or small storms is responsible for the differences in

Table 3-3. Surface storage or closely related quantities for grass.

Species	Description of plant	Term used	Storage cm	(in)	Source
Big bluestem *(Andropogon gerardi)*	Dense stand, 2 ft high	Maximum interception capacity	0.23	(0.092)	Clark (1940)
Buffalo grass *(Buchloe dactyloides)*	8 in high, in bloom, many stolons	Maximum interception capacity	0.17	(0.065)	Clark (1940)
Slough grass *(Spartina pectinata)*	Nearly complete cover, few stems, lower leaves shed	Maximum interception capacity	0.08	(0.033)	Clark (1940)
Bluegrass *Poa pratensis)*	Maximum vegetative development	Total interception[1]	0.01	(0.040)	Haynes
Range mix[2]	10 in high, (1,500 lb/acre, air dry when 16 in high)	Interception storage[3]	0.10 to 0.12	(0.041 to 0.048)	Burgy and Pomeroy (1958)
Annual ryegrass	Uniform stands 4 to 19 in high	Surface storage	0.04 to 0.28	(0.017 to 0.111)	Merriam (1961)
Blando brome *(Bromus mollis)*	Various stages of growth	Interception storage capacity	0.03 to 0.51	(0.01 to 0.02)	Corbett and Crouse (1968)
Annual ryegrass *(Lolium multiflorum)*	Various stages of growth	Interception storage capacity	0.03 to 0.51	(0.01 to 0.02)	Corbett and Crouse (1968)
Tall wheatgrass *(Agropyron elongatum)*	Various *stages of* growth	Interception *storage* capacity	0.03 *to* 0.51	(0.01 *to* 0.02)	Corbett and Crouse (1968)
Black grama *(Boutelona eriopoda)*	Sparse stand	Retention storage capacity	0 to 0.38[4]	(0 to 0.15)	Amisial, et al. (1969
Agropyron-Koeleria grass prairie June-July	Evenly distributed, almost complete ground cover	Interception storage capacity	0.19	(0.07)	Couturier and Ripley (1973)

[1]Average of four storms with a total of 2.08 cm (0.82 in) rain.
[2]Principal species with tall fescue and soft chess.
[3]Storage 0.03 cm (0.01 in) greater than that shown was achieved with application of misty rain.
[4]Determined from calibrated model.
Date from Merriam (1961) and others.

interception percentages from year to year. Couturier and Ripley (1973) state that where vegetation is green and amply supplied with water, net interception (Formula 3-3) will be very small, but under arid conditions and with the presence of standing dead material it may be quite large, particularly if much of the rain falls in small amounts.

Kittredge (1939) measured the water retained by organic matter at field

 Table 3-4. Distribution of annual interception losses for grass, chaparral, and grass litter, San Dimas Experimental Forest, California.

Interception (percent)	Number of years		
	Grass	Chaparral	Grass litter
2.0– 2.9	—	—	2
3.0– 3.9	—	—	7
4.0– 4.9	—	—	10
5.0– 5.9	5	—	7
6.0– 6.9	7	—	5
7.0– 7.9	6	—	0
8.0– 8.9	1	—	1
9.0– 9.92	1	—	1
10.0–10.9	1	4	—
11.0–11.9	2	5	—
12.0–12.9	5	6	—
13.0–13.9	3	4	—
14.0–14.9	—	2	—
15.0–15.9	—	3	—
16.0–16.9	—	1	—
17.0–17.9	—	3	—
18.0–18.9	—	1	—
19.0–19.9	—	2	—

Data from Corbett and Crouse (1968).

capacity under chaparral in the San Gabriel Mountains in California. The amounts ranged from 0.1 to 0.7 cm (0.04 to 0.29 inch). Garcia and Pase (1967) found the water-holding capacity of Pringle manzanita *(Arctostaphylos pringlei)* litter averaged 0.5cm (0.20 inch) and that of shrub liveoak *(Quercus turbinella)* litter 0.5 cm (0.19 inch) under dense, uniform canopies in Arizona.

Total water retention of the three layers (*L*,freshly fallen litter; *F*, partially decomposed litter still recognizable as to origin and age; and *H*,, completely decomposed litter) of undisturbed forest floor under ponderosa pine *(Pinus ponderosa)* in Arizona was reported by Clary and Ffolliott (1969). With an average forest floor weight of 21,136 kg/ha (9.3 t/ac) the average amount of water retained was:

> *L* layer, .008 cm (0.003 inch); *F* layer, 0.03 cm (0.012 inch); *H* layer, .19 cm (0.075 inch); and total forest floor, 0.22 cm (0.088 inch).

This suggests a moisture-holding capacity of about 0.13 cm (0.05 inch) of water for 2.5 cm (1 inch) of forest floor, and an additional 0.18 cm (0.07 inch) of water for additional 2.5 cm (1 inch) of forest floor. Helvey (1971) reports, based on limited data relating to moisture relationships in the forest floor of conifers, that field capacity of litter (water held against drainage) averages about 215% by weight and annual losses range from 2 to 17% of gross rainfall.

Other Interception Losses

The role of bryophytes and lichens in rainfall interception has not been explored on rangeland sites. Moul and Buell (1955) report that well-developed moss mats from frequently burned sites in the New Jersey pine barrens absorb an average of 1.6 cm (0.62 inch) of rainfall. Well-developed lichen mats absorb from 0.4 to 0.7 cm (0.2 to 0.28 inch), depending on the species.

Much more might be said about interception losses but unfortunately, data are not widely available to substantiate the importance of the interception component. In some recent efforts to develop hydrologic models, interception losses are not treated as a term separated from infiltration. Thus, the interception losses are lumped with the infiltration, and when subtracted from the gross precipitation, provide an estimate of the precipitation excess (Smith and Chery 1973).

Literature Cited

Amisial, R.A., J.P. Riley, K.G. Renard, and E.K. Israelson. 1969. Analog computer solution of the unsteady flow equation and its use in modeling the surface runoff process. Utah Water Res. Lab., Utah State Univ., Logan, Rep. PRWG 38-2.

Beall, H.W. 1934. Penetration of rainfall through hardwood and softwood forest canopy. Ecology 15: 412-415.

Burgy, R.H., and C.R. Pomeroy. 1958. Interception losses in grassy vegetation. Trans. Amer. Geophys. Union, 39: 1095-1100.

Clark, O.R. 1940. Interception of rainfall by prairie grasses, weeds, and certain crop plants. Ecol. Monogr. 10: 243-277.

Clary, W.P., and P.F. Ffolliott. 1969. Water holding capacity of ponderosa pine forest floor layers. J. Soil Water Conserv. 24: 22-23.

Collings, M.R. 1966. Throughfall for summer thunderstorms in a juniper and pinyon woodland, Cibecue Ridge, Arizona. U.S. Geol. Surv. Prof. Pap. 485B. 13 p.

Connoughton, C.A. 1934. The accumulation and rate of melting snow as influenced by vegetation. J. Forestry 33: 564-569.

Corbett, E.S., and R.P. Crouse. 1968. Rainfall interception by annual grass and chaparral. USDA, Forest Serv. Res. Pap. PSW-48. 12 p.

Couturier, D.E., and E.A. Ripley. 1973. Rainfall interception in mixed grass prairie. Can. J. Plant Sci. 53: 659-663.

Crouse, R.P., E.S. Corbett, and D.W. Seegrist. 1966. Methods of measuring and analyzing rainfall interception by grass. Int. Ass. Sci. Hydrol. Bull. 11: 110-120.

Dunford, E.G., and H.C. Niederhof. 1944. Influence of aspen, young lodgepole pine, and open grassland types upon factors affecting water yield. J. Forestry 42: 673-677.

Garcia, R.M., and C.P. Pase. 1967. Moisture-retention capacity of litter under two Arizona chaparral communities. USDA, Forest Serv. Res. Note RM-85, 2 p.

Goodell, B.C. 1963. A reappraisal of precipitation interception loss by plants and attendant of water loss. J. Soil Water Conserv. 18: 231-234.

Grah, R.M., and C.C. Wilson. 1944. Some components of rainfall interception. J. Forestry 42: 890-898.

Hamilton, E.L., and P.B. Rowe. 1949. Rainfall interception by chaparral in California. U.S. Dep. Agr. and State of California Dep. Natur. Resources Div. Forest. Unnumbered Pub. 43 p.

Haynes, J.L. 1940. Ground rainfall under vegetation canopy of crops. J. Amer. Soc. Agron. 32: 176-184.

Helvey, J.D. 1971. A summary of rainfall interception by certain conifers of North America. p. 103-113. *In:* Proc. Third International Seminar for Hydrol. Prof., Biological Effects in the Hydrol. Cycle, Purdue Univ., West Lafayette, Ind., July 18-30.

Helvey, J.D., and J.H. Patric. 1965. Canopy and litter interception of rainfall by hardwoods of eastern United States. Water Resour. Res. 1: 193-206.

Hewlett, J.D., and W.L. Nutter. 1969. An outline of forest hydrology. Univ. of Georgia Press, Athens. 137 p.

Hicks, W.I. 1943. Interception tests on mountain-desert shrubs. Rep. to Storm Drain Div., Los Angeles City Eng. 4 p. (Original not seen, cited by Zinke (1967)).

Hull, A.C., Jr., and G.J. Klomp. 1974. Yield of crested wheatgrass under four densities of big sagebrush in southern Idaho. U.S. Dep. Agr. Tech. Bull. 1438. 38 p.

Kittredge, J., Jr. 1939. The forest floor of the chaparral in San Gabriel Mountains, California. J. Agr. Res. 58: 521-535.

Kittredge, J., Jr., H.S. Loughead, and A. Mazurak. 1941. Interception and stemflow in a pine plantation. J. Forestry 39: 505-522.

plantation. J. Forestry 39: 505-522.

McMillan, W.D., and R.H. Burgy 1960. Interception loss from grass. Trans. Amer. Geophys. Union Res. 65: 2389-2394.

Merriam, R.A. 1961. Surface water storage on annual rye grass. Trans. Amer. Geophys. Union Res. 66: 1833-1838.

Miner, N.H., and J.M. Trappe. 1957. Snow interception, accumulation, and melt in lodgepole pine forests in the Blue Mountains of eastern Oregon. USDA, Forest Serv. Res. Note PNW. 4 p.

Moul, E.T., and M.F. Buell. 1955. Moss cover and rainfall interception in frequently burned sites in the New Jersey pine barrens. Torrey Bot. Club Bull. 82: 155-162.

Murphy, C.E., Jr., and K.R. Knoerr. 1975. The evaporation of intercepted rainfall from a forest stand: an analysis by simulation. Water Resour. Res. 11: 273-280.

Rowe, P.B. 1948. Influence of woodland chaparral on water and soil in central California. U.S. Dep. Agr. and California Dep. Natur. Resources, Div. Forest unnumbered Pub. 70 p.

Rowe, P.B., and E.A. Colman. 1951. Disposition of rainfall in two mountain areas of California. U.S. Dep. Agr. Tech. Bull 1048.

Rowe, P.B., and T.M. Hendrix. 1961. Interception of rain and snow by second-growth ponderosa pine. Trans. Amer. Geophys. Union, 32: 903-908.

Skau, C.M. 1964. Interception, throughfall and stemflow in Utah and alligator juniper cover types of northern Arizona. Forest Sci. 10: 283-287.

Smith, R.E., and D.L. Chery. 1973. Rainfall excess model from soil moisture flow theory J. Hydrol. Div. ASCE 99(HY9): 1337-1351.

Townsend, T.W. 1966. Plant characteristics in relation to the desirability of rehabilitating the *Arctostaphylos patula-Ceanothus velutinus-Ceanothus prostratus* association on the east slope of the Sierra Nevada. M.S. Thesis, Univ. Nevada 85 p.

West, N.E., and G.F. Gifford. 197 . Rainfall interception by cool desert shrubs. J. Range Manage. 29: 171-172 .

Wilm, H.G. 1943. The application and measurement of artificial rainfall types of FA and F infiltrometers. Trans. Amer. Geophys. Union, 24, Pt. II: 480-487.

Wood, O.M. 1937. Interception of precipitation in oak-pine forest. Ecology 18: 251-254.

Zinke, P.J. 1967. Forest interception studies in the United States. p. 137-161. *In:* Int. Symp. on Forest Hydrol. (Penn. State Univ.) Proc. Aug. 29-Sept. 10, 1965. p. 137-161.

Chapter 4

Infiltration

Once a raindrop reaches the soil surface, it must ultimately infiltrate the soil, evaporate, or become a part of overland flow and eventual runoff. The actual role of infiltration in the hydrologic cycle was probably first recognized by Horton (1933). He referred to infiltration as the movement or passage of water through the soil surface. Once in the soil, water movement is defined as percolation. The rate at which water can enter the soil is dependent on many factors, among which may be litter and rock cover, vegetal canopy coverage, surface crusting, rainfall energy, quantity of coarse material in soil surface, degree of dispersion of surface soil by sodium particle coatings, slope, soil texture, bulk density, and capillary force patterns. Other factors are percent organic matter, parent material, microbiological activity, wind action, temperature of soil, air, and water, shrinking and swelling of colloids, season of year, successional state and age of vegetative cover, exchangeable ions present, silicasesquioxide ratio, degree of aggregation, moisture content of soil, and soil structure (Fig. 4-1). These relationships are shown in Figure 4-1.

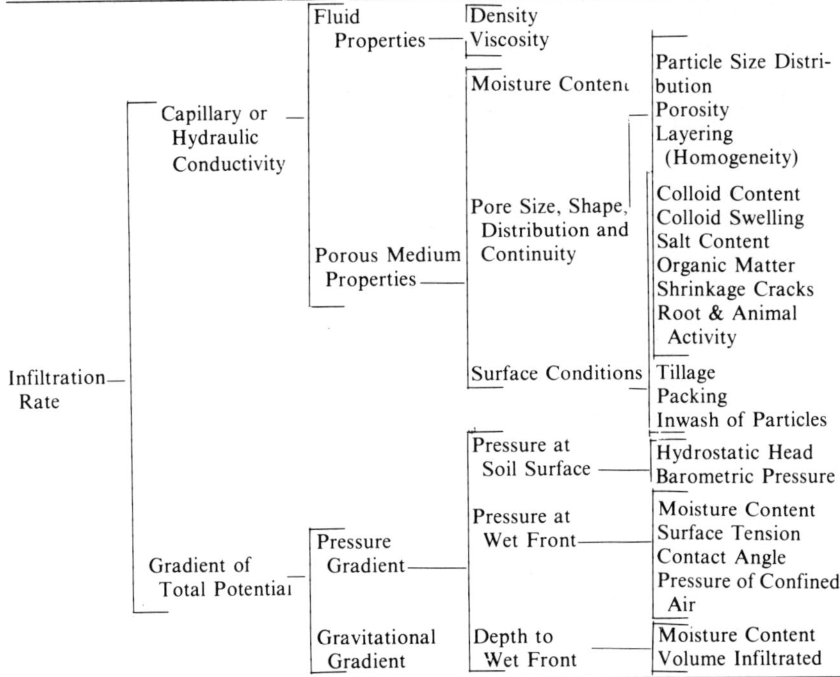

Fig. 4-1. Factors affecting the infiltration rate into unfrozen soil. (Data from Gray, Norum, and Murray 1969).

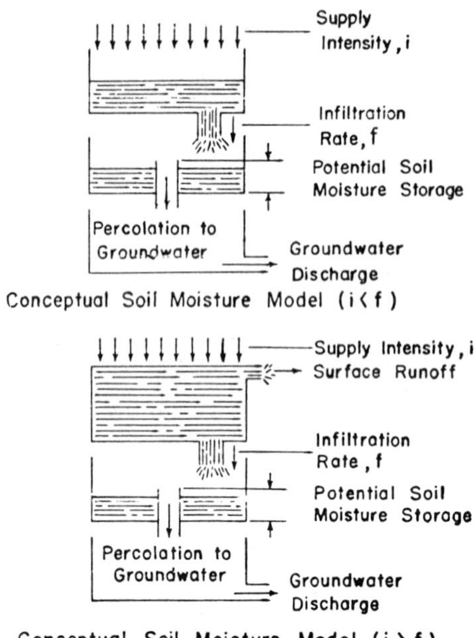

Fig. 4-2. (Top) The supply intensity, *i*, is less than the maximum rate at which the soil in its given condition can absorb water (*i*>f); hence, total supply goes to replenishing the soil moisture reservoir and recharging the ground water supply (neglecting evaporation and interception losses). (Bottom) In this case *i*>*f* and some water accumulates on the surface and appears as surface runoff (from Gray, Norum, and Murray 1969.

Without going into the physics of flow through porous media, suffice it to say that movement of water in the liquid phase in unsaturated soils takes place through soil moisture films, or in small pores through the combined action of gravitational and capillary forces. Therefore, as may be seen in Figure 4-2. any factor in the soil system that affects the magnitude of the capillary conductivity and the potential gradient (sum of the two terms gravitational potential and capillary potential) will likewise affect the infiltration rate.

Horton recognized a maximum and minimum infiltration capacity. The maximum capacity for any given rainstorm occurs at the beginning of the storm. This rate (quantity of water absorbed by the soil per unit of time) decreases rapidly at first because of changes in the structure of the surface soil and because of increases in surface soil moisture. Minimum infiltration approaches the percolation rate of the soil profile (Fig. 4-3).

The shape of the infiltration rate curve may be illustrated by the work of Beutner, Gaebe, and Horton (1940), who used a sprinkler apparatus on a variety of Arizona desert soils to study infiltration for different slopes, soils, and durations of rainfall. All of the infiltration curves had similar shapes, which began with high values and declined rapidly for the first 10 minutes, after which they continued to decline slowly until a nearly constant infiltration rate was reached. Final infiltration rates on "dry" runs (rainfall applied to soil in dry condition) varied from 0.28 to 2.22 inches/hr (71 to 564 mm/hr)

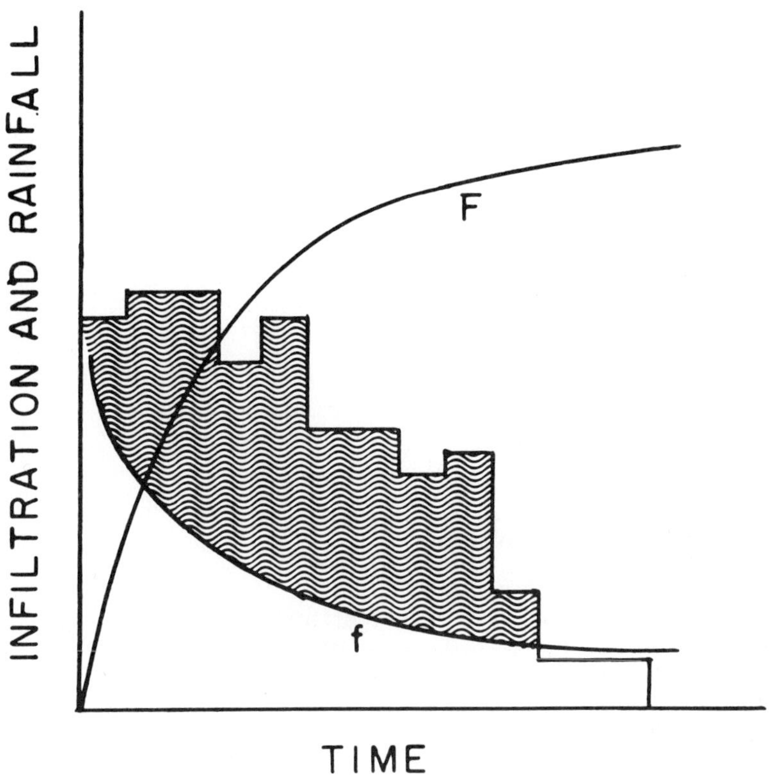

Fig. 4-3. ***F*** **represents the cumultive quantity of water entering the soil, and** ***f*** **represents infiltration rate that becomes relatively constant after the soil profile has become wetted. The wavy hatched area represents the excess precipitation that is available for runoff.**

and on "wet" runs (soil at field capacity at the start of the run) from 0.18 to 1.28 inches/hr (46 to 325 mm/hr).

During infiltration, distribution of water content with depth in the soil is referred to as the moisture profile, which possesses the features shown in Figure 4-4 (Bodman and Colman 1943). Figure 4-4 is rather idealistic with regard to many rangeland situations in that it applies where the rate of water entry through the soil surface is not limiting under a ponded water surface. However, the concept is useful. In essence the saturation and transition zone is the zone in which the soil is saturated. The transmission zone is an ever-lengthening unsaturated zone of fairly uniform water content. The wetting zone joins the wetting front and the transmission zone. The wetting front is a rather sharp line indicating where the soil changes from wet to dry. The wetting front marks the lower boundary of the wetting zone.

Infiltration under Field Conditions

How various factors interact to control the infiltration process under

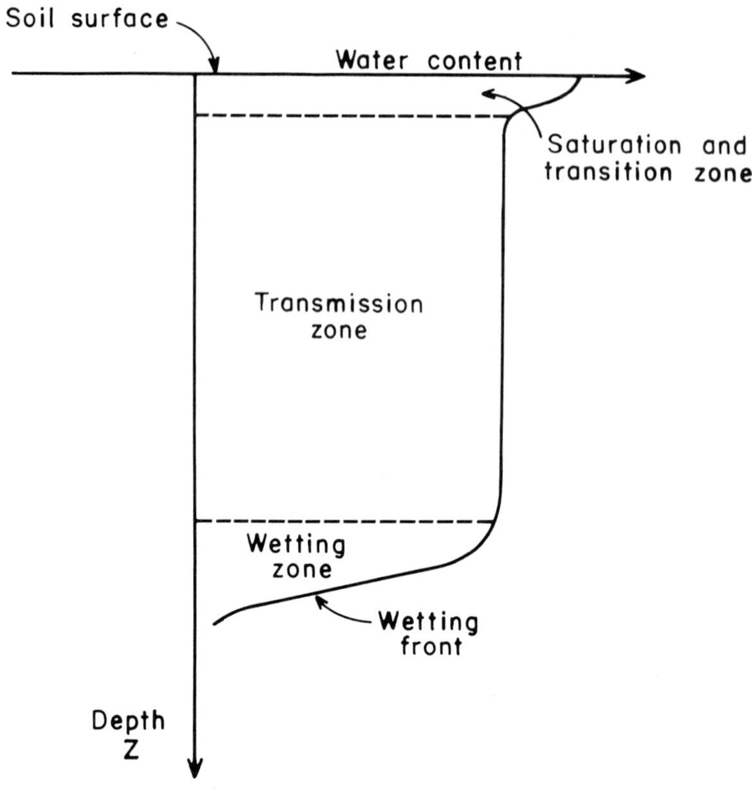

Fig. 4-4. Zones of the moisture profile during infiltration (from Bodman and Colman 1943).

rangeland conditions is not well understood. One idea, as an example, for controlling water infiltration into soils is the air-earth interface (AEI) concept which holds that interfacial roughness and openness control the rates and routes of water infiltration by governing the flow of air and water in underlying macropore and micropore systems (Dixon 1975). Roughness refers to the microrelief that produces depression storage, whereas openness refers to the macroporosity that is visible at the soil surface. Macropores include those voids produced by clay shrinkage, tillage, earthworms, roots, internal erosion, ice lenses, pebble dissolution, and entrapped gas. In contrast, the micropore systems includes the spaces within and between individual soil aggregates (textural and structural pores or simple and compound packing voids) that fill and drain largely by capillarity (Figure 4-5).

The AEI concept embodies six physical interfacial models (Figure 4-6) representing two degrees of surface roughness and three degrees of surface openness. The subterranean part of the macropore system is depicted as a single U-shaped tube to graphically reflect its infiltration role as a water-intake air-exhaust circuit. Models RO, RP, and RC represent rough interfaces with open, partly open (unstable), and closed macropore interfacial

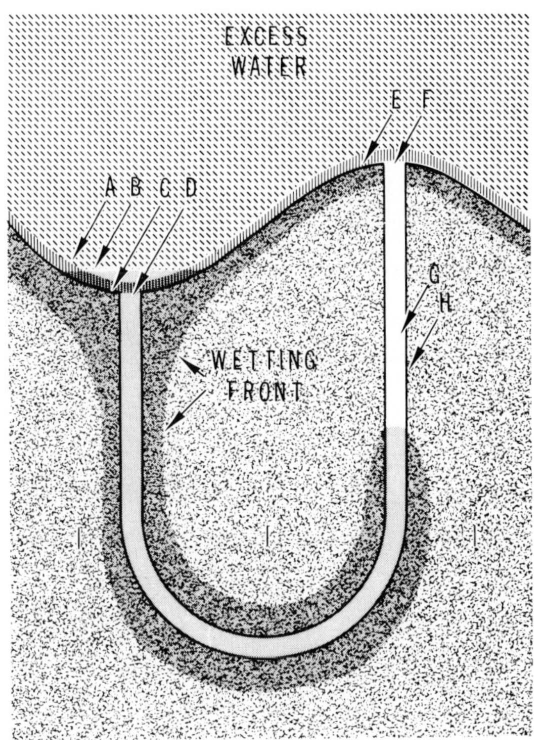

Fig. 4-5. Soil model containing a micropore system and a macropore system. A = plant residue cover on air-earth interface; B=free water surface; C=microdepression in air-earth interface; F=soil air exhaust port of macropore; G=interface; D=water intake port of macropore; E=microelevation in air-earth interface; F=soil air exhaust port of macropore; G=macropore space; H=macropore wall; and I=micropore space (from Dixon and Peterson 1971).

openings or *macroports,* respectively; whereas models, SO, SP, and SC represent smooth interfaces containing open, partly open, and closed macroports.

Under the rough open surface of model RO, interfacial exchange of air and water occurs freely and water infiltrates rapidly via the relatively short broad straight paths of the macropore system, whereas under the smooth closed surface of model SC, surface exchange of air and water is greatly impeded and water infiltrates slowly via the relatively long narrow tortuous paths of the micropore system. Infiltration under these model extremes often differs by more than an order of magnitude. The generaly hydraulic behavior of these models may be deduced by ranking them with respect to various properties or characteristics. By definition, interfacial roughness and depression storage rank in the order RO=SO>RP=SP>RC=SC. Also by definition, interfacial openness or physical continuity of the interface and macropores rank in the same order.

According to Dixon (1975), the major significance of the AEI concept lies

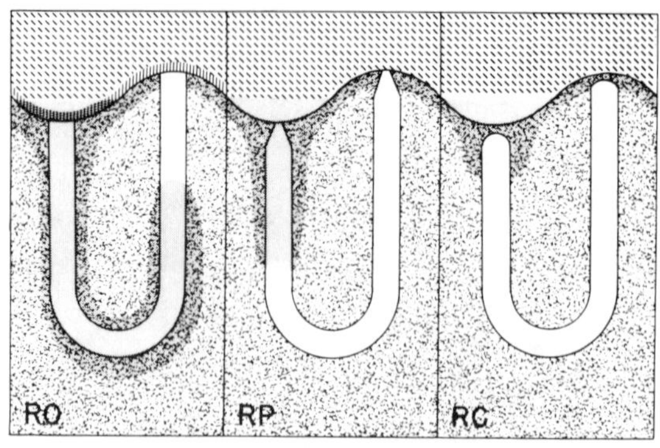

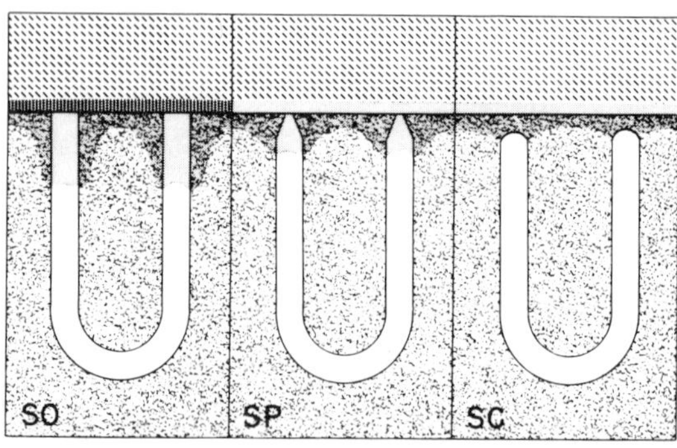

Fig. 4-6. Air-earth interface models and associated U-shaped macropore for water infiltration into soils. Models RO, RP and RC represent rough interfaces containing open, partly open (unstable) and closed macroports, respectively; whereas models SO, SP and SC represent smooth interfaces containing open, partly open (unstable) and closed macroports. (from Dixon and Peterson 1971).

in its potential for practical field application. Since the soil surface controls the rate and route of water movement into, within, and through the soil, soil and water management practices which appropriately alter this surface can be used to control various infiltration-related problems. Surface management practices can be directed to changing the existing interface into the desired one by means of the *transformation* processes shown in Figure 4-7. For example, interface RO is changed to SC by the *exposing-smoothing-sealing* sequence of processes. Although the transformation processes often occur naturally, their rates may be controlled by appropriate cultural practices. For instance, *exposing* of the soil surface may occur slowly through biological decomposition of plant residue, or very rapidly via cultural practices such as burning and moldboard plowing. Similarly *covering* of a barren soil in a semiarid region may be achieved rapidly with combinations of cultural practices such as irrigation, fertilization, and mulching. The transformation processes are general processes which include many specific processes such as

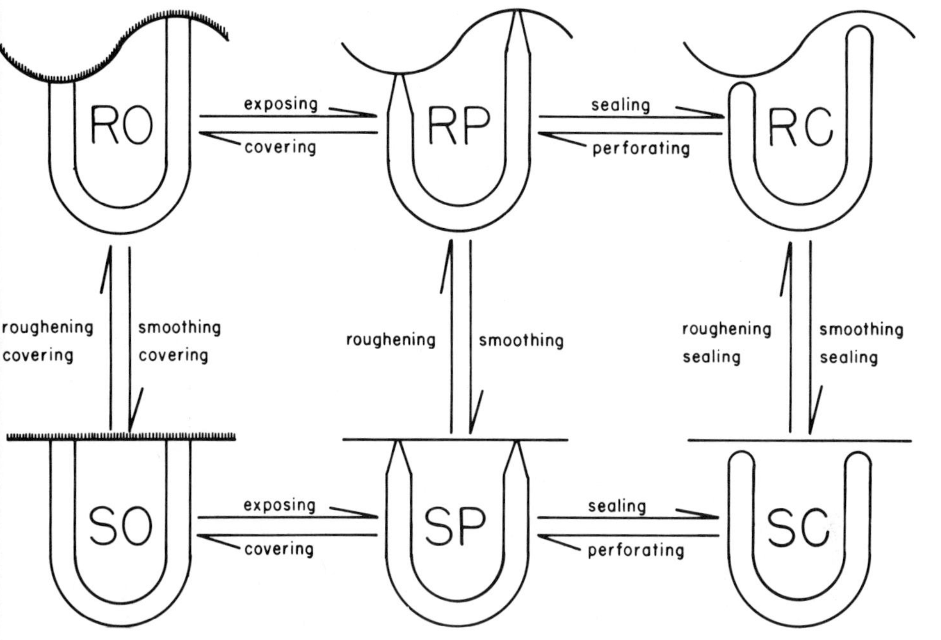

Fig. 4-7. Air-earth interface cycles and transformation processes (from Dixon and Peterson 1971).

physical, chemical, and biological processes and man-imposed processes (cultural practices).

In essence then, infiltration rates as a function of various natural or modified plant-soil complexes are of direct concern to watershed managers. Theoretical analyses of infiltration of water into soil have not generally been attempted for conditions as found in most rangeland situations. Such analyses are usually made for physical conditions represented in terms of a vertical column of unsaturated porous material to which water is applied at the top, usually by ponding; but differences in soil surface conditions, initial soil water content, and natural variation between field plots make approximate infiltration theory equations (Philip 1957; Green and Ampt 1911) or empirical equations (Horton 1940; Holtan, Kirkham, and Nielsen 1967) difficult to apply under wildland situations. Several infiltration equations and general conditions to which they may apply are shown in Figure 4-8. Figure 4-9 is an example of an attempt to apply three of the equations to infiltrometer data collected from a variety of pinyon-juniper (*Pinus sp. -Juniperus* sp.) sites in central and southern Utah. As may be seen, Horton's equation best explained the variance associated with various infiltration measures on given sites, yet potential usefulness of this equation may be questioned. Because of the extremely complex profiles found on natural watersheds, it may well be that arbitrary decay functions subject to intuitively reasonable constrains and boundary conditions are one answer to overcoming the problem of character-

	Horton	Kostiakov	Holtan-Overton	Philip	Natural Conditions
Formula	$f = f_c + (f_o - f_c)e^{-K_f t}$	$f = abt^{(a-1)}$	$f = f_c + aF_p^2$	$f = \frac{1}{2}St^{-12} + A$	
Type	Empirical	Empirical	Empirical	Analytic	
Schematic flow field	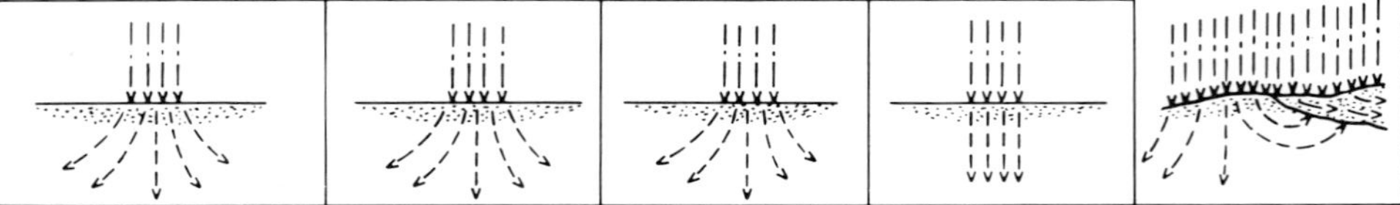				
Area of inflow	Small	Small	Small	Unlimited	Large
Medium	Relatively Homogeneous	Relatively Homogeneous	Relatively Homogeneous	Homogeneous	Heterogeneous
Flow field	Approximately Axi-symmetric Three-dimensional	Approximately Axi-symmetric Three-dimensional	Approximately Axi-symmetric Three-dimensional	Uni-dimensional	Complex Three-dimensional

Fig. 4-8. A qualitative comparison between the domain of application of various infiltration equations and conditions likely to occur in a natural basin (from Amorocho 1967). In the formulae:

A = function depending on the hydrologic conductivity of the soils at time O (initial conductivity) and time t, a = constant depending on the soil vegetation complex, alpha = a constant which may vary approximately between 0.2 and 0.8,

f = infiltration rate at time t, fc = ultimate infiltration rate (t = oo),

f_o = initial infiltration rate, F=potential infiltration rate at time t,

K_f and b are positive constants depending on soil characteristics,

t = time since the beginning of infiltration, and S=sorbtivity.

izing watershed infiltration (Amorocho 1967). For further information the reader is referred to the recent review by Amerman et al. (1975).

Methods for Measuring Infiltration

Many methods have been used to measure infiltration rates (Parr and Bertrand 1960). These may be categorized as plot studies (lysimeters, small flooded plots in confined areas, and sprinkling devices) and watershed studies. All of these methods have produced useful information, but extreme care must be exercised in attempts to use infiltration rate estimates in predicting runoff or flood flows from watersheds. An infiltration (or runoff) study might involve one of three approaches: (1) a sprinkler type rainfall applicator; (2) an artificial flooding or ponding approach; or (3) infiltration under natural rainfall. If a sprinkler is used, it might be any one of a dozen different kinds which produce various drop sizes, intensities, and areal coverage; the

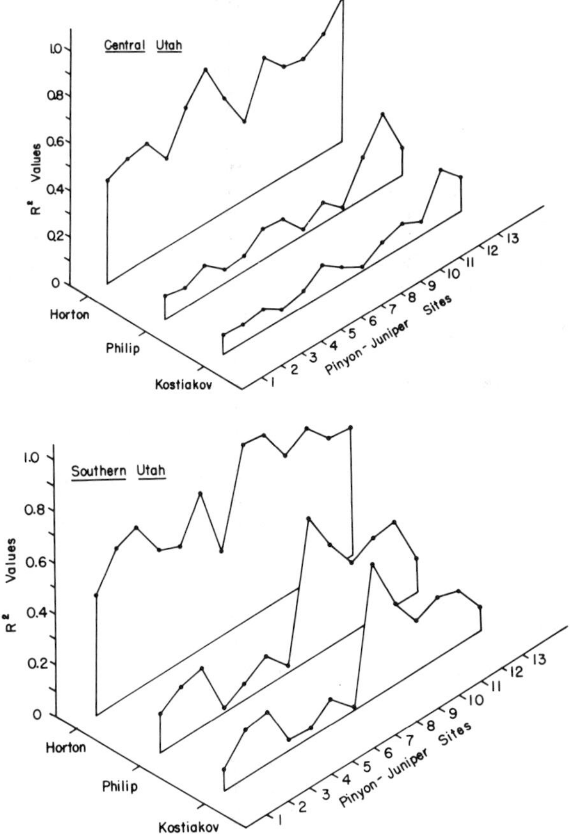

Fig. 4-9. ***R^2*** **values indicating amount of variance explained by three infiltration equations using infiltrometer data from 13 pinyon-juniper sites in southern Utah (bottom) and 13 pinyon-juniper sites in central Utah (top). Each "set" of data to which the equations were fit represented from 8 to 30 replications (from Gifford 1976).** ***R^2*** **= coefficient of determination.**

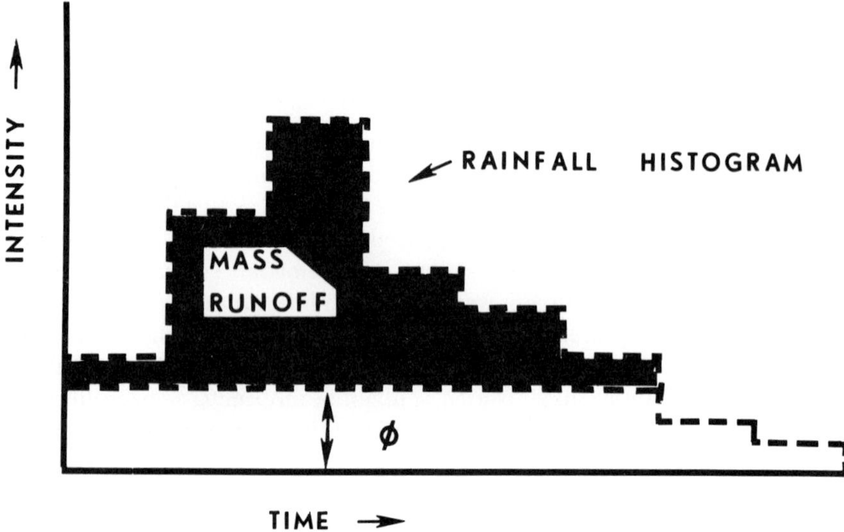

Fig. 4-10. A diagram of the phi (ϕ) index of infiltration which is the average rainfall intensity above which runoff occurs.

latter often affects size of plot used. Sprinkler type infiltrometers are used in an effort to duplicate natural falling raindrops from a particular type storm and thereby obtain a more realistic measure of infiltration.

The flooding or ponding approach, e.g., with ring infiltrometers, is generally a poor technique if some approximation to infiltration under natural rainfall is desired. Comparative differences, however, can be obtained by this method and the technique is probably suited for instances where natural flooding does occur, as in flood irrigation or within contour trenches.

Infiltration rates under natural rainfall can be derived from data on small plots or natural watersheds. Probably the most meaningful estimates of infiltration rates for flood prediction are obtained from watershed studies where rainfall intensity above which runoff occurs is considered an index of infiltration rate (Langbein et al. 1947). Two elements not considered in this index (call the phi [ϕ] index) are interception quantities and depression storage. Figure 4-10 shows the relationship between mass runoff and the phi index. The phi index is not a constant, rather it varies with rainfall duration and intensity but approaches an *f*-average value for storms of long duration.

Infiltration measurements are useful in the evaluation of magnitudes of certain causative factors such as range condition, soil texture, and vegetation types. However, limited attempts to predict runoff from infiltration data have been largely unsuccessful. This statement is not surprising when one considers that precipitation intensity and amount represent only two of many factors affecting quantities of water entering soils in a watershed. One such additional factor is slope shape, i.e., convexity or concavity of the slope. A convex slope will have less infiltration than a concave slope receiving the same precipitation because water runs off the convex slope and runs into the concave slope or moves more slowly across it. Duration of flow over a site (run-in vs. runoff sites) also affects the quantity of water entering the soil and results in greater water storage at the foot of a long slope than at the top, even though infiltration rates may be similar or the same for upper and lower slopes. Alluvial sites in lowlands may have low infiltration rates because of small soil particle size, but may store moisture in amounts that exceed precipitation applied because of duration of flow over them. These points are illustrated in Figure 7-10 for 14 plant communities in the Willow Creek watershed in northern Montana; the lowland western wheatgrass community had one of the lowest infiltration rates but was the wettest site because of duration of flow.

Numerous devices have been designed to study infiltration under natural and artificial conditions (Pearse 1936; Pearse and Bertelson 1937; Arend and Knight 1940; Rowe 1940; Wilm 1943; Evanko 1950; Friedrich 1951; Dortignac 1951; Bondurant 1957; Barnes and Costel 1957; Adams, Kirkham, and Nielsen 1957; Krimbold and Shahori 1957; Packer 1957; Meyer and McCune 1958; Rauzi 1960; Rhoades 1961; Johnson 1963; McQueen 1963; Chow and Harbaugh 1965; Steinhardt and Hillel 1966; Morin 1967; Costin and Gilmour 1970; Meeuwig 1971; Blackburn, Meeuwig, and Skau 1974; and Burton 1976). Because of space limitations it is impossible to discuss all the various pieces of infiltrometer equipment. However, equipment used should duplicate as closely as possible conditions that actually occur in nature during the period of the year when sampling takes place.

Infiltration in Natural Rangeland Plant Communities

On the arid and semiarid rangelands of the western United States, it is difficult to discuss rangeland plant communities without considering the influence of soil characteristics and parent material on a particular plant community. Pearse and Wooley (1936) indicated a close relationship between the absorption rate of water by the soil and the presence and character of the plant cover on various soils in the Boise River watershed. In general, plots bearing a single plant absorbed water 0.03 inch/min (0.76 mm) faster than the adjacent barren plots (same soil type), an increase of 71%. An increase in infiltration of 127% occurred when fibrous-rooted species were present, but there was only a 51% increase with taprooted species.

Data from four watersheds near Newell, South Dakota, showed that infiltration rates were correlated with range sites (delineated primarily by soil type) as mapped by the Soil Conservation Service where range condition class was comparable. In this study, all sites were classified in good condition. Water intake during the first 15-minute period of a 1-hour test was high (1.52 to 2.04 inches/hr 386 to 518 mm/hr), even on thin or fine-textured soils. However, the intake rate declined more rapidly on such sites in later periods of a test except on thick well-structured clay which maintained a rate comparable to a sandy loam. Average infiltration rates after 1 hour for various sites were: sandy, 1.52 inches/hr (386 mm/hr); panspots, 0.36 inch/hr; (91 mm/hr); dense clay, 0.40 inch/hr (102 mm/hr); dense clay, 0.40 inch/hr (102

Table 4-1. Average water losses (percent of water applied[1]) on various sites by range conditions, (from Osborn 1950).

Site	Range condition: Excellent (%)	Good (%)	Fair (%)	Poor (%)	Bare (%)	All conditions (%)
Deep soils						
Grand Prairie[2]	15	35	17	28		23
Edwards Plateau[2]		30	48	65	46	49
Trans-Pecos[3]			33	84	79	52
Trans-Pecos[4]			3	59	81	58
Rolling Red Plains[5]		44	88		88	62
Rolling Red Plains[6]			53	70	78	63
Rio Grande Plain[7]			40	59	53	54
Trans-Pecos[8]			60	80	71	68
High Plains[8]		26	80	70	72	63
Rolling Red Plains[9]	8	54	66	62	79	56
Shallow soils						
Rolling Red Plains[10]	74	76	81	75	73	76
Cross Timbers[11]	50	56	66	69	67	60

[1]Approximately 2.00 inches (508 mm) in 20 minutes
[2]Deep upland site.
[3]Clay-loam flats site.
[4]Clay-loam upland site.
[5]Red shale hills, draws site.
[6]Heavy deep upland site.
[7]Red sandyland site.
[8]Sandy loam upland site.
[9]Deep sand site.
[10]Red shale hills, thin upland site.
[11]Shallow sandy loam upland site.

mm/hr); and clayey, 1.26 inches/hr (320 mm/hr) (Rauzi and Kuhlman 1961). Elsewhere Osborn (1950) summarized the findings of the first year of various field studies in Texas and Oklahoma. Table 4-1 summarizes average water losses by site and range condition as determined with a mobile raindrop infiltrometer.

Rauzi, Fly, and Dyksterhuis (1968) used a mobile raindrop infiltrometer and 2 ft² (30.5 cm²) plots to sample many sites within both the shortgrass plains and the tall-grass prairie. Analysis of results was based on the second 30-minute period of 1-hour infiltrometer runs. Water-intake rates at their extremes were lowest on the range sites characterized by fine-textured dispersed soils, and highest on the range sites characterized by coarse-textured soils (Table 4-2). Simple correlations between water-intake rates during the second 30-minute period showed that soil structure of the first horizon was highly correlated with water intake. Texture of the second horizon and nature of the boundary of the first horizon were of lesser importance. Among all variables measured, however, the amount of both new and old vegetation showed greatest general correlation with water-intake rate.

Other investigators, using various methods of measuring infiltration and working on different range types, have concluded that the quantity of living plant material and associated litter is more significantly correlated with infiltration than any other variables that have been measured (Duley and Kelly 1939; Duley and Domingo 1949; Osborn 1952; Meeuwig 1970). The degree of correlation for a given site may vary however, depending on season

Table 4-2. Rate of water intake, amount of vegetal cover, percentage of bare ground, and number of observations for various range-soil groups in short grass plains and tall grass prairie.

Range sites	Rate of water intake, second 30-minutes (inches/hr)	Vegetal cover (air dry) Midgrass (lb/acre)	Herbage (lb/acre)	Mulch (lb/acre)	% Bare ground	Number of observations
All range-soil groups	1.35	929	1,319	1,139	38	670
Alkali upland	.29		683	346	70	59
Dense clay	.50	1,003	1,108	445	59	59
Panspot	.90	768	1,317	528	46	29
Shallow complex	1.26	901	1,379	826	47	69
Clayey	1.45	736	1,154	993	32	138
Silty	1.46	1,136	1,487	1,665	26	218
Sandy	1.67	885	1,319	1,119	35	83
Overflow	2.31	2,905	2,942	2,406	14	11
Sand	3.13	997	1,785	1,412	36	23

Data from Rauzi, Fly, and Dyksterhuis (1968). (Multiply lb/acre by 1.12 to obtain kg/ha).

of year and length of time since beginning of rainfall during a given storm (Gifford 1972). Where the potential for plant growth is limited, as for instance on semiarid rangeland sites, the relationship between plant cover (and/or litter) and infiltration rates is not so well established (Gifford 1968; Williams and Gifford 1969; Blackburn 1975). In these xeric plant communities the relatively sparse plant cover interacts with numerous other factors in determining ultimate infiltration rates.

The physical and biological conditions of the soil that promote good cover conditions also influence the infiltration rate (Osborn 1952). At the Davis County Experimental Watershed in Utah, Meeuwig (1970) found that in addition to plant and litter cover, bulk density of the surface 4 inches (10 cm) of soil, particles and aggregates larger than 0.5 mm in the surface inch of soil, initial moisture content of the surface 2 inches of soil, and air-dry weight of vegetation accounted for 80% of the variance in the amount of water retained on his Rocky Mountain infiltrometers plots. Soil properties were particularly important for predicting infiltration rates on a deteriorated subalpine range with decomposed granitic soils in central Idaho (Meeuwig 1969).

Lyford and Qashu (1969) used a double-ring infiltrometer to evaluate the influence of paloverde *(Cercidium microphyllum)* and creosotebush *(Larrea tridentata)* on infiltration rates at different radial distances from plant centers. Infiltration of water averaged nearly three times greater under these plants than in the openings. Bulk density was lower and organic matter content was higher in topsoil under the plants than in the openings.

Several investigators have compared infiltration rates between plant communities. Dee, Box, and Robertson (1966) found that infiltration rates in plant communities on the same soil series increased with increasing position of a plant in the successional scale, the state of succession of the community, the amount of standing vegetation, and accumulated litter from previous years. Using ring infiltrometers these investigators found that soil under blue grama *(Bouteloua gracilis)* absorbed the most water, followed in order by soils under windmillgrass *(Chloris verticellata),* annual forbs and buffalograss *(Buchloe dactyloides).* In Texas, Thomas and Young (1954) found that initial infiltration rates were highest on tobosagrass *(Hilaria mutica)* sod, followed by buffalograss sod, curlymesquite sod *(Hilaria belangeri)*, and bare ground. However, the total amount of infiltration at the end of 2 hours was greatest on tobosagrass sod and bare ground and was significantly lower on buffalograss and curlymesquite sod. Infiltration was greater for a buffalograss community than a western wheatgrass *(Agropyron smithii)* community on soil derived from Pierre Shale near Denver (Branson, Miller, and McQueen 1965).

Smith and Leopold (1942) using a North Fork sprinkling infiltrometer measured the following infiltration rates for wet runs: 1.55 inches (394 mm)/hr for 24 grasslands, 1.24 inches (3.5 mm)/hr for 39 woodlands (pinyon-juniper), and 0.71 inches (180 mm)/hr for 63 desert shrub plots. The rates for desert shrub plots differed significantly from woodland and grassland.

Using a type-F sprinkling infiltrometer on selected plant-soil complexes in Utah, Woodward (1944) concluded that, in general, browse areas exhibited the greatest infiltration capacities, followed by sagebrush-pinyon-juniper and herbaceous rangeland areas. Box (1961) tested four plant communities occurring on three soil types in south Texas with the ring infiltrometer. Mesquite and chaparral communities occurred on clay, a prickly-pear community on clay loam, and a bunchgrass community on fine sand. Infiltration rates varied between communities and between conditions within a community. All vegetation improved water intake on the clay soil, but infiltration increases with vegetation increases were more rapid for grass than shrubs.

Infiltration rates may also vary with the time of year when measurements are made. In western Colorado on salt desert shrub range, seasonal variations were attributed to loosening and heaving of the soil surface by frost action in winter, followed by compaction of the soil surface in the spring and summer by raindrop impact (Schumm and Lusby 1963).

The physical effects of forest and meadow vegetation on soil freezing and snowpack characteristics, and the hydrologic effects of simulated winter rainstorms on snowmelt, infiltration, overland flow, and soil movement were studied by Haupt (1967). The plots were located 17 miles (27 km) northwest of Reno, Nevada, near Dog Creek, a tributary of the Truckee River. Results indicated that plant, litter, and snow cover generally dissipate raindrop energy and increase infiltration, but exposed rock usually accelerates overland flow and erosion. Soil losses from snowcovered plots, regardless of vegetative cover, were practically nil. Stalactite soil frost promoted infiltration while porous concrete frost usually reduced infiltration on burned or sparsely vegetated sites, but neither type of frost impaired infiltration where plant and litter cover were appreciable. A rapidly melting snowpack over soil containing dense frost may accelerate runoff.

Infiltration as Influenced by Grazing

Grazing animals remove protective plant material and compact the soil surface. Both of these actions affect the infiltration rate. In many instances grazing intensity and range condition are related. Previous grazing intensities are the cause of present range conditions, but present grazing intensity may vary from heavy to light for any range condition. Figures 4-10 and 4-11 are a graphic representation of data from Leithead (1959), Knoll and Hopkins (1959), Johnson (1962), Branson Miller, and McQueen (1962), Rhoades et al. (1964), Sharp et al. (1964), Rauzi Fly and Dyksterhuis (1968), Rauzi and Smith (1973), and Gifford (1978). The linear regression equations provide a means for quantifying the impact of one level of grazing (as based either on grazing intensity or range condition) relative to other grazing levels. Data used to construct the figures cover a wide range of both soils and vegetation types, and each point represents one point in time. Studies defining how long it takes to reach the conditions shown in Figure 4-11 are not available.

Elsewhere, trampling by animals reduced the intake rate on bluegrass land in Nebraska from a normal of 2.02 inches (51.3 mm)/hr to 0.14 inches (3.6 mm)/hr under "dry" conditions, and from a normal of 0.85 inches (216 mm)/hr to 0.13 inch (3.3 mm)/hr under "wet" conditions (Duley and Domingo 1949). Hendricks (1942) found that more effective use of rainfall can be had on semiarid rangelands if grazing is managed to allow for the accumulation of grass litter. Retarding runoff affords greater opportunity for rainwater infiltration.

At Badger Wash in western Colorado, average infiltration rates were slightly higher on grazed than ungrazed plots for the last 20 minutes of the infiltrometer runs (Lusby et al. 1971). This indicates that under certain conditions grazing has no appreciable effect on infiltration during the latter stages of extended rains. However, the initial quantity of water absorbed on ungrazed plots before runoff began was significantly higher than on grazed plots (Thompson 1968).

Tromble, Renard, and Thatcher (1974) found in southeastern Arizona that infiltration on selected rangeland sites was greater for brush-dominated plots than for either grazed plots or grass plots without grazing. Grazing intensities, however, were not defined nor were range condition ratings given. Recovery of infiltration rates following grazing pressures has received little attention. At the Manitou Experimental Forest (Colorado) protection from cattle grazing resulted in increased infiltration rates; the recovery period coverd 6 years on ponderosa pine-grass sites and 13 years on grassland sites. Infiltration rates of the soils on grassland and pine-grass sites could be estimated by measuring the quantity of dead organic material and noncapillary pores in the surface soil (Dortignac and Love 1961). In southwestern Wisconsin Sartz and Tolsted (1974) found that runoff from moderately grazed sloping pastures was reduced sharply within 3 years after grazing was stopped. Busby[1] has found in southeastern Utah on sandy loam soils in the pinyon-juniper type

---

[1]Busby, F.E., Unpublished data.

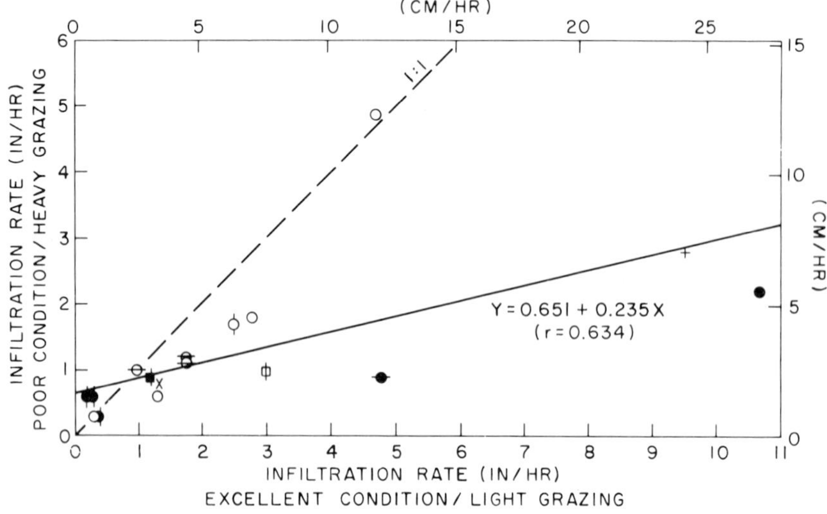

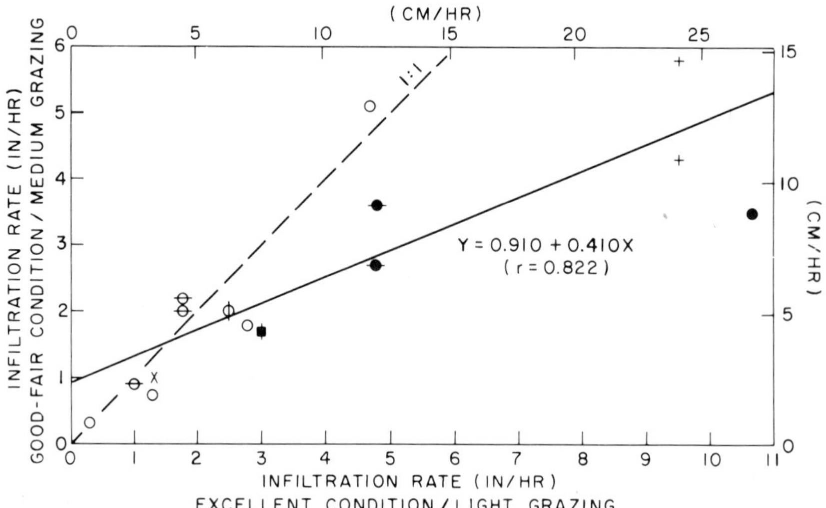

Fig. 4-11. Linear regression lines for infiltration as related to grazing intensity and range condition. Each symbol represents a different data source. Linear regression line for infiltration as related to grazing intensity and range condition.

that 4 years of protection is sufficient time for recovery of infiltration rates. Following complete protection, one grazing period is sufficient to again reduce infiltration rates (Buckhouse and Gifford 1976).

Effects of Fire and Water Repellency on Infiltration

Problems associated with water repellent soils have become increasingly important during the past few years because of the possible effects on infiltra-

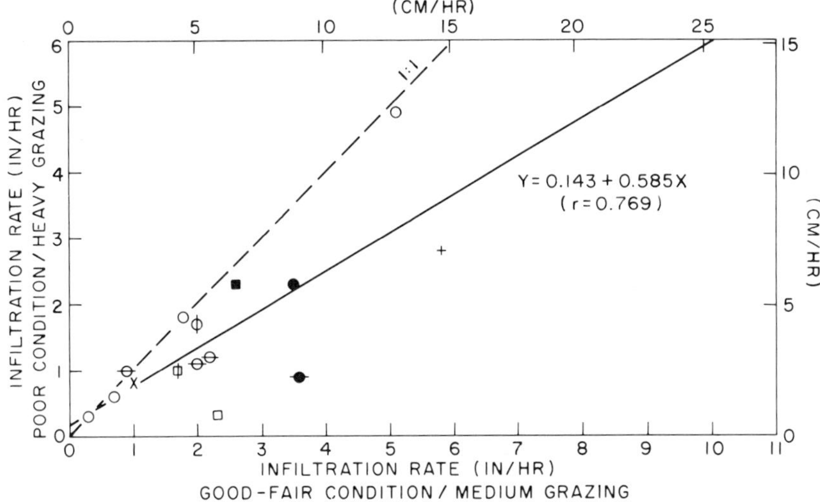

Fig. 4-11. Continued.

tion. Soil nonwettability is a widespread phenomenon that has been recognized on rangelands in the west and on burned brushland soils of southern California. Water repellent soils have been found under a variety of plant types and under widely different climatic conditions. In forested areas they have been observed under ponderosa pine *(Pinus ponderosa)*, subalpine fir *(Abies lasiocarpa)*, mountain hemlock *(Tsuga mertensiana)*, lodgepole pine, *(Pinus contorta)*, western white pine *(P. monticola)*, Jeffrey pine *(P. jeffreyi)*, spruce (*Picea* sp.), larch (*Larix* sp.), and Douglas fir *(Pseudotsuga taxifolia)*. In more arid climates they have been found under various species of oak (*Quercus* sp.), chaparral communities, and under juniper (*Juniperus* spp.), pinyon (*Pinus* spp.), sagebrush (*Artemisia* spp.), and rabbitbrush (*Chrysothamnus* spp.) (DeBano 1969).

DeBano and Rice (1973) proposed that both plants and microorganisms produce waxy organic substances that can make soils water-repellent. This condition has been documented in Australia as well as in many areas of the United States. These substances collect on the soil surface forming a moderately repellent surface, but because vegetation and litter are also present, infiltration rates are moderate. Eventually fire removes this protective layer of litter and vaporizes the hydrophobic substances. Some is lost upward, and some is distilled downward where it condenses on cooler soil particles forming a new, more intense water-repellent layer. Coarse-textured soils become more water-repellent than fine-textured ones since little coating material is needed for the relatively small surface area per unit volume.

Infiltration tests on pine *(Pinus jeffreyi)*, manzanita *(Arctostaphylos patula)*, and snowbush ceanthus *(Ceanothus velutinus)* sites near Reno, Nevada, revealed that apparent soil nonwettability was caused by pine litter, but the litter had compensating importance in physically restraining soil erosion and increasing the moisture holding capacity (Hussain, Skau and Meeuwig 1968). Chemical analysis of ethanol extractions from the above plants indicates that nonwettability may be associated with unsaturated aliphatic carboxylic acids with some simple sugar side-chain attachments composed chiefly of pentose and hexose sugars (lignin-type structure) (Morris and Natalino 1969). Savage et al. (1972) found essentially the same substances in their study of fire-induced water repellency.

Meeuwig (1971) studied water repellency with an infiltrometer on granitic soils near Lake Tahoe in Nevada. Wetting patterns of plots could be classified into eight general types, depending on size, continuity, and location of water-repellent zones (Figure 4-12).

For additional information on soil wettability problems the reader is referred to the brief literature review given in Scholl (1975).

Considering the length of time that fire has been used as a tool in manage-

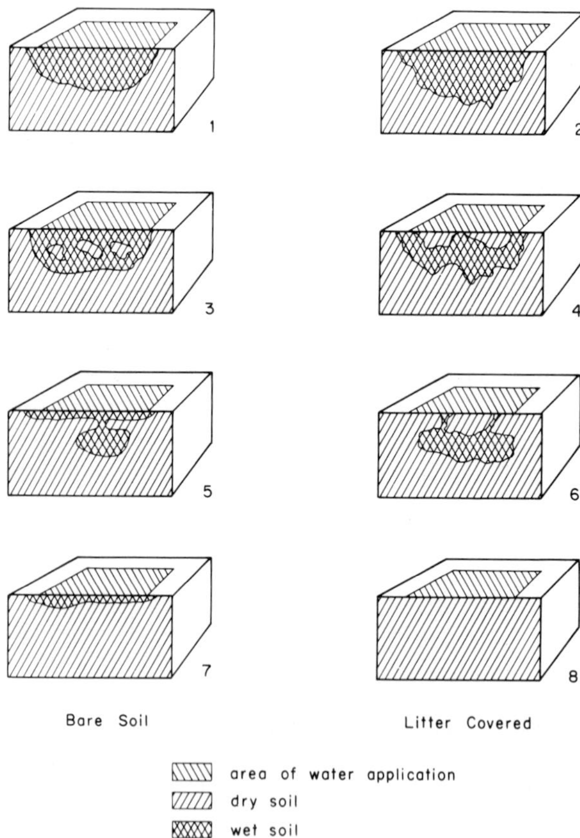

Fig. 4-12. Eight generalized wetting patterns. There was little or no runoff from the plots with wetting patterns 1, 2, or 3. Runoff from plots with pattern was generally slight because the wettable portions of the soil surface absorbed the runoff from the water-repellent portions. All plots with wetting patterns 5 through 8 produced appreciable runoff from Meeuwig 1971).

ment of rangeland resources, it is interesting that so little information is available regarding the hydrologic impact of this activity within range plant communities. Fire will influence infiltration by changing variables upon which infiltration seems dependent: characteristics of accumulated litter, soil surface area, structure, porosity, apparent liquid-solid contact angle, and solute concentration of the infiltrating water. Where fire is of high intensity and the organic covering of the soil is completely consumed or if moderate intensity fires are repeated, infiltration is generally decreased. Therefore, the effect of fire on infiltration is the difference between changes in factors that decrease water entry and those that increase it.

There are indications that initial site characteristics are extremely important to the ultimate hydrologic response to fire (Settergren 1967). This point is amplified by variable results from infiltration studies on burned sites. As examples, Scott (1956) found that relative infiltration rates were higher on burned areas immediately after the fire, and that this condition persisted one year later in all but one case. Other studies in California by Scott and Burgy (1956), Burgy and Scott (1952) and Veihmeyer and Johnson (1944) on different soils showed no decrease in infiltration rates as a result of burning. In contrast, Rowe (1948) found that repeated annual burning in woodland chaparall reduced the average infiltration rate under natural rainfall by 95%. Working in the Flint Hills region of Kansas, Hanks and Anderson (1957) reported that plots burned at various times of the year stored an average of 40% of the water from a 11.3-centimeter storm as compared to the unburned control plot, which stored 83% of the water from the same storm.

Settergren (1967) has found that when infiltration for burned sites versus that for unburned sites is plotted on logarithmic paper, a trend appears (Figure 4-13). In stratifying this data by soil textural type some trend toward a lessening effect of fire is apparent for the lighter soils (Figure 4-14). Data included in these two figures come from both forest and range plant communities. Settergren found that the response to fire of the infiltration capacity of soils varied as a function of normal infiltration rates, litter depth, the number of annual burns, and soil texture.

Effect of Other Range Improvement Practices on Infiltration

Plowing

Plowing is a mechanical technique used for eliminating undesirable plant species and also for seedbed preparation for desirable species.

Gifford and Skau (1967) looked at the first-year impact of plowing (moldboard plow) with drilling and plowing with contour deep furrow drilling on infiltration rates and potential sediment production rates on two big sagebrush *(Artemisia tridentata)* sites in Eastgate Basin, Nevada, about 192 km (119 miles) east of Reno. In general infiltration rates (measured with a sprinkling infiltrometer) for the two plowing treatments were significantly less than rates measured under undisturbed conditions. Sediment production potentials on the plowing treatments, however, were not significantly different from the natural sagebrush community. Sediment yields were highly

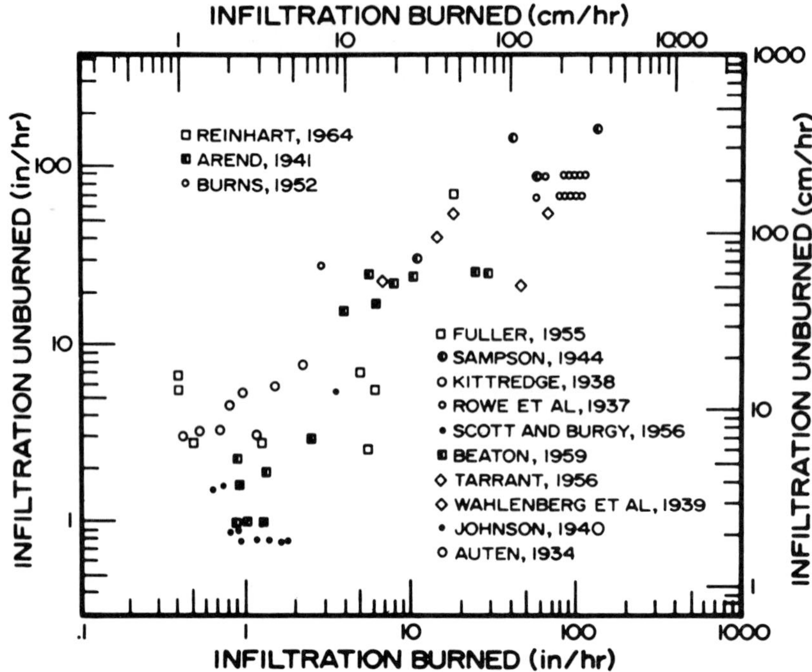

Fig. 4-13. Infiltration rates on soils which have had the vegetation burned versus that for the same soils without burning (From Settergren 1967).

variable in both instances.

In another study on silty loam soils in southern Idaho, Gifford (1972) and Gifford and Busby (1974) conducted intensive infiltrometer studies on a plowed big sagebrush site over a 4-year period. Results of the study indicate there was a natural decay in absorptive capabilities of surface soils due to the plowing treatment. The apparent result of grazing (no grazing for 2 years following plowing and seeding) was not to reduce the minimal infiltration capacities measured on the respective site, but rather to eliminate seasonal trends so that infiltration rates were at the low end of the scale throughout the year. Grazing did not increase sediment production potentials beyond the increases expected as a result of mechanical disturbance associated with plowing (Figure 4-15).

Blackburn and Skau (1974) studied infiltration and sediment production rates in two big sagebrush communities just south of Ely, Nevada, that had been plowed and seeded 5 to 6 years prior to the study. They found that infiltration rates for the sandy loam surface soils were not significantly different from their undisturbed counterparts. The same was true with respect to sediment production. Orr (1957) and Meeuwig (1965) had similar results on a high elevation range in central Utah 3 years and 7 years, respectively, following disking and seeding to grass. Assuming the previous studies already discussed to adequately reflect what is happening during the first few years following a plowing treatment, it would appear that time and subsequent

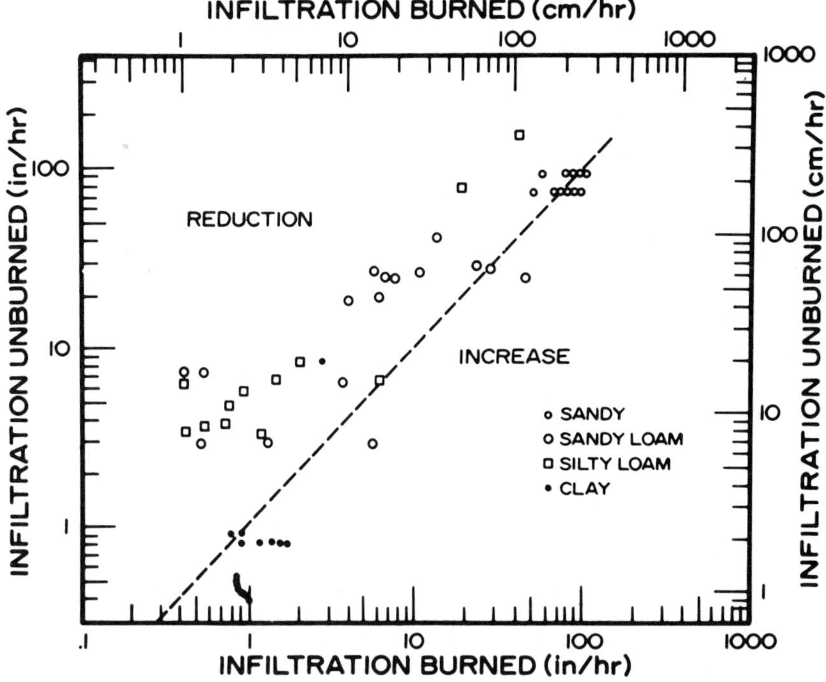

Fig. 4-14. Change in infiltration following fire by soil textural class (from Settergren 1967). The dashed line represents the division between a reduction and an increase in the infiltrate rate following fire.

grazing management interact with surface soil characteristics to play an important part in eventual *recovery (not improvement)* of desirable hydrologic characteristics.

Chaining

Gifford (1975) has recently provided an in-depth review of the impacts of pinyon-juniper manipulation on watershed values. Another recent review, on a more limited scale, is that of Clary et al. (1974). Therefore, rather than reiterate what has already recently been summarized, summary statements from Gifford (1975) regarding infiltration and runoff are presented as follows.

1. Studies completed to date indicate that infiltration rates have been only slightly affected when comparing chained sites to the undisturbed woodland. Coarse-textured soils probably account for much of this. If there is a decrease in infiltration rates due to chaining activities, then the decrease will probably occur on chained-with-windrowing treatments, this being the result of rather severe mechanical disturbance of surface soils during the windrowing process. The mechanical disturbance may actually increase permeability of surface soils but infiltration at the soil-air interface is decreased.

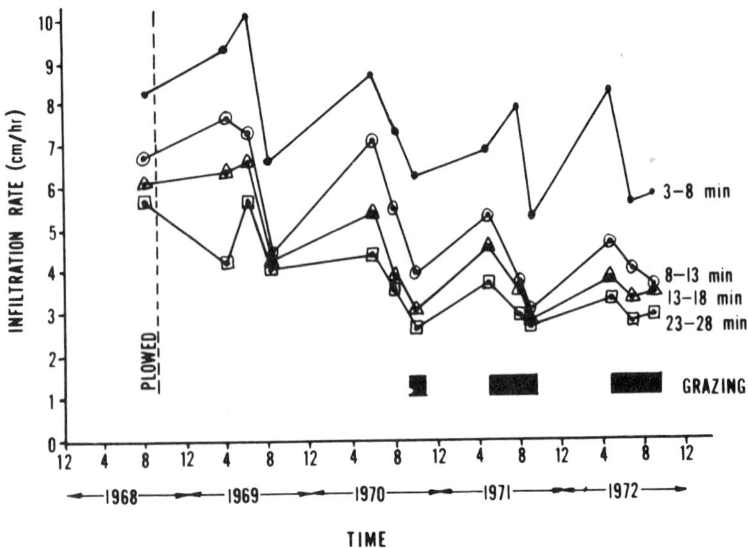

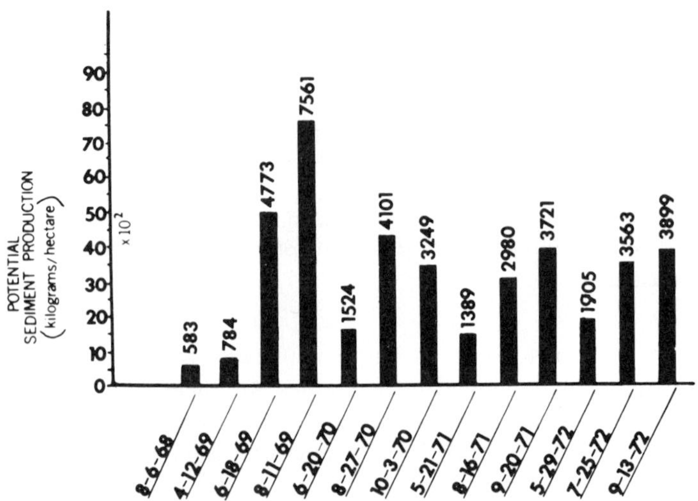

Fig. 4-15. Average infiltration rates by 5-minute intervals and potential sediment production for various sampling dates on a big sagebrush site near Holbrook, Idaho. The first bar on the left represents potential sediment production prior to plowing (from Gifford and Busby 1974). (Multiply kg/ha by 1.12 to obtain lb/ac).

2. Because of the variability in pinyon-juniper site characteristics, it is difficult to pinpoint exactly those factors that consistently influence infiltration rates.

3. The impact of grazing on infiltration rates appear accumulative (up to

some undefined point), and effects of even a single grazing season can be detected. Complete protection for four years of a grazed site on sandy loam soils restored infiltration capacities to a maximum.

4. Burning of debris appears to depress infiltration rates.

Literature Cited

Adams, J.E., D. Kirkham, and D.R. Nielsen. 1957. A portable rainfall-simulator infiltrometer and physical measurements of soil in place. Soil Sci. Soc. Amer. Proc. 21: 473-477.

Amerman, C.R., A. Klute, R.W. Skaggs, and R.E. Smith. 1975. Soil water. Reviews of Geophysics and Space Physics 13: 451-454.

Amorocho, J. 1967. Role of infiltration in nonlinear watershed analyses processes. Amer. Soc. Agr. Eng., Trans. 10: 396-399, 404 p.

Arend, J.L., and D.M. Knight. 1940. A portable infiltrometer and unit system for determing relative infiltration rates. USDA, Forest Serv., Central States Forest Exp. Sta. Tech. Note No. 11: 4 p.

Barnes, O.L., and G. Costel. 1957. A mobile infiltrometer. Agron. J. 49: 105-107.
British Columbia, Parts 1 and 2. Can. J. Soil Sci. 39: 1-11.

Beutner, E.L., R.R. Gaebe, and R.E. Horton. 1940. Sprinkled plot runoff and infiltration experiments of Arizona desert soils. Trans. Amer. Geophys. Union, Pt. 2: 550-558.

Blackburn, W.H. 1975. Factors influencing infiltration and sediment production of semiarid rangelands in Nevada. Water Resour. Res. 11: 929-937.

Blackburn, W.H., R.O. Meeuwig, and C.M. Skau. 1974. A mobile infiltrometer for use on rangeland. J. Range Manage. 27: 322-323.

Blackburn, W.H., and C.M. Skau. 1974. Infiltration rates and sediment production of selected plant communities in Nevada. J. Range Manage. 27: 476-480.

Bodam, G.B., and E.A. Colman. 1943. Moisture and energy conditions during downward entry of water into soils. Soil Soc. Amer., Proc. 8: 116-122.

Bondurant, J.A. 1957. Developing a furrow infiltrometer. Agr. Eng. 38: 602-604.

Box, T.W. 1961. Relationships between plants and soils of four range plant communities in south Texas. Ecology 42: 794-810.

Branson, F.A., R.F. Miller, and I.S. McQueen. 1962. Effects of contour furrowing, grazing intensities, and soils on infiltration rates, soil moisture and vegetation near Fort Peck, Montana. J. Range Manage. 15: 151-158.

Branson, F.A., R.F. Miller, and I.S. McQueen. 1965. Plant communities and soil moisture relationships near Denver, Colorado. Ecology 46: 311-319.

Buckhouse, J.C., and G.F. Gifford. 1976. Sediment production and infiltration rates as affected by grazing and debris burning on chained and seeded pinyon-juniper. J. Range Manage. 29: 83-85.

Burgy, R.H., and V.H. Scott. 1952. Some effects of fire and ash on the infiltration capacity of soils. Trans. Amer. Geophys. Union. 33: 405-416.

Burton, T.A. 1976. An approach to the classification of Utah mine spoils and tailings based on surface hydrology and erosion. M.S. Thesis, Utah State Univ., Logan. 92 p.

Chow, V.T., and T. Harbaugh. 1965. Raindrop production for laboratory watershed experimentation. J. Geophys. Res. 70: 6111-6119.

Clary, W.P., M.B. Baker, Jr., P.F. O'Connell, T.N. Johnsen, Jr., and R.E. Campbell. 1974. Effects of pinyon-juniper removal on natural resource products and uses in Arizona. USDA, Forest Serv. Res. Pap. RM-128. 28 p.

Costin, A.B., and D.A. Gilmour. 1970. Portable rainfall simulator and plot unit for use in field studies of infiltration, runoff, and erosion. J. Appl. Ecol. 7: 192-200.

DeBano, L.F. 1969. Observations on water-repellent soils in western United States. Symp. on Water-Repellent Soils (Univ. California, Riverside), Proc. May 6-10, 1969, p. 17-29.

DeBano, L.F., and R.M. Rice. 1973. Water-repellent soils: their implications in forestry. J. Forestry 71: 220-223.

Dee, R.F., T.W. Box, and E. Robertson, Jr. 1966. Influence of grass vegetation on water intake of Pullman silty clay loam. J. Range Manage. 19: 77-79.

Dixon, R.M. 1975. Infiltration control through soil surface management. p. 543-567. *In:* Proc., A.S.C.E. Symp. Watershed Management, Utah St. University, Logan, August 11-13: 543-567.

Dixon, R.M., and A.E. Peterson. 1971. Water infiltration control: a channel system concept. Soil Sci. Soc. Amer. Proc. 35: 968-973.

Dortignac, E.J. 1951. Design and operation of Rocky Mountain infiltrometer. USDA, Forest Serv., Rocky Mt. Forest and Range Exp. Sta. Pap. No. 5. 68 p.

Dortignac, E.J., and L.D. Love. 1961. Infiltration studies on ponderosa pine ranges of Colorado. USDA, Forest Service, Rocky Mt. Forest and Range Exp. Sta. Pap. 59: 34 p.

Duley, F.L., and C.E. Domingo. 1949. Effect of grass on intake of water. Univ. Nebraska Agr. Exp. Sta. Res. Bull. 159. 15 p.

Duley, F.L., and L.L. Kelly. 1939. Effect of soil type, slope, and surface conditions on intake of water. Univ Nebraska Agr. Exp. Sta. Res. Bull. 112. 16 p.

Evanko, A.B. 1950. A tin can infiltrometer with improvised baffle. USDA, Forest Ser., Northern Rocky Mt. Forest and Range Exp. Sta. Res. Note No. 76: 3 p.

Friedrich, C.A. 1951. Suggestions for manufacture and operation of a six-inch infiltrometer ring. USDA, Forest Serv., Northern Rocky Mt. Forest and Range Exp. Sta. Res. Note No. 104: 3 p.

Gifford, F.G. 1968. Rangeland waterhsed management-a review. Univ. Nevada Agr. Exp. Sta. Pap. R52. 50 p.

Gifford, G.F. 1972. Infiltration rate and sediment production trends on a plowed big sagebrush site. J. Range Manage. 25: 53-55.

Gifford, G.F. 1975. Impacts of pinyon-juniper manipulation on watershed values. p. 127-140. *In:* Proc., Pinyon-Juniper Ecosystem—A Symposium, Utah State Univ., Logan, May 1-2.

Gifford, G.F. 1976. Applicability of some infiltration formulae to rangeland infiltrometer data. J. Hydrol. 28: 1-11.

Gifford G.F. 1977. Infilrometer studies in rangeland plant communities of the Northern Territory, Australia. I. Infiltration studies. Australian Range J. 1: 142-149. **Gifford, G.F., and F.E. Busby. 1974.** Intensive infiltrometer studies on a plowed big sagebrush site. J. Hydrol. 21: 81-90.

Gifford, G.F., and C.M. Skau. 1967. Influence of various rangeland cultural treatments on runoff and sediment production from the big sagebrush type, Eastgate Basin, Nevada. Third Amer. Water Resour. Conf. Proc., (San Francisco), Nov. 8-10 p. 137-148.

Gray, D.M., D.I. Norum, and J.M. Murray. 1969. Infiltration characteristics of prairie soils. Paper presented at Amer. Soc. Range Manage. Annual Meeting (Calgary, Alberta). 16 p.

Green, W.H., and G.A. Ampt. 1911. Studies on soil physics: I. Flow of air and water through soils. J. Agr. Sci. 4: 1-24.

Hanks, R.J., and K.L. Anderson. 1957. Pasture burning and moisture conservation. J. Soil Water Conserv. 12: 228-229.

Jaupt, H.F. 1967. Infiltration, overland flow, and soil movement on frozen and snow-covered plots. Water Resour. Res. 3: 145-161.

Hendricks, B.A. 1942. Effect of grass litter on infiltration of rainfall on granitic coils in a semidesert shrub grass area. USDA, Forest Serv., Southwest Forest and Range Exp. Sta. Res. Note 96. 3 p.

Holtan, H.N., C.B. England, and V.O. Shanholtz. 1967. Concepts in hydrologic soil groupings. Amer. Soc. Agr. Eng., Trans. 10: 407-410.

Horton, R.E. 1933. The role of infiltration in the hydrologic cycle. Trans. Amer. Geophys. Union. 14: 446-460.

Horton, R.E. 1940. An approach toward physical interpretation of infiltration capacity. Soil Sci. Soc. Amer. Proc. 5: 399-417.

Hussain, S.B., C.M. Skau, and R.O. Meeuwig. 1968. Infiltrometer studies on non-wettable soils on East Sierra Nevada. Univ. Nevada Center for Water Resour. Res. Proj. Rep. 11: 19-28.

Johnson A. 1962. Effects of grazing intensity and cover on the water intake rate of fescue grassland. J. Range Manage. 15: 79-82.

Johnson, A.I. 1963. A field method for measurement of infiltration. U.S. Geol. Surv. Water-Supply Pap. No. 1544-F: 27 p.

Knoll, G., and H.H. Hopkins. 1959. The effects of grazing and trampling upon certain soil properties. Trans. Kansas Acad. Sci. 62: 221-231.

Krimbold, D.A., and A. Shahori. 1957. Design rain—aid in erosion control studies. Agr. Eng. 38: 740-743.

Langbein, W.B., and others. 1947. Major winter and non-winter floods in selected basins in New York and Pennsylvania. U.S. Geol. Surv. Water-Supply Pap. 915. 139 p.

Leithead, H.L. 1959. Runoff in relation to range condition i the Big Bend-Davis Mountain section of Texas. J. Range Manage. 12:83-87.

Lubsy, G.C., V.H. Reid, and O.D. Knipe. 1971. Effects of grazing on the hydrology and biology of the Badger Wash basin in western Colorado, 1953-1966. U.S. Geol. Surv. Water-Supply Pap. 1532D 90 p.

Lyford, F.P., and H.K. Qashu. 1969. Infiltration rates as affected by desert vegetation. Water Resour. Res. 5: 1373-1376.

McQueen, I.S. 1963. Development of a hand portable rainfall simulator infiltrometer. U.S. Dep. Interior, U.S. Geol. Surv. circ. 482. 16 p.

Meeuwig, R.O. 1965. Effects of seeding and grazing on infiltration capacity and soil stability of a subalpine range in central Utah. J. Range Manage. 18: 173-180.

Meeuwig, R.O. 1969. Infiltration and soil erosion on Coolwater Ridge, Idaho. USDA, Forest Serv. Res. Note INT-103: 5p.

Meeuwig, R.O. 1970. Infiltration and soil erosion as influenced by vegetation and soil in northern Utah. J. Range Manage. 23: 183-188.

Meeuwig, R.O. 1971. Infiltration and water repellency in granitic soils. USDA, Forest Serv. Res. Paper INT-11: 20 p.

Meyer, L.D., and D.L. McCune. 1958. Rainfall simulator for runoff plots. Agr. Eng. 39:

Morin, J. 1967. A rainfall simulator with a rotating disk. Trans. Amer. Soc. Agr. Eng. 10: 74-79.

Morris, R.J., and M. Natalino. 1969. The chemical nature of the organic matrix believed to limit water penetration in granitic soils. Univ. Nevada Desert Res. Inst. Proj. Rep. No. 13: 13 p.

Orr, H.G. 1957. Effects of plowing and seeding on some forage production and hydrologic characterisics of a subalpine range in central Utah. USDA, Forest Serv., Intermountain Forest and Range Exp. Sta. Res. Pap. 47: 23 p.

Osborn, B. 1950. Range cover tames the raindrop—a summary of range cover evaluations, 1949. USDA, Soil Conserv. Serv. Unnumbered Publ. (mimeo): 92 p.

Osborn, B. 1952. Range soil conditions influence water intake. J. Soil Water Conserv. 7: 128-132.

Packer, P.E. 1957. Intermountain infiltrometer. USDA, Forest Serv., Intermountain Forest and Range Exp. Sta. Misc. Pub. No. 14:41 p.

Parr, J.F., and A.R. Bertrand. 1960. Water infiltration into soils. Adv. Agron. 12: 311-363.

Pearse, C.K. 1936. Specifications for the construction and operation of a portable apparatus for measuring runoff and erosion. USDA, Forest Serv., Intermountain Forest and Range Exp. Sta., Mimeo: 39 p.

Pearse, C.K., and F.D. Bertelson. 1937. The construction and operation of an apparatus for the measurement of adsorption of surface water by soils. USDA, Forest Serv., Intermountain Forest and Range Expt. Sta., Mimeo: 13 p.

Pearse, C.K., and S.B. Wooley. 1936. The influence of range plant cover on the rate of absorption of surface water by soils. J. Forest. 34: 844-847.

Philip, J.R. 1957. The theory of infiltration: 4. Sorptivity and algebraic infiltration equations. Soil Sci. 84: 257-264.

Rauzi, F. 1960. Water-intake studies on range soils at three locations in the Northern Plains. J. Range Manage. 13: 179-184.

Rauzi, R., and A.R. Kuhlman. 1961. Water intake as affected by soil and vegetation on certain western South Dakota rangelands. J. Range Manage. 14: 267-271.

Rauzi, F.C., C.L. Fly, and E.J. Dyksterhuis. 1968. Water intake on Midcontinental rangelands as influenced by soil and plant cover. U.S. Dep. Agr. Tech. Bull. 1390: 58 p.

Rauzi, F.C., and F.M. Smith. 1973. Infiltration rates: three soils with grazing levels in northeastern Colorado. J. Range Manage. 26: 126-129.

Rhoades, E.D. 1961. A mobile double-tower sprinkling infiltrometer. J. Soil and Water Conserv. 16: 181-182.

Rhoades, E.D., L.F. Locke, H.M. Taylor, and E.H. McIlvain. 1964. Water intake on a sandy range as affected by 20 years of differential cattle stocking rates. J. Range Manage. 17: 185-190.

Rowe, P.B. 1940. The construction, operation, and use of the North Fork infiltrometer. USDA, Forest Serv., California Forest and Range Exp. Sta. Misc. Pub. No. 1:64 p.

Rowe, P.B. 1948. Influence of woodland chaparral on water and soil in central California. U.S. Dep. Agr. and State of California Dep. Natur. Resour. Div. Forest. Unnumbered Pub. 70 p.

Sartz, R.S., and D.N. Tolsted. 1974. Effect of grazing on runoff from two small watersheds in southwestern Wisconsin. Water Resour. Res. 10: 354-356.

Savage, S.M., J.F. Osborn, J. Letey, and C. Heaton. 1972. Substances contributing to fire-induced water repellency in soils. Soil Sci. Soc. Amer. Proc. 36: 674-678.

Scholl, D.G. 1975. Soil wettability and fire in Arizona chaparral. Soil Sci. Soc. Amer. proc. 39: 356-361.

Schumm, S.A., and G.C. Lusby. 1963. Seasonal variation of infiltration capacity and runoff on hillslopes in western Colorado. J. Geophys. Res. 68: 3655-3666.

Scott, V.H. 1956. Relative infiltration rates of burned and unburned upland soils. Trans. Amer. Geophys. Union. 37: 67-69.

Scott, V.H., and R.H. Burgy. 1956. Effects of heat and brush burning on the physical properties of certain upland soils that influence intiltration. Soil Sci. 82: 63-70.

Settergren, C.D. 1967. Reanalyses of past research on effects of fire on wildland hydrology. Missouri Agr. Exp. Sta. Res. Bull. 954, 16 p.

Sharp. A.L., J.J. Bond, J. W. Neuberger, A.R. Kuhlman, and J.K. Lewis. 1964. Runoff as affected by intensity of grazing on rangeland. J. Soil and Water Conserv. 19: 103-106.

Smith, H.L., and L.B. Leopold. 1942. Infiltration studies in the Pecos River watershed, New Mexico and Texas Soil Sci. 53: 195-204.

Steinhardt, R., and D. Hillel. 1966. A portable low-intensity rain simulator for field and laboratory use. Soil Sci. Soc. Amer. Proc. 30: 661-662.

Thomas, G.W., and V.A. Young. 1954. Relation of soils, rainfall and grazing management to vegetation, western Edwards Plateau of Texas. Texas Agr. Exp. Sta. Bull. 786, 22 p.

Thompson, J.R. 1968. Effect of grazing on infiltration in a western watershed. J. Soil Water Conserv. 23: 63-65.

Tromble, J.M., K.G. Renard, and A.P. Thatcher. 1974. Infiltration for three rangeland soil-vegetation complexes. J. of Range Manage. 27: 318-321.

Veihmeyer, F.J. and C.N. Johnson. 1944. Soil moisture records from burned and unburned plots in certain grazing areas of California. Trans. Amer. Geophys. Union. Part I: 72-88.

Williams, G., and G.F. Gifford. 1969. Analysis of hydrologic, edaphic, and vegetative factors affecting infiltration and erosion on certain treated and untreated pinyon-juniper sites. Prog. Rep., Utah Agr. Exp. Sta. (Logan), Proj. 728, Contract 14-11-0008-2837 (Bureau of Land Manag.) 162 p.

Wilm, H.G. 1943. The application and measurement of artificial rainfall on types FA and F infiltrometers. Trans. Amer. Geophys. Union. 24: 480-487.

Woodward, L. 1944. Infiltration capacities on some plant-soil complexes on Utah range watershed lands. Trans. Amer. Geophys. Union. Pt. II: 468-475.

Chapter 5

Runoff and Water Harvesting

Runoff from rangeland watersheds is important from several aspects. Runoff provides the water used for livestock when it is collected in storage impoundments. It is also the water source for downstream or offsite domestic, industrial, or agricultural purposes. Runoff is also important because it is the primary force in initiating erosion and transporting sediment and dissolved solids into rivers and reservoirs (Figure 5-1). Thus runoff is the mechanism that produces much of the pollution encountered in water courses.

Runoff may be divided into three major components. *Surface runoff* or *overland flow* is that part of runoff that travels over the soil surface to the nearest stream channel. *Storm seepage* is that part of precipitation which infiltrates into the soil to a relatively impermeable layer, and as a result spreads out and flows laterally a short distance below the soil surface to the stream. *Groundwater runoff* is precipitation which percolates through the soil mantle to the groundwater table and is eventually discharged into a stream.

Surface runoff is the most common kind of runoff on rangelands. This is

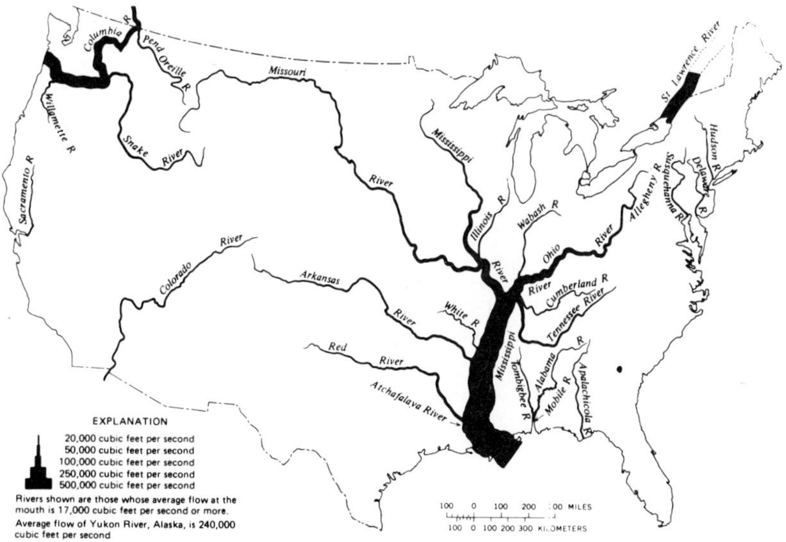

Fig. 5.1 Large rivers in the United States. The outflow of all streams from the contiguous United States into the oceans or across its boundaries is about 185,800 m^3 per second (2,000,000 ft^3 per second). Of this total more than 75% is accounted for by 11 independent streams. The remaining 25% is discharged through a host of comparatively minor coastal streams (from Iseri and Langbein 1974). (Multiply ft^3/sec by 0.0283 to obtain m^3/sec.)

indicated by the ephemeral nature of most stream channels found on western rangelands. They discharge water for only brief periods as a result of snow melting in the spring or following rainfall when the intensity exceeds the capacity of the soil to absorb it.

Surface runoff on rangelands is initiated when water from precipitation occupies all available soil and vegetation surface storage and/or the infiltration rate of the soil is exceeded. Additional rainfall causes overland flow and runoff. The total volume of annual runoff is affected by climatic factors such as potential evapotranspiration and seasonal distribution of rainfall as well as watershed features such as the contributing watershed area. Timing of periods of runoff and the volume from an individual storm from a given watershed depends on the type of precipitation, its intensity, duration, and distribution; watershed topography; geology; soil type; vegetal cover characteristics; and antecedent soil moisture conditions. Many of these relationships are discussed more fully in Chapter 2 as they are affected by precipitation.

The Hydrograph and the Water-yielding Process

The relationship of the volume and timing of surface and groundwater

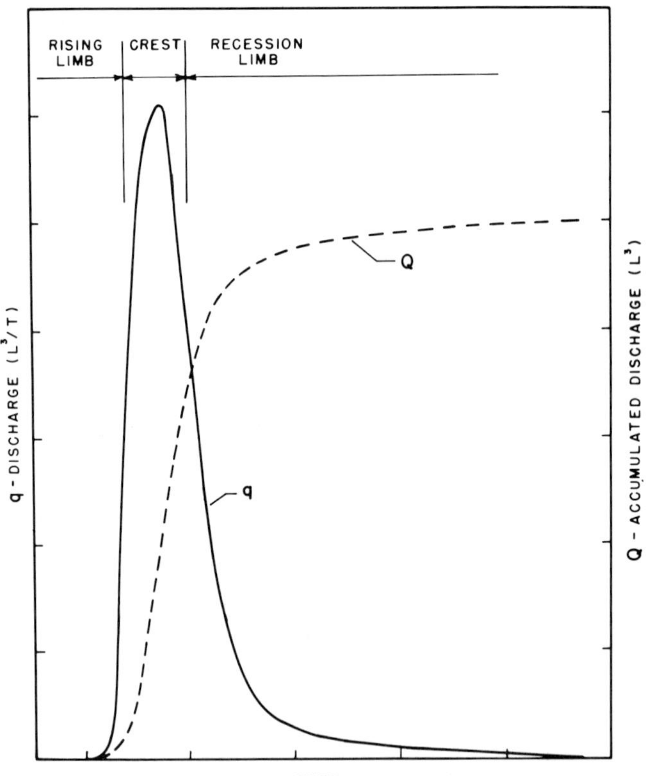

Fig. 5-2. Component parts of a simple hydrograph. Q is the volume of runoff as related to time, and q is the discharge rate at any point in time.

runoff combine to shape the stream hydrograph, or the graph of runoff plotted against time. In Figure 5-2 the relationship of mass or total runoff (Q) and runoff rate (q) is shown.

Slatyer and Mabbutt (1964) discuss runoff under arid and semiarid conditions. They state the following:

> Much of the rain falling on the desert surfaces is absorbed by the porous mantle of soil and rock waste or is lost by evaporation. Since desert rainfall is naturally infrequent and short-lived, these initial losses are proportionately greater than in humid regions, so that in general, percentage runoff tends to decrease with annual rainfall and is less than 10 percent in most arid zones.
>
> The intensity of the rainfall may be more important than the total amount although runoff will tend to increase with duration of the fall. . .
>
> Because of high evaporation losses, the degree of slope exercises a stronger control on the amount of runoff than in humid areas. On hill slopes percentage runoff may be as high as in humid uplands, the initial losses being compensated by steeper gradients, bare rocky surfaces, and low wastage through vegetation, but rain falling on gently sloping plains will be less effective in producing runoff than in humid areas.
>
> In the absence of dense vegetation, the nature of the surface is an important factor in controlling the amount and character of runoff. On desert hill slopes there is little retention of rainfall by the slope mantle and runoff is very rapid. There is also a deficiency of fine material on such slopes, so that surface runoff is underloaded and the length of overland flow is reduced. As a result, a fine meshwork of slope rills is formed and these sink up into a channel network incised to varying degrees, depending on rock resistance. . . .
>
> Because of the thinness of soils and sparsity of vegetation, these surfaces offer little storage for rainfall; such runoff as does occur will ensue rapidly, and surface flow will be less restricted than on similar slopes in humid areas. For the same reasons, there can be little feedback from the surface layer, so that runoff will rapidly cease when the rain stops. This leads to a concentration of runoff into short flood phases.

Rangelands at higher elevations or in areas of greater precipitation differ from semiarid rangelands in rainfall-induced runoff characteristics. Rainfall-induced runoff from these areas is closely related to the four general runoff types recognized by Horton (1933) (Figure 5-3). For runoff "Type O," rainfall does not exceed infiltration capacity of the soil, so no surface runoff occurs. The rainfall is not sufficient to satisfy the soil moisture deficit, so an increase in groundwater runoff does not occur. Therefore, the stream hydrograph follows the form of the normal dry weather depletion curve, with groundwater runoff gradually decreasing from A to B as the groundwater reserves are used up.

In runoff "Type 1," rainfall intensity again does not exceed infiltration capacity of the soil. However, water infiltrated exceeds the soil moisture deficit, so that some percolation to the groundwater reservoir takes place. The result is an increase in groundwater runoff XZ, after which the normal groundwater depletion hydrograph is resumed at the higher ZB_1.

For runoff "Type 2," rainfall intensity exceeds soil infiltration capacities and surface runoff occurs. The runoff causes an increase in streamflow XY, but any infiltration which occurs does not exceed the soil moisture deficit. Therefore, once the surface runoff peak has passed, the hydrograph rejoins the normal ZB depletion curve.

Runoff "Type 3" occurs as rainfall intensity exceeds the infiltration capacity and total amount of infiltration exceeds the soil moisture deficit. Surface

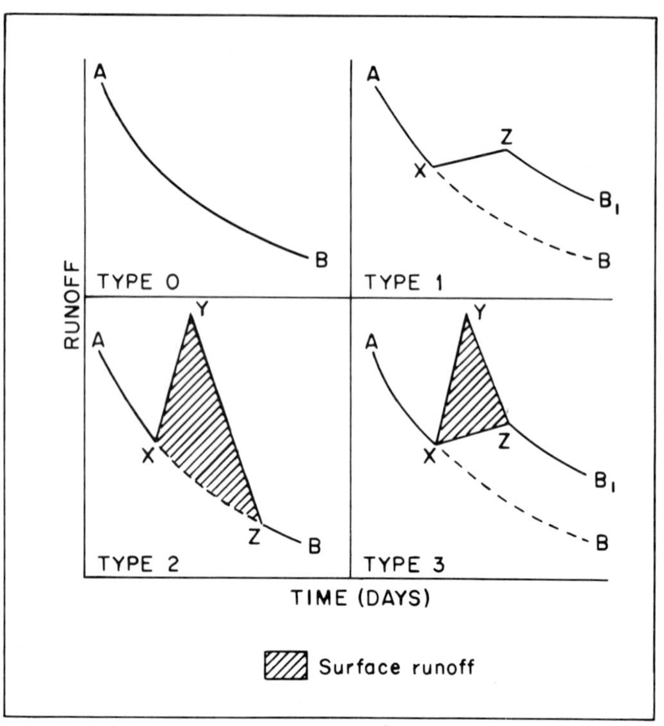

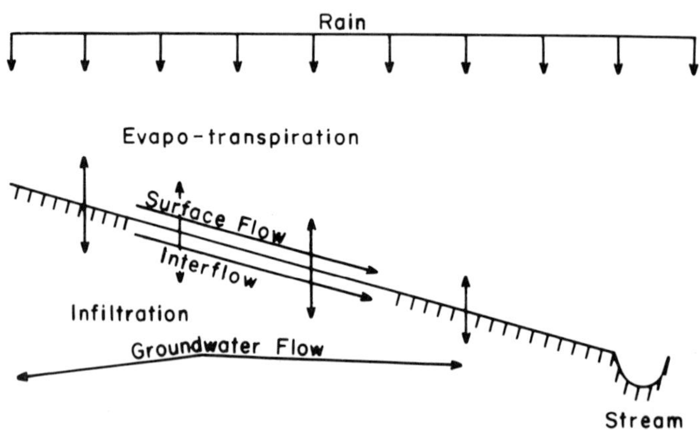

Fig. 5-3. Stream hydrographs showing the main types of runoff increase and schematic presentation of the rainfall-runoff process (from Horton 1933). See text for explanation of graph notations.

runoff increase is represented by the line XY and the increase in groundwater runoff by the line XZ. After the surface runoff peak has passed, they hydrograph follows a new groundwater depletion curve ZB_1.

This early work of Horton did much to provide the foundation for the

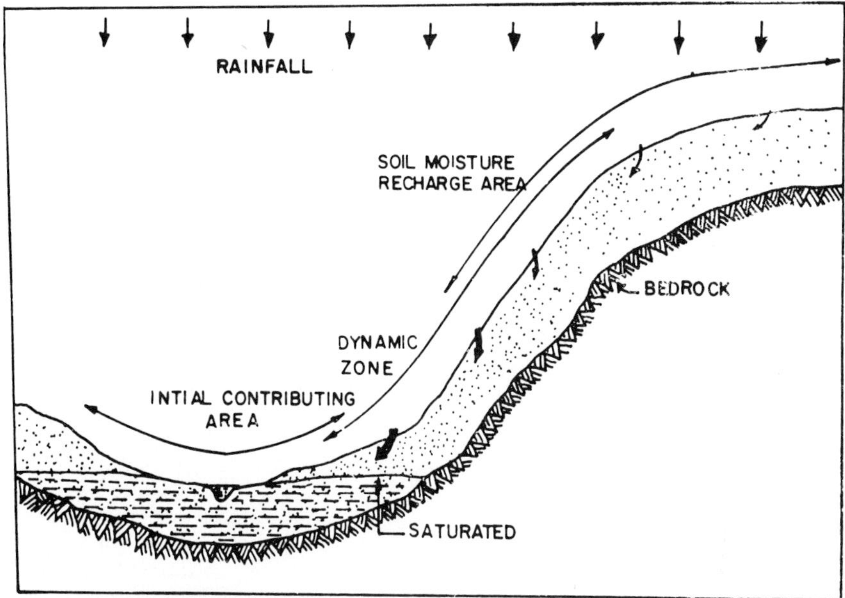

Fig. 5-4. Schematic diagram of the Partial Area or Variable Source Area Concept (from Tennessee Valley Authority, 1964).

science of hydrology as we recognize it today. For this reason Mr. Horton is often referred to as the father of present day hydrology.

The "Variable Source Area" Concept

Not all parts of a watershed contribute equally to the runoff process. As pointed out by Hewlett (1974), the origin of the "variable source area concept" of streamflow generation has been the subject of reports by Hursh (1936), Hursh and Brater (1941), Hursh and Fletcher (1942), Roessel (1950), U.S. Forest Service (1961), Hewlett (1961), Hewlett and Hibbert (1963, 1967), Hewlett and Nutter (1970), Troendle and Homeyer (1971), Nutter (1973), Engman (1974), and Engman and Rogowski (1974). In essence, the concept assumes that certain regions within a watershed contribute runoff to the storm hydrograph while other areas act as recharge or storage zones. Important factors to consider in determining whether an area contributes to runoff (or to the groundwater) include its physical position with respect to the channel, its soil properties, and the storm characteristics. Valley bottoms are generally considered to be the areas that contribute to streamflow while ridge tops constitute recharge areas. The area in between the valley bottoms and the ridge tops, often referred to as the dynamic zone, may be either contributing or recharging, depending upon the storm size and temporal characteristics, antecedent soil water content, and soil properties (Figure 5-4). Figure 5-5 shows a time-lapse view of a basin showing expansion of the source area and the channel system during a storm. An example of the usefulness of being able to delinate a watershed into contributing and noncontributing areas is shown in Figure 5-6. The watershed manager would be able to indicate areas that

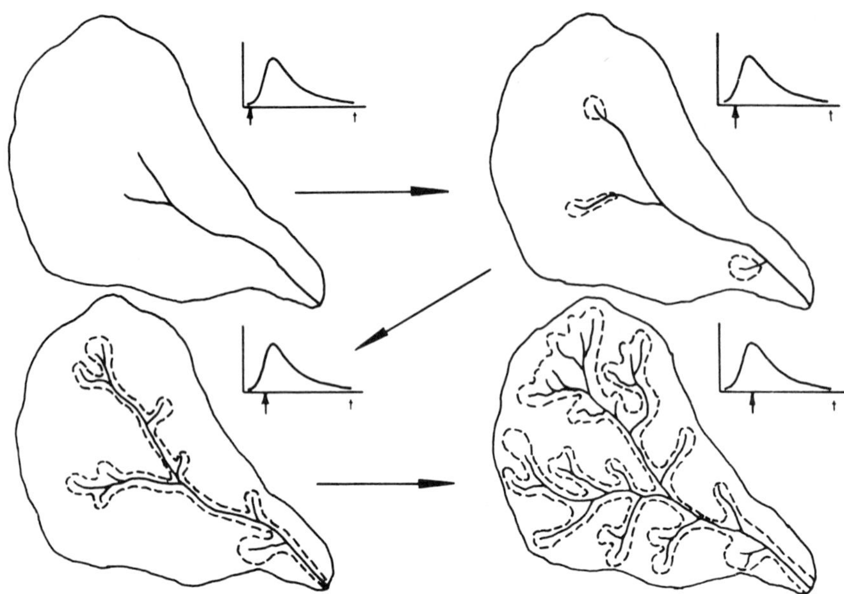

Fig. 5-5. A time-lapse view of a basin showing expansion of the source area and the channel system during a storm (from Hewlett and Nutter 1970).

could safely utilize heavy applications of fertilizers and areas where it would be unsafe. In a similar way, areas safe for land disposal of sewage effluent, agricultural wastes, and landfill could be readily selected.

Streamflow characteristics of rangelands can be illustrated by comparing

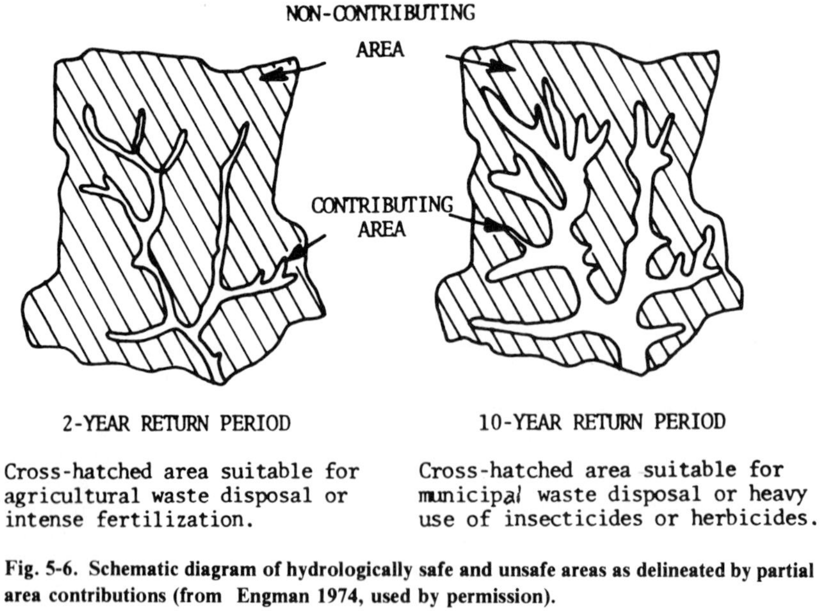

NON-CONTRIBUTING
AREA

CONTRIBUTING
AREA

2-YEAR RETURN PERIOD

Cross-hatched area suitable for agricultural waste disposal or intense fertilization.

10-YEAR RETURN PERIOD

Cross-hatched area suitable for municipal waste disposal or heavy use of insecticides or herbicides.

Fig. 5-6. Schematic diagram of hydrologically safe and unsafe areas as delineated by partial area contributions (from Engman 1974, used by permission).

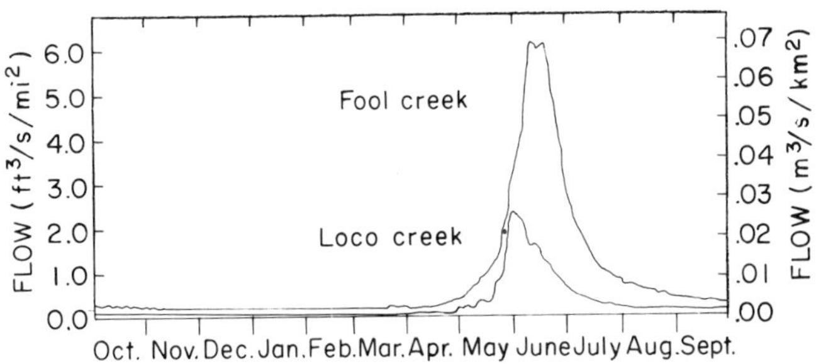

Fig. 5-7. Base flow, peak flow during snowmelt and volume of snowmelt runoff are greater on forested subalpine watersheds (Fool Creek) than on sagebrush watersheds (Loco Creek) (from Sturges 1975). (Multiply ft^3/sec by 0.0283 to obtain m^3/sec.)

flow regimes of sagebrush lands with those of a timbered subalpine watershed. Figure 5-7 compares yearly hydrograph of Loco Creek on the Stratton Experimental Area with Fool Creek on the Fraser Experimental Forest near Fraser, Colorado. Sturges (1975) points out that gross differences between the hydrographs are readily apparent, and he states the following:

> . . .Base flow on the timbered watershed is more than twice as great as from Loco Creek and peak flow rates during snowmelt are almost three times as great. Fool Creek yields 42 percent of annual precipitation as runoff, while the yield on Loco Creek is 22 percent. Snowmelt runoff persists longer on the forest watershed. The timing of snowmelt, and snowmelt discharge rates, probably vary more in the sagebrush type than on timberedd land. . . The form of the snowmelt hydrograph is similar for Loco Creek and Fool Creek even though runoff volume, flow rate, and runoff duration were greater for the timbered watershed.

However, as pointed out by Branson (1976), of the approximately 40 percent of the world's land surface that is classified as rangelands, more than 80 percent is within the arid and semiarid zones. Runoff as a product of these lands has often been overemphasized and the much larger component, evapotranspiration, under-emphasized (Figure 7-2).

Channel Transmission Losses

Transmission losses as water moves downstream, are an important factor in evaluating the response of ephemeral streams to precipitation. Stream channels in many of the streams in the rangeland areas of the western United States are dry except for brief periods following thunderstorms or when prolonged snowmelt runoff occurs. Many researchers have documented transmission losses (Babcock and Cushing 1942; Allis, Dragoun, and Sharp 1964; Keppel and Renard 1962; Lane, Diakin, and Renard 1970; Burkham 1970), but the success of predicting their magnitude and their effect on the downstream hydrograph shape has been limited (Lane 1972; Smith 1972).

Figure 5-8 (a and b) is an example that illustrates the transmission losses measured on the Walnut Gulch Experimental Watershed, Arizona. Precipitation for the storm event of August 5, 1968, was concentrated in one of the subwatersheds with 2.1 inches measured near the storm center (and subwatershed center) in 36 minutes. Runoff from this watershed was measured as 27.4 acre-feet (33,784 m^3) with a peak discharge of almost 1,100 cfs (31,152

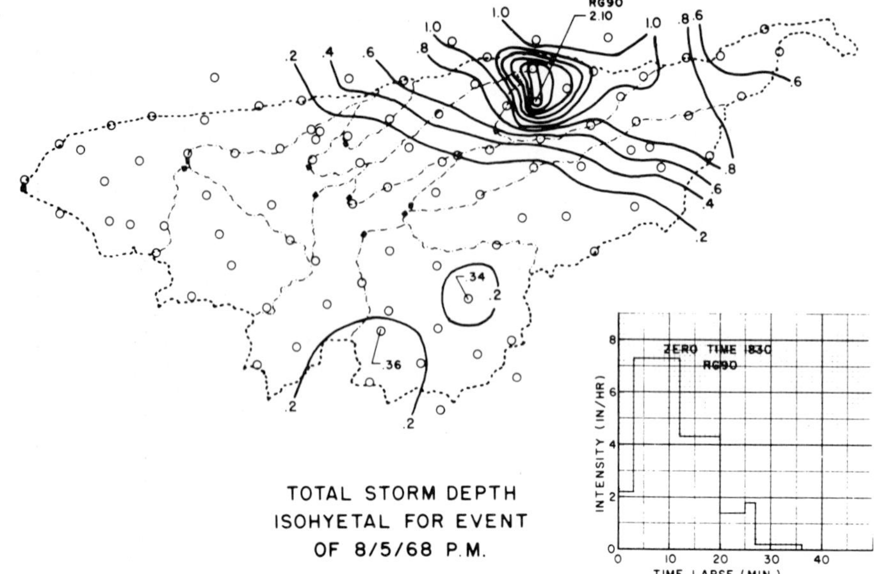

Fig. 5-8a. Isohyetal map of total storm depth in inches for the event on August 5, 1968, on the Walnut Gulch watershed, Arizona. Also shown are rainfall intensities as computed from a recording rain gage (RG 90). Circles indicate locations of rain gages in the watershed.

l/s). In traversing the intervening 4.0 miles (6.4) km) between measuring points at flumes 11 and 8 (Figure 5-8b), much of the storm volume (14.55 acre-feet) was absorbed by the alluvial channel. Traversing the 0.9 mile (1.4

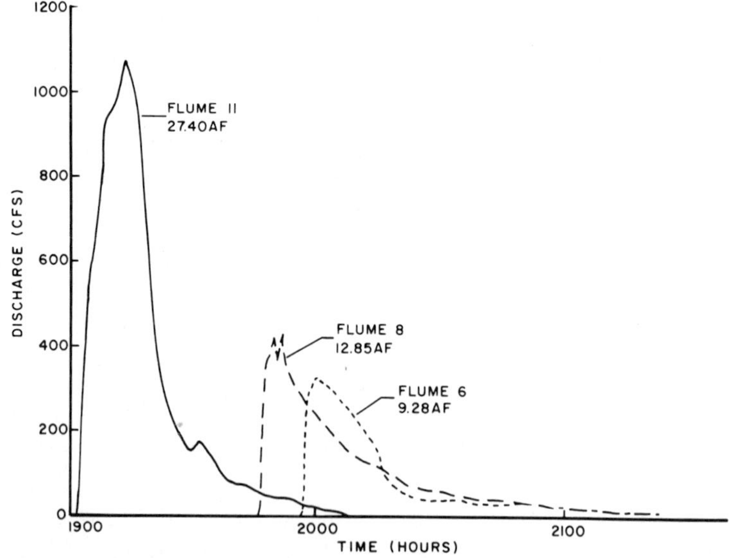

Fig. 5-8b. Runoff hydrographs for selected flumes on the Walnut Gulch watershed for the event of August 5, 1968 (Renard 1970). (Multiply ft³/s by 0.0283 to obtain m³/s.)

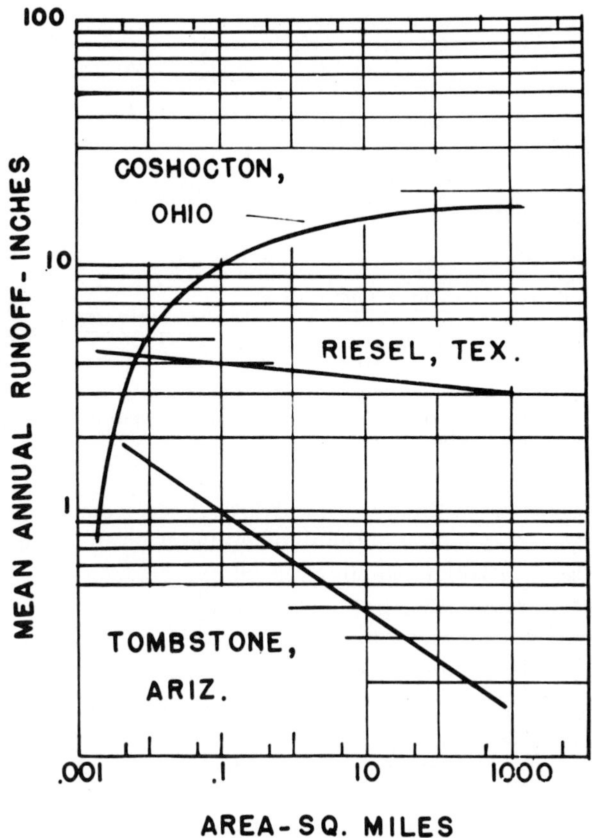

Fig. 5-9. Relationship between mean annual runoff and area (from Glymph and Holtan 1969). (Multiply square miles by 2.59 to obtain km².)

km) from flume 8 to flume 6 caused a further reduction in flow. Events such as this are common in thunderstorm areas. This phenomena leads to the observation that in such areas the water yield per unit area actually decreases with increasing watershed size. In other areas, the yield may increase or be independent of watershed size (Figure 5-9). In such ephemeral streams as Walnut Gulch, the hydrologic balance as shown Figure 5-10 illustrates the magnitude of these flow reductions.

Rangeland Runoff Studies from Natural Plant Communities

Studies of runoff relationships on arid and semiarid rangelands are complicated because of the infrequent runoff events and the highly variable nature of precipitation. Runoff varies with the kind of vegetation as well as the quantity of vegetation. For instance, runoff as percent of precipitation for different vegetation zones of the Rio Grande Basin is shown in Figure 5-11. These data indicate that 25 cm (10 inches) or more precipitation must occur before there is a significant amount of runoff. As pointed out by Branson (1976), it is

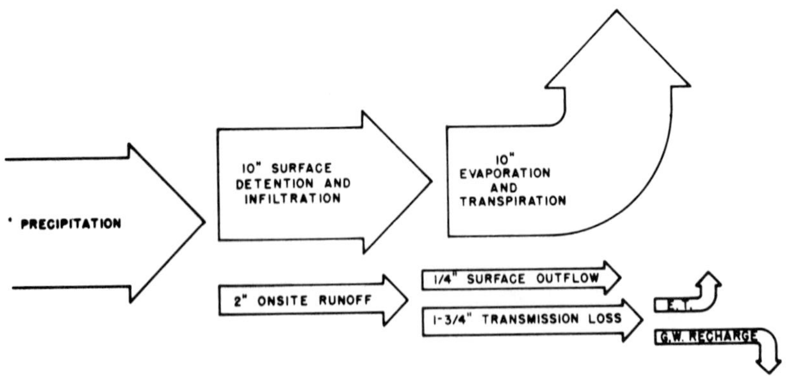

Fig. 5-10. Water balance of Walnut Gulch watershed for an input of 12 inches (30.5 cm) precipitation (from Renard 1970).

undoubtedly incorrect to think of these data as runoff from different vegetation types. Rather they represent runoff quantities from environmental complexes with major variables including vegetation, soil, temperature, altitude, and precipitation zones.

The nature of precipitation on arid and semiarid lands has been discussed earlier, and thus will be only briefly reviewed here.

For very small watershed, runoff can be directly correlated to point rainfall values by simple regression equations (Osborn and Lane 1969). About 70 to 80 percent of the variability in peak discharge and runoff volume was explained by the variability in either the maximum 15-minute or total rain depth for 4 very small watersheds 0.5 to 11 acres (0.2 to 4.5 ha). In general, thunderstorm rainfall so dominated the rainfall-runoff relationships that no other single variable appreciably improved the initial regression equation (5-1 and 5-2).

$$Q = a + bP \quad \textbf{(5-1)}$$

and

$$V = c + dP \quad \textbf{(5-2)}$$

where

Q = peak discharge in cubic feet or cubic liters per second;
V = volume of runoff in cubic feet or liters;
P = total storm rainfall in inches;
(a, b, c, and d are constants).

For larger watersheds (over 1 square mile or 2.6 km^2), runoff was best correlated to the maximum 30-minute rainfall for both peak discharge and total storm runoff. Simple correlations between the maximum 15- and 30-minute point rainfall values and peak (Q) and total discharge (V) are shown in Table 5-1. As would be expected, the best correlations were for the very small watersheds.

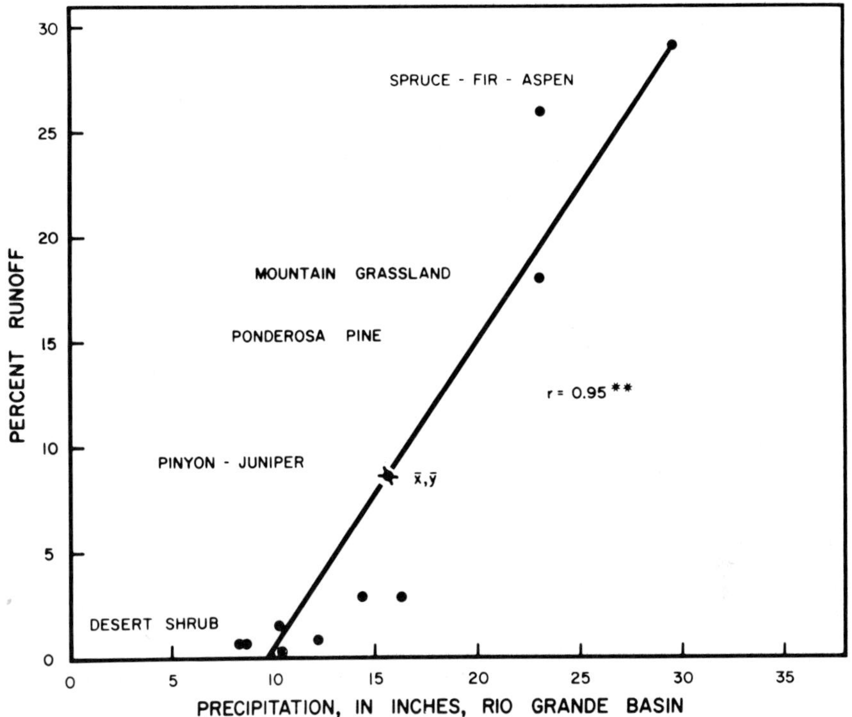

Fig. 5-11. Runoff as percent of precipitation for different vegetation and precipitation zones of the Rio Grande Basin of Colorado and New Mexico (computed from data by Dortignac 1956). ** = significant at 0.01. The solid dots are the available data points.

For watersheds less than 120 acres (49 hectares), point rainfall from a single gage gave good rainfall-runoff correlations. For the 560-acre (227 hectares) watershed, W-4, with a 3-gage network, the correlations were poor, and the differences between using the central gage and average rainfall were not significant.

Correlations between point rainfall values and runoff for four larger watersheds, W-11, W-5, W-6, and W-1 are also shown in Table 5-1. For each of these stations, 20 to 26 events were chosen from the period 1961-1968. In general, except for the occasional "double" event, antecedent rainfall and pre-storm channel conditions account for less than 10 percent of the variability in runoff. Rainfall variability was by far the most significant variable for explaining runoff.

The correlations between rainfall and runoff for W-11, W-6, W-5, and W-1 could be rated as good, poor, and nonexistent for the 4-watersheds, respectively. The most significant point here was the difference in correlation between W-6 and W-1. There was obvious correlation, although admittedly poor, between rainfall and runoff for W-6, and just as obvious in lack of correlation on W-1. The amount of difference would appear to rule out chance. Probably, the combined effect of the limited areal extent of the

Table 5-1. Coefficient of determination for rainfall and runoff on eight Walnut Creek Gulch Watersheds (Osborn and Renard 1970).

Water-shed	Area (acres)[3]	Runoff variable[2]	Coefficient of determination (R^2)					
			P15			P30		
			Avg.	Max.	Central	Avg.	Max.	Central
LH-5	0.5	V/A			0.81[3]			0.77[3]
		Q/A			0.85[3]			0.74[3]
LH-3	8.5	V/A			0.85[3]			0.85
		Q/A			0.88[3]			0.86[3]
K-1	120.	V/A	0.61	0.60	0.60			
		Q/A	0.72	0.72	0.71			
W-4	560.	V/A	0.48	0.37	0.41			
		Q/A	0.48	0.37	0.43			
W-11	2,030.	V/A				0.71	0.67	[4]
		Q/A				0.76	0.66	[4]
W-5	5,500	V/A				0.35	0.23	0.33
		Q/A				0.33	0.33	0.28
W-6	23,500	V/A				0.60	0.48	0.53
		Q/A				0.50	0.46	0.49
W-1	36,900	V/A				0.24	0.15	0.34
		Q/A				0.24	0.35	0.27

[1]Acres × 0.405 = hectares.
[2]V/A = total storm runoff per acre; Q/A = peak discharge per acre; P15 = 15 minute rainfall; and P30 = 30 minute rainfall.
[3]Only one rain gage for the very small watersheds.
[4]No "central" gage on the W-11 watershed until 1966.

runoff-producing thunderstorm and the larger channel abstraction between Station W-6 and W-1 cause the difference.

In another study at Walnut Gulch a multiple regression analysis of 34 storms revealed that precipitation quantity, crown spread of vegetation, and antecedent soil moisture accounted for 72 percent, 3 percent, and 0.5 percent, respectively, of the prediction variance for runoff from five 6×12-ft plots (Schreiber and Kincaid 1967). These examples illustrate the relative importance of different variables in storm runoff prediction. Clearly, precipitation accounts for the most variance in runoff.

In many runoff studies, comparable amounts of precipitation on study plots are obtained either by artificial application of rainfall to small areas or by comparison of runoff amounts from adjacent small areas. In the case case it is assumed that rainfall is uniform over the study areas. At the present stage of technolgy, watershed managers cannot control or predict precipitation with any degree of accuracy. However, for a given storm, differences in runoff do occur from site to site. This is due to differences in soils, plant type, and range condition, and the watershed manager does exercise some control over these factors (Bailey 1945).

Branson and Owen (1970) found that runoff on salt desert shrub watersheds in western Colorado was directly related to the percentage of bare soil within a watershed. These investigators used mean annul runoff for 15 years of record and 1 to 6 years of vegetation measurements. Their regression

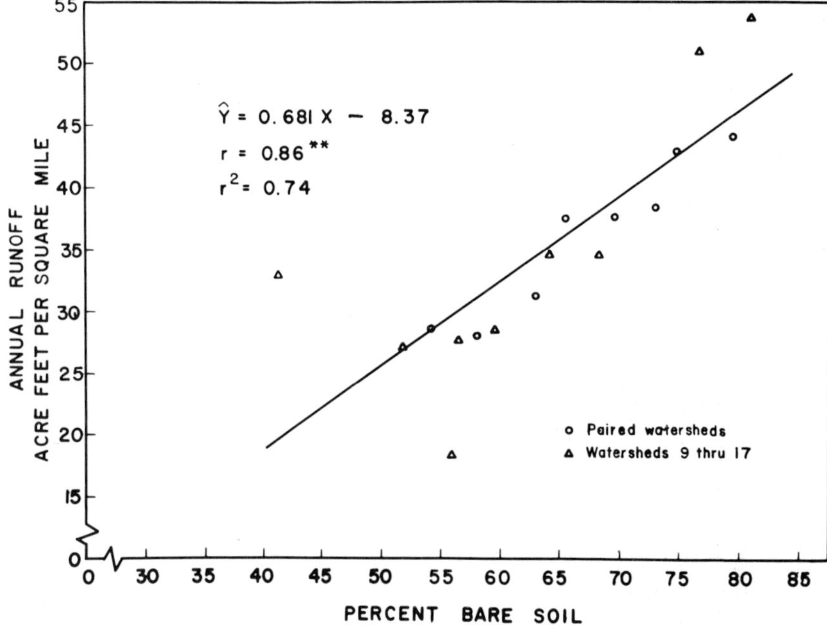

Fig. 5-12. Regression line for percent bare soil versus 15-year average annual runoff for 17 watersheds (after Branson and Owen 1970). Four of the "paired watersheds" were proteted from grazing by livestock, the remaining 13 watersheds were moderately grazed during the 15 years of study. Data points for grazed and ungrazed watersheds were equally distributed below and above the regression line (acre-ft/square mile × 0.48 P **mm).**

curve is shown in Figure 5-12. Similar results were found in predominately grass-covered watersheds (11 inches or 279 mm annual precipitation) in north central New Mexico (Shown 1971). In this instance vegetation plus mulch gave a higher correlation coefficient ($r = 0.89$) with runoff than did bare ground. Similar results are reported by Kincaid and Williams (1966) and Schreiber and Kincaid (1967) at Walnut Gulch.

Marston (1952) analyzed 23 summer storms that produced runoff in an aspen community. His data show that rainfall at a rate of 1.5 inches/hr (38 mm/hr) may result in 20 percent runoff from bare soil, but less than 5 percent runoff when 5 percent or more of the ground is covered. At precipitation rates of 3.0 inches/hr (76 mm/hr), as much as 45 percent of the rainfall occurred as runoff from bare ground, while ground cover (live vegetation plus litter) of at least 65 percent was required to keep runoff at 5 percent of the precipitation.

Influence of Grazing on Runoff

One of the earliest studies of the effects of grazing on runoff was at the Manitou Experimental Forest near Colorado Springs, Colorado. It is typical of many ponderosa pine-bunchgrass ranges. Soils at Manitou are permeable, gravelly alluvium which developed from Pikes Peak granite. The study included control plots which were not grazed and others which were grazed

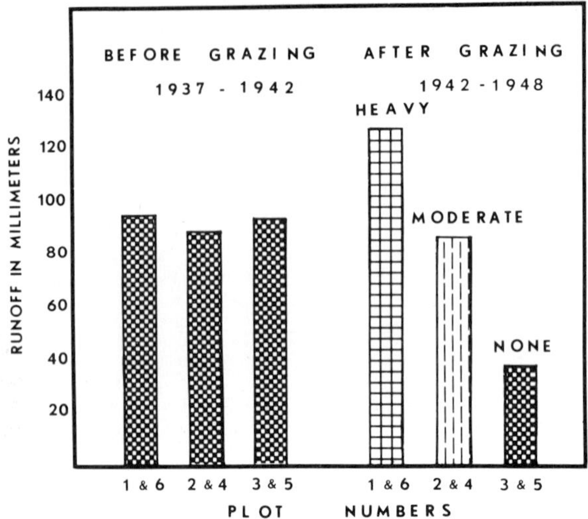

Fig. 5-13. Average runoff, June to September, for bunchgrass rangeland subjected to different intensities of grazing for the precalibration period (1937-1942) and the treatment period (1942-1948) (after Dunford 1949) (mm ÷ 25.4 × inch).

heavily and moderately.

The most runoff was from plots (0.01 ac, 0.004 ha) that were heavily grazed, while the least runoff was from the protected plots (Figure 5-13). The results also showed that although there was an increase in runoff from moderately grazed plots, this was not accompanied by increased soil loss. The investigators concluded, therefore, that moderate grazing was permissible if the resulting loss of water does not cause a critical shortage of moisture for plant development (Dunford 1949, 1954).

Similar runoff results are reported by Hanson et al. (1970) on mixed shortgrass and midgrass prairie near Cottonwood, South Dakota. Average seasonal runoff from the small experimental watersheds was 0.79, 0.56, and 0.42 inch (20, 14, and 11 mm) for watersheds that were grazed heavily, moderately, and lightly, respectively. Their data indicate that heavily grazed watersheds produce runoff from short intense storms as well as from storms of long duration, whereas the lightly grazed watersheds produced runoff mainly from long duration storms that followed periods of antecedent precipitation. If long-duration storms follow a wet period, runoff from lightly grazed watersheds may be as much as from heavily or moderately grazed watersheds.

A third study was carried out on the salt desert shrub rangelands at Badger Wash in western Colorado, an important winter-spring range for sheep and cattle of the intermountain area. In this study the effect of winter-spring grazing by both sheep and cattle on four experimental watersheds was tested against four paired watersheds from which livestock were excluded. Grazing intensity was held constant. After the initial 2 years of the study, there was 30 percent less runoff from the ungrazed watersheds (Figure 5-14) (Lusby 1970).

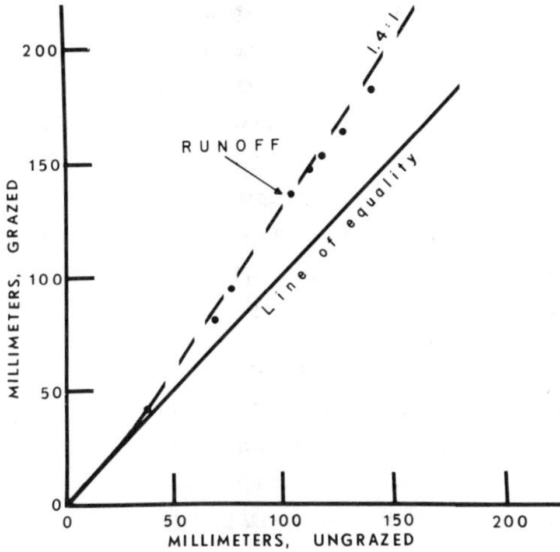

Fig. 5-14. Mass diagram of runoff from grazed and ungrazed watersheds in western Colorado. The data are for 13 years of measurement of 4 grazed and 4 ungrazed watersheds (adapted from Lusby 1970). (Divide mm by 25.4 to obtain inches.)

Additional studies of the effect of grazing on runoff are for the chaparral rangelands found at the Sierra Ancha Experimental Forest in Arizona. In this investigation Rich and Reynolds (1963) found that spring and fall grazing by horses and cattle at intensities of 40% and 80% removal of perennial grasses did not increase runoff. Removal of 80% of the growth of perennial grasses

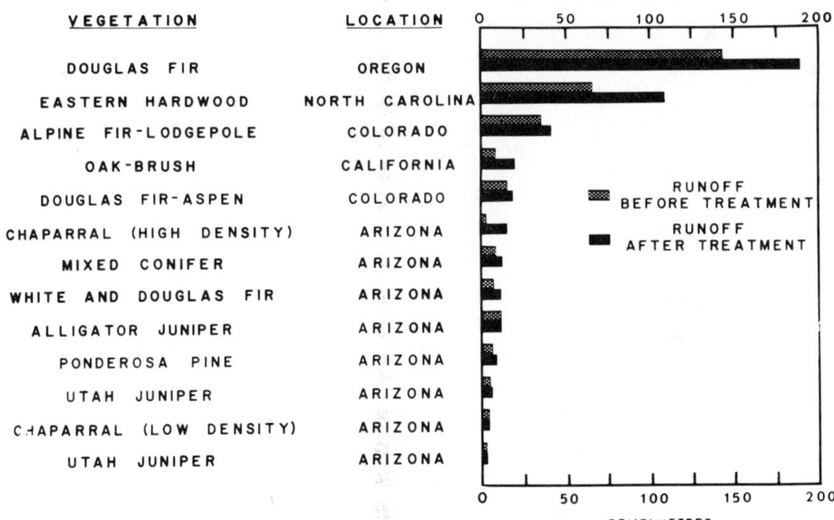

Fig. 5-15. Runoff before and after vegetation conversions, clearing or replacement of woody with herbaceous vegetation, for different vegetation types (from Branson 1976, data adapted from Rothacher 1970, Hoover 1944, Lewis 1966, Coleman 1953, Brown 1970, Rich 1961).

 reduced their basal cover. However, from an earlier study by Martin and Rich (1948) it was found that winter runoff was independent of the amount of perennial grass vegetation. In this area, runoff from winter precipitation accounts for 82% to 88% of the total runoff. This is in contrast to the first three study areas in which runoff comes mainly from summer rainfall. A lysimeter study at Sierra Ancha (Rich 1959) showed that both runoff and sediment yields increased as grazing intensities increased.

It is interesting to note the general lack of studies which adequately define the hydrologic impact of grazing. Various grazing systems (especially rest rotation systems) which are being proposed for much of our public lands have not been studied at all from the standpoint of hydrologic impacts (Gifford and Hawkins 1976). For more information see Chapters 4 and 6 and review papers of Gifford (1975) and Smeins (1975).

Vegetation Conversion Effects on Runoff

Changes in vegetal type or modification of cover on a watershed may result in a corresponding change in the hydrologic regimen (Figure 5-15). These changes may be beneficial or disastrous depending on the circumstances and objectives. The effects wildfires and vegetation conversions have on runoff are discussed here. For additional information, see the chapters on *Infiltration* and *Erosion, Yield, and Water Quality*.

In the case of wildfires, the land manager's objective is usually to get replacement vegetation started on the burned area as soon as possible after the fire in order to minimize flooding and erosion. Vegetation conversions on the other hand, are controlled efforts to replace one vegetal type of low value with one that is more beneficial. From a watershed viewpoint, increased on-site water use efficiency or increased runoff are the usual objectives (Ffolliott and Thorud 1974).

During a 9-year period from 1929 to 1938, runoff was measured on plots that were annually burned, twice burned, and on undisturbed woodland-chaparral (Rowe 1948). Results of this California study show that surface runoff from the plots burned annually averaged 4.76 inches/yr (1,209 mm/yr) compared to 0.63 inch/yr (16 mm/yr) from the twice burned (during the 8-year study) plots, and only a trace of runoff from undisturbed plots. Yearly records revealed that woodland-chaparral was highly effective in minimizing runoff during years of high, as well as low, precipitation (Rowe 1948). Not only can total annual water yields be increased by burning, but yields and peak flows from individual storms may be increased (Figure 5-16).

Runoff from summer storms returned to nearly normal prefire conditions at the end of 4 years following a wildfire in a second-growth ponderosa pine forest near Deadwood, South Dakota (Orr 1970). The area was seeded with grass-legume mixture immediately following the fire. The consequent decline in runoff from rainfall closely followed the increase in live vegetation and litter, and runoff approached stability when a ground cover of about 60% was reached. Runoff from snowmelt was not related to ground cover, but rather to frozen ground and to the amount and distribution of the snowpack (Orr

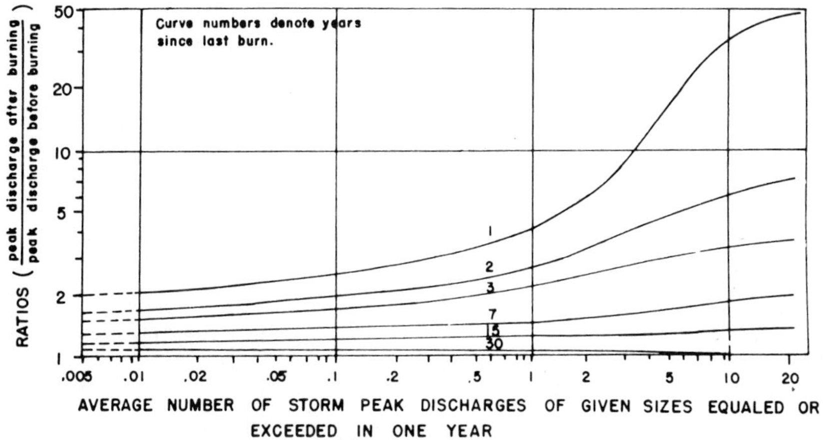

Fig. 5-16. Average effect of fire on peak discharge by frequency classes and number of years following burning in southern California. The greatest impact obviously occurs on the smaller or more common flows (from Rowe, Countryman, and Storey 1954).

1970).

Following a wildfire near Roosevelt Dam in Arizona, which burned over previously calibrated watersheds, two small watersheds of 32 ha (80 acres) and 38 ha (95 acres) were selected, one for studying the effects of natural shrub recovery and the other for studying the conversion to grass on runoff and sediment production. Before the fire the area supported a dense stand of chaparral (*Quercus turbinella, Q. emoryi,* and *Cerocarpus betuloides*). Runoff was highest on both watersheds during the first year following the fire, but by the end of the fourth year runoff from the shrub-covered area was approaching prefire conditions. On the watershed converted to grass, however, runoff had increased from an annual 2.5 cm/yr (1 inch/yr) to nearly 22 cm/yr (8.6 inches/yr) (Pase and Ingebo 1965, Brown 1970) during the sixth the ninth year after treatment. When annual precipitation is less than 40 centimeters (16 inches), increase in water yield resulting from chaparral conversions is likely to be less than 5 centimeters (2 inches). However, the efficiency of the conversion for producing extra water improves with rainfall, at least up to 86.7 centimeters (34 inches) (Hibbert 1971; Hibbert, Davis, and Brown 1975) (Figure 5-17). When low density chaparral was converted to grassland, however, it did not result in any measurable increase in runoff (Brown 1970).

In the pinyon-juniper type in Arizona, Collings and Myrick (1966) compared the runoff from Corduroy Creek Basin (213 square miles) (552 km^2) to Carrizo Creek Basin (237 square miles) (614 km^2). Folliwng 6 years of calibration, 38% of the Corduroy Creek Basin was converted to grass while the Carrizo Creek Basin was left undisturbed. Five years of post-treatment data showed a slight increase in runoff from the treated watershed; however, the variability of the data was great and the investigations concluded that the

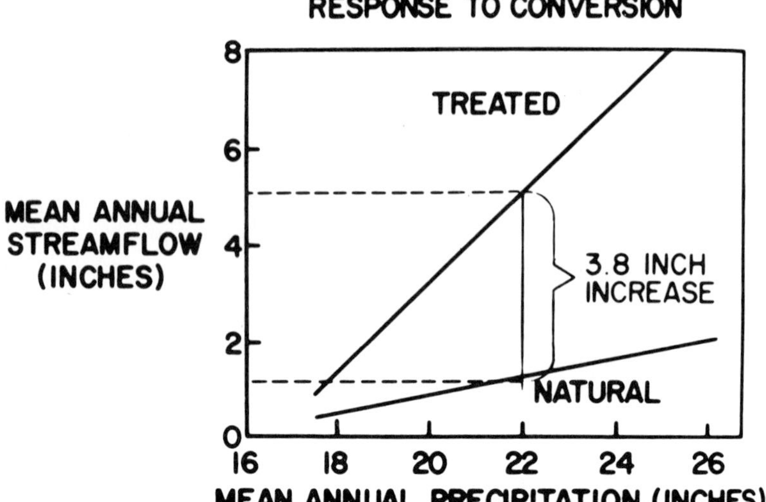

Fig. 5-17. Average water yield from natural and converted chaparral as a function of precipitation. Difference between lines is attributed to treatment (from Hibbert, Davis, and Brown 1975) (inches × 25.4 = mm).

increase in runoff could have been due to chance alone.

Replacing Utah juniper with grass at Beaver Creek in Arizona resulted in a 10% increase in runoff according to 3 years of post-treatment data; this was not, however, a statistically significant increase (Wilm 1966). The soils at Beaver Creek are shallow and have developed from Tertiary and Cretaceous volcanics, which tend to be more porous and permeable than the sedimentary shale with interbedded sandstone of the Supai Formation found at Cibecue Ridge, Arizona. Clary (1975) discussed these same experiments with additional years of data showing similar results. He discussed the data on the basis of three types of treatment (cabling, herbicide, and felling) and concluded that only the herbicide-treated areas had statistically significant increases in water yield. In a similar study at Cibecue Ridge, runoff was greater from the converted (by chaining followed by burning and hand seeding) watershed (than was predicted using a model developed from pretreatment data) during the first two years following treatment, but for the last three years of post-treatment measurements, runoff was considerably lower than predicted. As can be seen in Figure 5-18, runoff was closely related to amount of bare soil following conversion of the woodland to grass. Prediction error in the model was about 30% but changes in recorded water yield exceeded 30% during the five years of post-treatment records (Myrick 1971). He concluded that his work plus that of others indicate that areas receiving less than 20 in (508 mm) annual precipitation have limited potential for water yield increases regardless of vegetation type. Gifford (1975) summarized the relationship of pinyon-juniper manipulation on runoff as follows:

1. Given a runoff event due to high intensity rainfall, least runoff may be

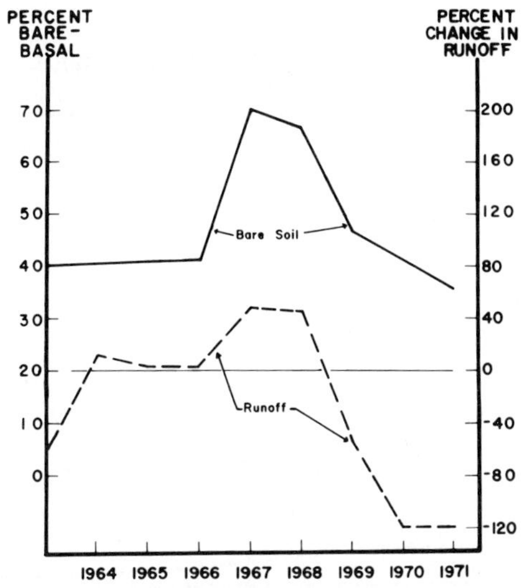

Fig. 5-18. Runoff as related to bare soil (measured by basal point quadrat method) for four pretreatment (1963-1966) (1967-1971) and five post-treatment years on a watershed converted from pinyon-juniper woodland to grassland. Runoff was greater than predictated during the first two years following treatment but considerably less than predicted during the last three years. (Runoff data from R.M. Myrick published data except for last year and unpublished bare soil data by F.A. Branson.)

expected from sites chained with debris-left-in-place, followed very closely by the natural woodland and also sites which have simply been sprayed to kill the trees. Greatest runoff will occur on sites chained with debris windrowed.

2. Where water yield is important, spraying (but not tree removal) is most effective in the Utah juniper type. At higher elevations in Arizona where alligator juniper is found, tree removal may result in a slight increase in water yield. Where only select areas of a watershed are treated or where tree densities are low, increases in water yield should not be expected.

Baker (1975) showed that appreciable increases in water yield could be obtained by thinning pine watersheds in Arizona. Table 5-2 illustrates the magnitude of the water yield increase that might be obtained by various thinning treatments. Results of his computer simulation are shown in Figure

Table 5-2. Summary of effects of thinning pine watersheds on winter streamflow (inches) (Baker 1975). (inches × 25.4 = mm)

Treatment	Mean winter streamflow untreated condition	Mean increase	Percent difference
Clearcut	5.64	1.84	33**
3/4 thin	7.12	1.34	19*
1/3 stripcut	6.30	0.98	16*
1/3 stripcut and thin	4.42	0.82	19*

**Statistically significant at the 1% level.
*Statistically significant at the 5% level.

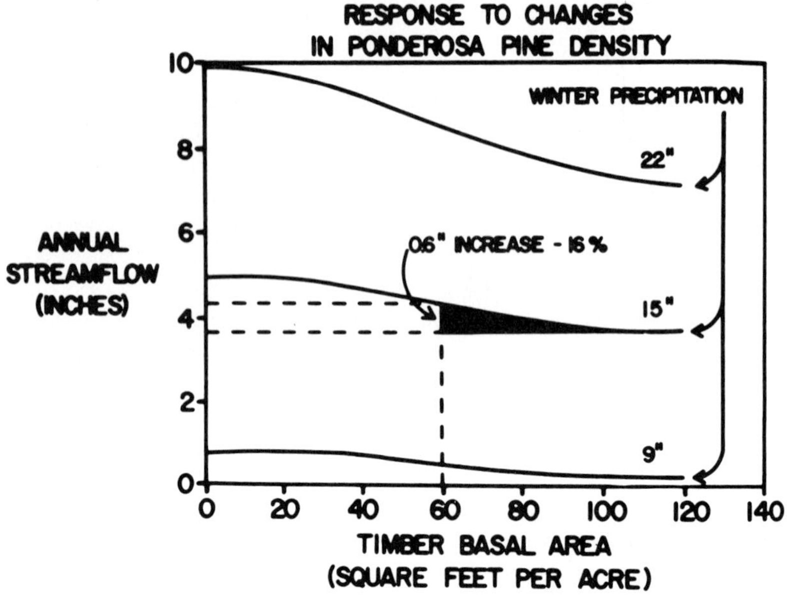

Fig. 5-19. Predictions of annual streamflow by regression analysis based on winter precipitation, insolation, and timber basal area (from Baker 1975). (Multiply inches by 25.4 to obtain mm. Multiply ft²/acre by 0.23 to obtain m²/ha.)

5-19 for various precipitation amounts. It is important to note that these drastic treatments may come under severe criticism from environmentalists if such widespread operations were begun.

There have been many conversions of sagebrush to grassland in order to increase forage production for livestock. A few studies have been made on the hydrologic effects of converting this type to grassland, but it does not seem likely that they will result in increased runoff. Runoff from plowed and seeded watersheds from one study in Colorado has been lower than from untreated big sagebrush watersheds during 3 years following treatment (Branson 1975).

For a discussion of the effects of forest cutting practices on water yield in the west the reader is referred to Bates and Henry (1928), Hoover (1944), Love (1960), Brown (1970), and Rothacher (1970).

At the Central Great Plains Experimental Watershed in Nebraska, four small watersheds were selected for study to determine the effects of reseeding cultivated fields to perennial prairie grasses. Three of the watersheds (two of which were reseeded) were broken out of virgin prairie 50 to 60 years ago and had been cultivated continuously since that time. The remaining watershed had not been plowed or pastured since 1937. From 6 years of poast-treatment data, Dragoun (1969) concluded that plant cover on the watersheds seeded to perennial native grasses resulted in approximately 90% reduction in surface runoff the second year after planting. After the third year, there was little surface runoff and no measurable soil erosion. Upon conversion of a native prairie grass area to cultivation, surface runoff increased substantially the first year and after 4 years runoff began to approach that of continuously cultivated areas.

Impact of Other Range Improvements on Runoff

Though hydrologic impacts of range improvement practices have been of interest for many years, relatively little information is available for evaluating these practices. Most of the data which have been published are derived from something other than small watershed studies (infiltrometer plots, runoff plots, etc.). However, this cannot be considered a serious criticism since most range improvement practices are applied only to selected parts of watersheds and therefore any hydrologic changes which do occur are reasonably localized, at least with respect to the total watershed area. As a result, total watershed studies may not reflect specific on-site hydrologic advantages or disadvantages of a particular range improvement practice.

Range improvements are generally considered to be special treatments, developments, and structures used to improve range forage resources or to facilitate their use by grazing animals (Vallentine 1971). In many cases only broad generalizations regarding hydrologic impact are possible because several improvement practices are generally used together (for instance, pitting + seeding + grazing management), thus confounding the impact of a single treatment by itself. Individual storm characteristics (intensity, duration, quantity) following applications of treatments as well as soil characteristics, will also have a big impact on future behavior and life of projects.

Pitting

Pitting on rangelands generally consists of forming small basins or pits in the soil in an effort to catch and hold rain and runoff water. Capacities of the pits range from about 2 to 15 mm (0.08 to 0.6 in) of precipitation and/or runoff. Hickey and Dortignac (1964) evaluated soil pitting over a 3-year period on easily eroded shale-derived soils near Cuba, New Mexico. They

Table 5-3. Kilograms of herbage, mulch per hectare, and infiltration rates on native shortgrass rangeland, Archer Substation, Cheyenne, Wyoming (from Rauzi, Lange, and Becker 1962).[1]

Pasture treatment	Average air-dry weight		Infiltration rates	
	Herbage (kg/ha)	Mulch (kg/ha)	Second 30-minutes (cm/hr)	Total 1-hr (cm/hr)
Pitted	619[b]	793[ab]	4.3[a]	5.1[a]
Wyoming range seeder	1709[a]	1053[a]	2.4[b]	3.6[b]
Sod drill[2]	950[b]	1007[a]	1.2[b]	2.1[c]
Moderately grazed[3]	813[b]	585[b]	1.2[b]	1.9[c]

[1]Means with the same letter or letters superscript are not statistically different from each other at the .05 level of significance. Figures in parenthesis represent 3-year (1958-60) averages for herbage yields. Treatments were applied in 1955 and were tested in 1960.

[2]Furrow openers were 1.5 in (3.75 cm) wide spaced 10 in (25.4 cm) apart. Penetration was 2 in (7.5) to 4 in (10 cm).

[3]Blue grama leaf length following grazing was 0.9 in (2.25 cm) (multiply kg/ha by 2.26 to obtain lb/acre.)

Table 5-4. Effect of soil-surface modification on ovendry units of forage produced per unit of precipitation received (from Wight and Siddoway 1972) and forage yields as related to checks or controls.

Treatment	Soil texture	Duration of study (years)	Precipitation use efficiency (kg/ha/cm)	Percent of check
Rotary subsoil[1]	Sandy loam	2	34	97
Scalping[2]	Sandy loam	3	42	121
Contour furrowing	Clay	3	15	213
Pitting				
Sidney, Montana	Sandy loam	3	42	121
Havre, Montana	Loam	4	42	176
Glasgow, Montana	Loam	3	41	152
Archer, Wyoming	Fine sandy loam	2	21	151

[1]Rotary subsoil punched holes in a 9 ft (2.7-m) by 9 ft (2.7-m) grid pattern.
[2]Scalped stripes approximately 22 in (56 cm) wide and 3 in (7.6 cm) deep on 15 ft (4.6-m) centers along the countour. (Multiply kg/ha/cm by 2.26 to obtain lb/acre/inch.)

found the surface pits caused reductions of only 12 to 24% of surface runoff and 16% in erosion the first year after treatment. At the end of 3 years, surface runoff was reduced only 10% and erosion was about the same from treated and check plots.

Rauzi, Lang, and Becker (1962) found in Wyoming on sandy loam soils that pitting shortgrass range did increase infiltration rates, even after 5 years following pitting. Summary data from their study are given in Table 5-3. They found that during the second 30-minute period of 1-hour tests, the test plots on the pitted pastures absorbed almost twice as much water as did the test plots on the pastures treated with the Wyoming range seeder, and almost 4 times as much as the pastures moderately grazed or these treated with the sod drill. Also, Rauzi (1956) reported higher water infiltration on pitted rangeland, also in Wyoming. Life of pits in northeastern Wyoming was estimated to be at least 10 years (Barnes, Anderson, and Heerwagen, 1958).

Branson, Miller, and McQueen (1966) reported that pits were obliterated at the end of 8 years on three sites (unidentified) studied in the western United States. Studies in Montana (Ryerson et al. 1970, cited by Wight and Siddoway 1972) and South Dakota (Nichols 1969) indicate the life of pits may approximate 6 years or less.

The fact that certain range improvement practices do increase soil moisture availability has been noted in several studies, an example of which is given in Table 5-4. Obviously the impact of some of the improvement practices has been to reduce runoff and possibly reduce erosion, but exact figures are lacking.

Ripping (also chiseling, subsoiling).

Ripping (to a depth of from 30 to 90 cm, 1 to 3 ft) is used to break or shatter compacted soil profile layers that may inhibit root development and/or moisture penetration. Dortignac and Hickey (1963) and Hickey and Dortignac (1964) studied runoff and erosion from ripping treatments for 3 years on shale-derived soils in New Mexico (see also discussion on pitting). For untreated soils, surface runoff was as high as 89 percent of storm rainfall, and annual erosion as high as 4,640 kg per ha (4,134 lbs/acre). Ripping (70 to 90 cm depth, 2.1 m apart) reduced surface runoff 96 percent and erosion 85 percent in the first year after treatment. Three years following treatment, the reductions amounted to 85 percent for runoff and 31 percent for erosion. Both erosion and runoff varied with topographic position. However, soil "piping" may have confounded these results somewhat.

Branson, Miller, and McQueen (1966) found that ripping treatments performed with the Jayhawk Soilsaver on six sites in the western United States actually decreased perennial grass production. They concluded that the minor soil surface modifications do not have a marked effect on runoff or water retention. Gifford and Skau (1967) indicate that shallow auger ripping (20 to 34 cm, 7.9 to 13.4 inch. penetration) did not increase infiltration rates on loam and sandy loam soils in Nevada during 30-minute-high-intensity simulated rainstorms. On sites where positive results may have resulted due to auger ripping, re-ripping at 3- to 5-year intervals may extend treatment effectiveness (Aldon 1972).

There are some indications that commericially available construction rippers (ripper teeth are 7.6 to 10 cm, 3 to 4 inches wide) have produced more satisfactory results than the relatively thin blades used.

Contour Furrows

Contour furrows are generally constructed with the Model B contour furrowing machine developed by the U.S. Forest Service. The machine makes two furrows 150 cm (5 ft) apart, 20 to 30 cm (8 to 12 inches) deep, 50 to 76 cm (20 to 30 inches) wide, and dammed at intervals of 1.2 to 1.5 m (4 to 5 ft). When newly constructed, the capacity of the furrows exceeds 5 cm (2 inches)

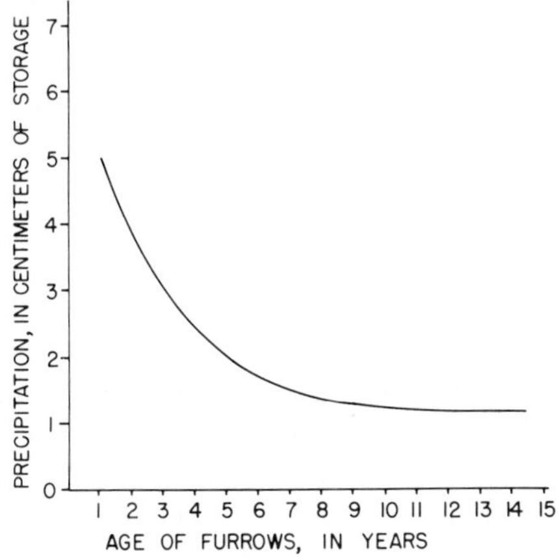

Fig. 5-20. Decrease in water storage of Model B contour furrows as affected by time. Note the rapid decline in storage during the first 5 years. Storage capacities were computed from field measurements of depth, width, and spacing of furrows (from Branson, Miller, and McQueen 1966). (cm

of precipitation. Because of the storage capacity of the furrows, little runoff or sediment would be expected to leave a treated site. Out of several mechanical treatments studied in the western U.S., Branson, Miller, and McQueen (1966) found broadbase furrowing the most productive and contour furrowing the second most productive of seven treatments sampled, with respect to perennial grass yields. They also found that storage capacity of furrows decreased rapidly during the first 5 years to less than 2.5 cm (1 inches) of precipitation, and thereafter stabilized at about 1.2 cm (0.5 inch) after 9 or more years (Figure 5-20). Neff (1973) found similar relationships for eastern Montana, though initial furrow storage capacities were smaller (2.5 cm, 1 inch). Furrow life in eastern Montana ranges from 20 to 35 years. Fisser, Mackay, and Nichols (1974) indicate that furrows on a site in Wyoming may have an influence on forage production for perhaps 35 years. However, where furrows are spaced more than 3 m (10 ft) apart and only 10 to 13 cm (4 to 5 inch) deep, they are ineffective (Hubbard and Smoliak 1953).

Soiseth, Wight, and Aase (1974) have shown that contour furrowing on a panspot (Solonetzic) range site in southeastern Montana improved infiltration, reduced the sodium hazard, and increased herbage production for at least 10 years. Increased infiltration, soil-moisture storage, and forage production were also measured in northeastern Montana on land contour furrowed 10 years prior to the year measurements were made (Branson, Miller, and McQueen 1962).

Branson, Miller, and McQueen (1966) report results from broadbase furrows at two sites. In broadbase furrow construction a motor patrol is used to push earth downdrainage to form dikes 45 to 60 cm (18 to 24 inch) in height, the dikes resembling waterspreaders but differing in that water is not diverted onto the systems from intermittent streams. Vegetation yield on medium to coarse-textured soils (1,750 kg/ha, 1,559 lbs/acre) for a 24 cm (9 inches) annual rainfall site was greater than that of other treatments studied.

Contour Trenches

Contour trenches in the western United States have evolved from small, handmade furrows, 30 to 60 cm (12 to 24 inches) deep, to large, bulldozed trenches, 90 to 120 cm (36 to 48 inches) deep. Since one of the objectives of these trenches is to completely contain all water (and sediment) on site, it is obvious that some impact on runoff and erosion (even if highly localized) should be forthcoming. Little information is available, however, on hydrologic impacts of contour trenching. A discussion of construction and design criteria for contour trenches has been given by Noble (1963).

Doty (1970, 1971, 1972) has looked at contour trenching effects on streamflow, snow accumulation, and soil water distribution on a watershed near Farmington, Utah. The trenches were constructed on a windswept southwest exposure in the upper 15 percent of the 186 ha (510 acres) drainage and were designed to hold 50 percent of the precipitation from a 5-cm (2-inch) storm lasting 1 hour, plus allowing an additional 45 cm (1.5 ft) free-board. Twelve years of streamflow records before trenching and 4 years of records after trenching were analyzed. In general trenching on the windswept site increased snow accumulation slightly, but this appeared to be more significant to revegetation than to water yield. Six years of soil water measurements indicated that the trenching did not alter the soil water conditions sufficiently to significantly change water yields from the catchment. And finally, neither annual water yields nor snowmelt runoff in spring and early summer were significantly altered in either volume or distribution over time as a result of trenching.

Bailey and Copeland (1961) studied two adjacent mountain watersheds in Utah; one had no history of watershed plant cover depletion or floods while the other produced damaging mudrock floods in 1930 following previous destruction of vegetation by overgrazing on the headwater lands. Vegetation was restored on the depleted and eroded areas by contour trenching and reseeding prior to the start of streamflow records. These rehabilitation measures effectively controlled summer storm floods from the previously poor condition watershed and were accompanied by a decrease of 2.70 in (686 mm) in annual runoff over the 22-year period.

In another study, DeByle (1970) has indicated that contour trenches probably reduce wet mantle (when soils were wet from winter storms) flood peaks on the east side of the Sierra Nevada range and may minimize downstream flood damage.

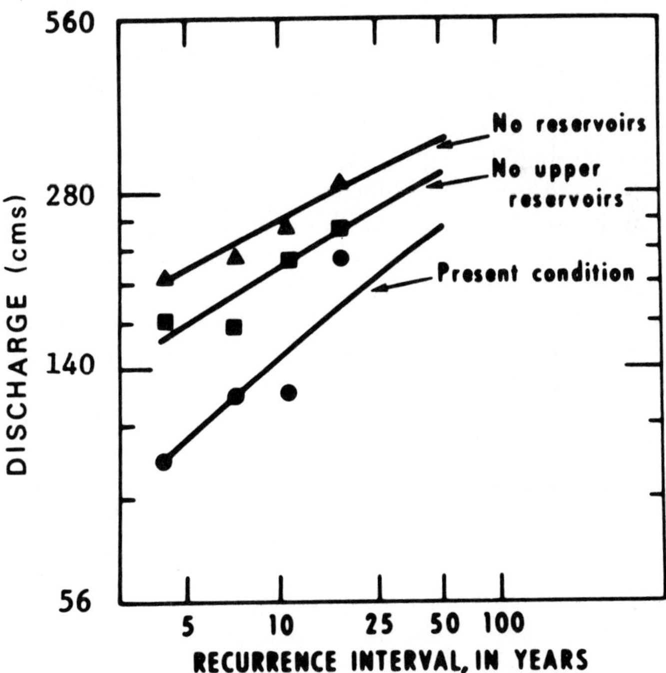

Fig. 5-21. Flood-peak frequency curves showing effects of reservoirs. (from Jennings, Hawkinson, and Bauer 1972). (Multiply m³/sec. by 35.2 to obtain ft³/sec.)

Branson, Miller, and McQueen (1966) studied 12 sites at various locations on which motor patrol trenches had been constructed. These contour trenches were relatively ineffective in improving grass production but appeared to be effective in reducing runoff and sediment yields. Lives of these structures are variable, ranging from 25 years in one instance in the above study to several tens of years in vary favorable instances elsewhere.

Small Reservoir Construction

Small floodwater-retarding structures with uncontrolled outlets and emergency overflow spillways are often used on upstream tributary watersheds. Some hydrologic impacts of these structures are given in the following three examples.

Jennings, Hawkinson, and Bauer (1972) attempted to model effects of floodwater retarding structures on small basin flood flows on Honey Creek, a tributary of East Fork Trinity River in north Texas, with a gaged drainage area of 101 kim^2 (39.0 mi^2). This small basin has 12 structures with a combined capacity of 9.7×10^6 m^3 (7,857 acre-ft) at emergency spillway crests which partially control 54.1 km^2 (20.9 sq. miles) of drainage area. The effectiveness of the Honey Creek reservoirs in reducing flood peak magnitudes was readily seen from various frequency analyses. For example, according to the frequency relation the flood peak of 224 m^3/s (8,000 ft^3/sec) has a recurrence interval of about 8 years under natural conditions but this flood magnitude occurs only once on the average of every 50 years under the present regulated conditions (Figure 5-21).

A second example from a central 2,927 km^2 (1,130 sq. miles) reach of the Washita River basin in south central Oklahoma provides information regarding sediment yield reductions on watersheds treated with flood-retarding structures (Allen and Welch 1971). Data are given for four small tributaries to the Washita River prior to and following construction of structures. Results of the study indicated that sediment yield dropped sharply on each of the four watersheds after flood-retarding structures were installed. The reductions ranged from 48 to 61 percent (Table 5-5).

The third example comes from hydrologic observations made in the 1,393-km^2 (538 sq. miles) Willow Creek basin in north-eastern Montana for the

Table 5-5. Summary of structural treatment, sediment yields, and percent sediment reduction after treatment (from Allen and Welch 1971).

	Structures			Sediment Yield[1]		
Watershed	Number	Dates constructed	Area controlled (%)	Before treatment (kg/ha/hr)	After treatment (kg/ha/yr)	Sediment reduction (%)
Sugar Creek	25	1962-63	40	3,515	1,361	54-61[1]
Winter Creek	9	1965-66	56	6,577	3,334	49[1]-60
Big Dry Creek	5	1966	63	3,107	1,361	51-56[1]
Little Dry Creek	1	1966	60	521	272	48[1]-51

[1]Determined by sediment transport curve method. (Multiply kg/ha/hr by 0.89 to obtain lb/ac/yr.)

period 1954-68 (Frickel 1972). During this study period numerous reservoirs and waterspreaders were constructed in the basin. By the end of the 1967, reservoirs providing storage capacity of about 5.9×10^7 m^3 (48,000 acre-ft) had been constructed and these reservoirs detained runoff and sediment from approximately 75 percent of Willow Creek Basin. This upstream regulation of flow has caused a decrease in both the channel dimensions and the suspended-sediment load at the gaging station near the mouth of the basin. The peak stream discharge recorded at the gaging station is estimated to be about 45% less than it would have been had no reservoirs or waterspreaders been in the basin. The annual runoff volume recorded at the gaging station is estimated to be 18% less than it would be if the basin were totally uncontrolled.

Other good exmples may be found in the symposium edited by Moore and Morgan (1969).

Influence of Cryptogams on Runoff

Little information is available on rangeland watershed values as influenced by cryptogam species. In an appraisal of research on surface materials of desert environments, H.D. Dregne (1968) has stated the following:

> Whether algal and lichen crusts are significant factors in soil stabilization. . .in upland soils is unresolved, as is the effect of such crusts on water penetration and runoff, to say nothing of their effect on soil development.

Soil algae have been recognized as an important agency in stabilizing surface crusts in areas denuded of macrovegetation by drought (Piercy 1917; Drouet 1937). Fletcher and Martin (1948) postulated from observational evidence near Tucson, Arizona, that the microfloral invasion of the raincrust improves infiltration, decreases erosion, and aids in establishment of plant seedlings under rigorous environments of desert communities. In contrast, some observers of the effect of algal crusts believe that they may cause considerable water runoff by sealing soil surfaces (Rozanov 1951).

Fletcher (1960) found in Arizona that a drying period following microfloral growth and crust formation results in incision of the crust and a marked increase in infiltration. If no drying period occurs (and no consequent crust incision) such growth may depress infiltration capacity. Soil type was also found to be in an important factor in determining effects of algal crusts (Figure 5-22). No detectable differences were found by Faust (1971) in runoff, infiltration, or sediment yields from soils covered with blue-green algae and soils denuded of blue-green algal growth.

Booth (1941) studied algal crusts in parts of Kansas, Oklahoma, and Texas and concluded that rate of infiltration is not slowed by the presence of an algal crust and sediment production is minimized. He postulated that algal crust resistance to erosion is apparently the result of binding the surface soil particles into a nonerodible layer, which is also very effective in breaking the force of falling water. Shields, Mitchell, and Drouet (1957) have provided similar findings.

 Ecologic studies in southeastern Utah have postulated the importance of cryptogams; following a comparative study of two grasslands with Canyonlands National Park, Kleiner (1968) commented:

> Because of the significant difference in cryptogamic components of the two parks [grasslands], it is the author's belief that this component of the vegetation may play a more important role in the stability of desert grasslands than has been generally recognized. Certainly the cryptogamic cover would tend to increase the stability of the highly erodible sandy soils of these grasslands during torrential summer rains and heavy wind storms. Many differences between the two parks [grasslands] seem attributable to the protection against erosion which the cryptogamic cover of Virginia (Park) affords. There is, for instance, considerable pedestalling of perennial bunchgrasses in Chesler. In contrast, Virginia shows little pedestalling. In the same vein, rapdily eroding secondary drainage channels cut far into the head of watersheds of Chesler while such channels are stable in Virginia. Lower organic matter, less available PO_4 and higher calcium content of surface soils in Chesler may be explained in terms of slow sheet erosion and loss of the weathered and organically enriched few centimeters of surface soil.

Despite this reasoning, a recent study (Loope and Gifford 1972) in

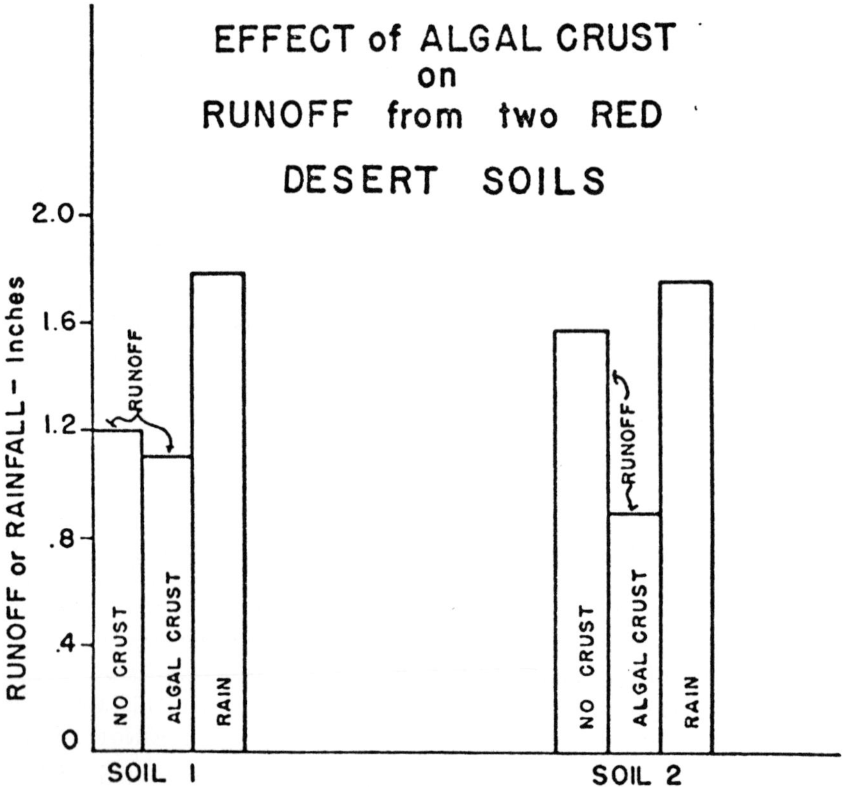

Fig. 5-22. The effect on runoff of removing microflora-infested crust from two soils near Tucson, Arizona. Soil No. 1 is gravelly granitic soil; soil No. 2 is an unstable recent alluvium. Infiltration into soil No. 1 was slightly affected but soil No. 2 was greatly affected by microflora removal (after Fletcher 1960).

southeastern Utah shows that the cryptogamic crust apparently has little effect on soil chemical properties. However, data obtained with the Rocky Mountain infiltrometer indicated that sites with cryptogamic cover had significantly higher infiltration rates than sites with no lichen cover. Patterns of sediment production indicated a potential for increased sediment once the crust had been disturbed.

Water Harvesting

Water harvesting is the process of collecting and storing precipitation from land that has been treated to increase the runoff of rainfall and snowmelt (Myers 1964). More recently, Courrier (1973) defined water harvesting as "the process of collecting natural precipitation from watersheds for beneficial use." Many of our western rangelands are classified either as arid or semiarid or water shortage areas and potential uses for water-harvesting techniques do exist. There are, in fact, many examples of water-harvesting techniques now being used by federal agencies, ranchers, and others for agricultural purposes, watering domestic livestock, or for use by wildlife. In the United States these developments or structures have received various names including rain traps, catchment basins, paved drainage basins, trick tanks, and guzzlers. Additional terms from Australia are roaded catchments and flat batter tanks.

The origin of the term "water harvesting" is not known, but it was probably first used by Geddes (1963) of the University of Sydney (Australia). Although the term "water harvesting" is relatively new, the practice is ancient. Shanon, Evenari and Tadmor (1969) reported that excavated runoff farms were used over 3000 years ago in what is now the Negev Desert of Israel. By the time of the Roman occupation, these runoff farms (rainwater collection and irrigation of lower lying areas) had evolved into relatively sophisticated systems covering a large portion of the Negev Highlands. There is also evidence that less complicated water harvesting systems were used by the Indians in the Four Corners areas of the southwestern United States about 700 to 900 years ago.

Cooley, Dedrick and Frasier (1975) discussed the "state-of-the-art" for water harvesting noting:

> As yet water harvesting is not accepted as a competitive method of providing water supplies, although over 3000 water harvesting systems have been installed around the world. Most catchments are the roaded catchment type and are used in Western Australia where private farms have supplied the capital for installation. In the U.S. catchments have been built almost exclusively on public lands by Government agencies or research organizations.

The methods used to increase runoff can be divided into four general categories: vegetation management, land alteration, chemical treatments, and soil covers.

As mentioned earlier, there have been numerous studies to indicate that runoff can be increased by vegetation management. Best results are obtained for areas of higher precipitation (>11 to 15 inches >279.4 to 381 mm) with the conversion efficiency for extra water improving as rainfall increases Cooley, Dedrick, and Frasier (1975) reported that the potential water yield increases from vegetation management depend upon the percent of the annual precipitation occurring as snow, the type, depth, and slope of the watershed soil, and the varieties of vegetation with their associated evapotranspiration rate, which can be managed considering all other constraints.

Land alteration is often the simplest and least expensive method of water harvesting when it involves nothing more than building walls or ditches to collect runoff from existing or manmade catchments such as rock outcrops, highways, airports, and parking lots. Treating soil surfaces to prevent infiltration is a most intriguing approach to water harvesting.

Water Harvesting Materials and Methods

Materials used to improve runoff from catchment surfaces are generally of two classes, either mechanical or chemical with a possible third category being a combination of mechanical and chemical. Many new substances and techniques will probably be used in the future but a classification of most that have been used is as follows:

A. Mechanical methods:
   1. Removal of stones and building of trenches to collect hillside runoff.
   2. Removal of vegetation plus shaping and compaction of water-collecting areas.
   3. Use of iron sheets and corrugated metal roofing. Other similar materials include plastic films or sheets, butyl rubber sheets, aluminum foil sheets, and heavy-weight roofing paper.
   4. Water collection and storage from naturally impermeable ("slickrock") areas.
   5. "Roaded catchments" and "flat batter tanks" in Australia—techniques which place subsurface clays from subsoils on the surface to improve runoff.

B. Chemical methods
   1. Some of the many chemicals used to decrease infiltration include: waxes such as paraffin, salts ($NaCl$, Na_2CO_3, $NNa_5P_3O_{10}$, etc.), latex, heavy oils, sodium rosinate, dialkyl quarternary ammonium chloride, metallic soaps, fatty amine acids, sodium methyl silanolate, and silicones.
   2. Various formulations of asphalt treated jute, asphalt-fibergrass, and asphaltic emulsions.
   3. Soil cement about 4 inches (10 cm) thick.

Water Harvesting Designs

Many designs have been used but nearly all have three components: (1) a water-collection area or apron, (2) a storage tank, earthen pit or storage bag, and (3) a watering trough with automatic float valve (Figure 5-23). A somewhat different design is referred to as a "trick tank" (Pearson, Morrison, and Wolke 1969). Trick tanks consist of an elevated roof which serves as both collection area and shade for livestock. Other features, storage tank and watering trough, are similar to that described above.

Collection areas should be large enough to provide the amount of water needed. Variables to consider include annual precipitation, permeability of the collection apron, and expected evaporation from collection apron or other collection surface, storage facility, and trough. In a 10-inch (254 mm) precipitation isohyet, 1-yd² (0.84 m²) should yield 56 gallons (212 liters) of water without losses. For arid lands the storage facility should be large enough to contain about 2-years water supply for carryover in dry seasons (McBride and Shiflet 1974).

Performance of Materials

All artificial catchment surfaces are subject to various kinds of damage and periodic maintenance efforts are required for optimum performance. If fences are not maintained, damage may result from large mammals, mainly puncturing of surfaces. Rodents may also create holes by gnawing and burrowing. Damage may also result from plant growth, insects, vandalism, wind action, and material failure due to oxidation, ozone attack, and tearing (Dedrick 1973). Ultra-violet radiation is damaging to both rubber and plastics. Possible solutions to these problems include: the use of soil sterilants before aprons are installed to control plant punctures, proper fencing, the use of open storage in snow areas to prevent storage bag failure due to snowpack, rodent repellents and barriers, reinforced materials to prevent tearing (ozone damage occurs only when material stretches), the uses of materials not subject to oxidation such as butyl rubber, and the use of gravel layers to protect plastic

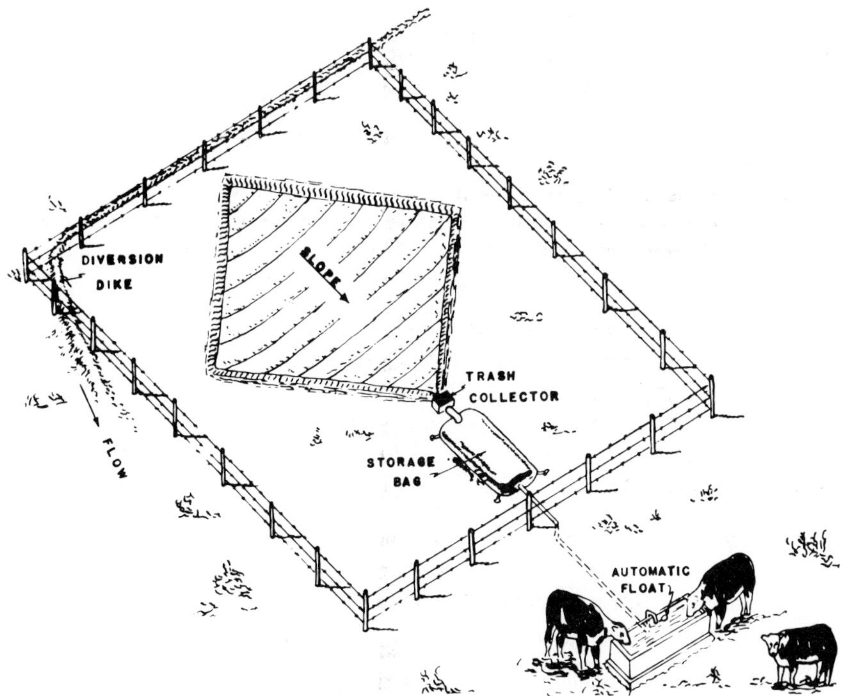

Fig. 5-23. Features of a water harvesting system (from Lauritzen 1961).

 films.

Estimated longevity of various materials is presented in Table 5-6. Although more expensive initially, more durable materials may be more economical if investment costs are discounted over time. Runoff as percent of precipitation may also be important because smaller catchments can be used when surfaces are less permeable.

Advantages and Disadvantages

Water harvesting has advantages and disadvantages. Some of these have been stated earlier but some of the advantages can be listed as follows:

1. It may be the only practical means of developing water in certain areas.
2. It does not require water rights for development in most areas.
3. It is usually less expensive and more convenient than hauling water.
4. It does not add to energy shortage or water pollution problems.
5. There are no "dry holes."

Disadvantages include high initial and maintenance costs of developments.

Additional Application of Water Harvesting

In Montana Hodder et al. (1970) used plastic-covered small basins for tree and shrub plantings. These served as "condensation traps," directing condensed water downward towards the seedlings and probably trapped precipitation in a similar manner. In New Mexico, Aldon and Springfield (1974) used 4-ft^2 (0.4 m^2) plots of wax and 4-mil black polyethylene to increase survival and growth of 4-wing saltbush seedlings. Runoff increase was 40% for paraffin and 34% for polyethylene.

Evaporation and Seepage Reduction

Reduction of water losses caused by evaporation may also be considered an important phase of water harvesting. Prevention of water losses has a monetary value equivalent to the cost of various methods of supplying water. Covered storage tanks and butyl rubber water storage bags control evaporation. Extensive tests with monomolecular films on stock reservoirs (Koberg, Curse, and Shrewsberg 1963) have not shown satisfactory results and have led to the use of more durable materials. Floating sheets of foam rubber, 3/16 in thick, have been used in Utah (Dedrick, Hansen, and Williamson 1973) to reduce evaporation loss from stock tanks. There were some problems encountered in the use of the material (pecking by birds, temporary clogging of bailing holes, and separation of the cover by ice), but none of these problems caused failure. The material is expensive (cost was $1.80 to $2.00 per 1,000 gallons (3785 liters of water saved) but may be justifiable when compared to alternate means of producing or saving water.

Seepage losses from reservoirs may equal or exceed evaporation losses. Many of the techniques used for rainwater harvesting can also be used to prevent seepage losses. Some of these include: soil compaction, chemical treatment of soil, and soil covers such as butyl rubber, sheet plastic, asphalt reinforced with plastic or fiberglass, and ferrocement (National Academy of Science 1974).

Table 5-6. Runoff efficiency, estimated longevity, and initial costs of methods and materials used in water harvesting procedures.

Material and geographic area	Runoff as percent of precipitation	Estimated life of treatment treatment years	Initial treatment cost in dollars/m^2	Source of data
Land clearing (Arizona)	20-30	5-10	0.01-0.02	Frasier 1974
Land clearing (Israel)	21	——	——	Tadmore and Shanan 1969
Soil smoothing (Arizona)	25-35	5-10	0.4- .06	Frasier 1974
Silicone water repellants (Arizona)	50-80	5-80	.0- .15	Frasier 1974
Paraffin wax (Arizona)	60-90	5- 8	.25- .33	Fraiser 1974
Paraffin was (Arizona)	92	——	.10- .15	Fink et al. 1973
Concrete (Arizona)	60-80	20	1.67-4.20	Frasier 1974
Gravel-covered sheeting (Arizona)	70-80	10-20	.42- .58	Frasier 1974
Gravel-covered sheeting (Wyoming and Colorado)	45	——	——	Rauzi et al. 1973
Asphalt-fiberglass (Arizona)	85-95	5-10	.83-1.67	Frasier 1974
Asphalt roofing (Wyoming and Colorado)	100	——	——	Rauzi et al. 1973
Butyl rubber (Arizona)	90-100	10-15	1.67-2.50	Frasier 1974
Butyl rubber (Arizona)	100	——	2.39-3.59	Fink et al. 1973
Sheet metal (Arizona)	90-100	20	1.67-2.50	Frasier 1974
Sheet metal (Colorado)	57	——	——	Mickelson 1974
Bentonite (Colorado)	19	——	——	Mickelson 1974
Salt treatment (NaCl) (Wyoming and Colorado)	25	——	——	Rauzi et al. 1973
Natural cover (Colorado)	5	——	——	Mickelson 1974
Natural cover (Arizona)	28	——	——	Fink et al. 1973
Natural cover (Israel)	7	——	——	Tadmor and Shanan 1969

Divide m^2 by 10.76 to obtain $ft.^2$

 Calcium aggregated clays sometimes cause reservoir seepage. Chemicals such as sodium carbonate and sodium tri-polyphosphate have been used to reduce seepage from reservoirs containing clay soils. Bentonite has been used to seal reservoirs having porous soils. If plastic or rubber sheeting is used, protection from livestock damage by fencing or soil covering is necessary.

Literature Cited

Aldon, E.F. 1972. Reactivating soil ripping treatments for runoff and erosion control in the southwestern U.S. Ann. of Arid Zone 11: 154-160.

Aldon, E.F., and H.W. Springfield. 1974. Using paraffin and polyethylene to harvest water for growing shrubs. *In:* Proc. Water Harvesting Symposium, Phoenix, Arizona, U.S. Dept. Agr., Res. Serv. ARS W-22, 329 p.

Allen, P.B., and N.H. Welch. 1971. Sediment yield reductions on watersheds treated with flood-retarding structures. Trans. Amer. Soc. Agr. Eng. 14: 841-871.

Allis, J.A., F.J. Dragoun, and A.L. Sharp. 1964. Tranmission losses in loessial watersheds. Amer. Soc. Agr. Engi. Trans. 7: 209-212.

Babcock, H.M., and E.M. Cushing. 1942. Recharge to ground water from floods in a typical desert wash, Pinal County, Arizona. Amer. Geophys. Union Trans., 23(1):49.

Bailey, R.W. 1945. Determining trend of range-watershed conditions essential to success in management. J. Forest. 43: 733-737.

Bailey, R.W., and O.L. Copeland. Jr. 1961. Low flow discharges and plant cover relations on two mountain watersheds in Utah. Int. Ass. Sci. Hydrol. Pub. 51: 267-278.

Baker, M.B., Jr. 1975. Modeling management of ponderosa pine forest resources. *In:* Watershed Management Symposium Proceedings, Amer. Soc. Civil Eng. p. 478-493.

Barnes, O.K. 1952. Pitting and other treatments on native range. Wyoming Agr. Exp. Sta. Bull. 318, 23 p.

Barnes, O.K., D. Anderson, and H. Heerwagen. 1958. Pitting for range improvement in the Great Plains and the southwest desert regions. USDA Production Res. Rep. 23: 17 p.

Bates, C.G., and A.J. Henry. 1928. Forest and stream flow experiment at Wagon Wheel Gap, Colorado. Monthly Weather Rev. Suppl. 30. 79 p.

Booth, W.E. 1941. Algae as pioneers in plant succession and their importance in erosion control. Ecology 22: 38-46.

Branson, F.A. 1975. Natural and modified plant communities as related to runoff and sediment yields. p. 157-172 *In:* Ecological Studies 10, Coupling of Land and Water Systems. Springer-Verlag New York, Inc.

Branson, F.A. 1976. Water use on rangelands p, 193-209. *In:* Proc. Fifth Workshop of U.S./Australia Rangelands Panel. Utah Water Res. Lab., Logan, Utah.

Branson, F.A., R.F. Miller, and I.S. McQueen. 1962. Effects of contour furrowing, grazing intensities, and soils on infiltration rates, soil moisture, and vegetation near Fort Peck, Montana. J. Range Manage. 15: 151-158.

Branson, F.A., R.F. Miller, and I.S. McQueen. 1966. Contour furrowing, pitting and ripping on rangelands of the western United States. J. Range Manage. 19: 182-190.

Branson, F.A., and J.R. Owen. 1970. Plant cover, runoff, and sediment yield relationships on Mancos Shale in Western Colorado. Water Resources Res. 6: 783-790.

Brown, H.E. 1970. Status of pilot watershed studies in Arizona. Amer. Soc. Civil Eng. Proc., J. Irrig. & Drainage Div. 96, Pap. 7129: 1-23.

Burkham, D.E. 1970. Depletion of streamflow by infiltration in the main channels of the Tucson basin, southeastern Arizona. U.S. Geol. Surv. Water-Supply Pap. 1939-B. 36 p.

Clary, W.P. 1975. Multiple use effects of manipulating pinyon-juniper. p. 469-477 *In:* Watershed Management Symposium Proceedings. Amer. Soc. Civil Eng.

Coleman, E.A. 1953. Vegetation and Watershed Management. The Ronald Press Co., New York. 412 p.

Collings, M.R., and R.M. Myrick. 1966. Effects of juniper and pinyon eradication on streamflow from Corduroy Creek Basin, Arizona. U.S. Geol. Surv. Prof. Pap. 491B. 12 p.

Cooley, K.R., A.R. Dedrick, and G.W. Frasier. 1975. Water harvesting: State of the art. p. 1-20. *In:* Watershed Management. Amer. Soc. Civil Eng. New York.

Courrier, W.F. 1973. Water harvesting by trick tanks, rain traps, and guzzlers. p. 169-181 *In: Water-Animal Relations Symposium, Twin Falls, Idaho.*

DeByle, N.V. 1970. Do contour trenches reduce wet-mantle flood peaks? USDA Forest Service, Res. Note INT-108: 7.

Dedrick, A.R. 1973. Raintrap performance on the Fishlake National Forest. J. Range Manage. 26(1): 9-12.

Dedrick, A.R., T.D. Hansen, and W.R. Williamson. 1973. Floating sheets of foam rubber for reducing stock tank evaporation. J. Range Manage. 26: 404-406.

Dortignac, E.J., and W.C. Hickey. 1963. Surface runoff and erosion as affected by soil ripping. USDA Misc. Pub. No. 970: 156-165.

Dortignac, E.J. 1956. Watershed resources and problems of the Upper Rio Grande Basin. U.S. Forest Serv., Rocky Mountain Forest and Range Exp. Sta. 107 p.

Doty, R.D. 1970. Influence of contour trenching on snow accumulation. J. Soil and Water Conserv. 25: 102-104.

Doty, R.D. 1971. Contour trenching effects on streamflow from a Utah watershed. USDA Forest Service Res. Pap. INT-95: 19 p.

Doty, R.D. 1972. Soil water distribution on a contour-trenched area. USDA Forest Service Res. Note INT-163: 6 p.

Dragoun, F.J. 1969. Effects of cultivation and grass on surface runoff. Water Resources Res. 5: 1078-1083.

Dregne, H.E. 1968. Appraisal of research on surface materials of desert environments, p. 287-316. *In:* Deserts of the World. Univ. Arizona Press, Tucson.

Drouet, F. 1937. The Brazilion Myxophycae I. Amer. J. Bot. 24: 598-608.

Dunford, E.G. 1949. Relation of grazing to runoff and erosion on bunchgrass ranges USDA Forest Service Rocky Mtn. Forest & Range Exp. Sta. Res. Note 7. 2 p.

Dunford, E.G. 1954. Surface runoff and erosion from pine grasslands of the Colorado Front Range. J. Forest. 59: 923-927.

Engman, E.T. 1974. Partial area hydrology and its application to water resources. Water Resour. Bull. 10: 512-521.

Engman, E.T., and A.S. Rogowski. 1974. A partial area model for storm flow synthesis.d Water Resources Res. 10: 464-472.

Faust, W.F. 1971. Blue-green algal effects on some hydrologic processes at the soil surface, p. 99-105. *In:* Hydrology and Water Resources in Arizona and the Southwest, Proc. 1971 Meetings of the Ariz. Amer. Water Resources Ass. and the Hydrol. Sec., Ariz. Acad. Sci. (Tempe), April 22-23.

Ffolliott, P.F., and D.B. Thorud. 1974. Vegetation management for increased water yield in Arizona. Ariz. Agr. Exp. Sta., Tech. Bull. 215. 38 p.

Fink, D.H., K.R. Cooley, and G.W. Frasier. 1973. Wax-treated soil for harvesting water. J. Range Manage. 26: 396-398.

Fisser, H.G., M.H. Mackay, and J.T. Nichols. 1974. Contour-furrowing and seeding on Nuttall saltbush rangeland of Wyoming. J. Range Manage. 27: 459-462.

Fletcher, J.E. 1960. Some effects of plant growth on infiltration in the southwest, p. 51-63. *In:* Water Yield in Relation to Environment in the Southwestern United States. Amer. Ass. Adv. Sci. Symp. (Texas).

Fletcher, J.E., and W.P. Martin. 1948. Some effects of algae and molds in the raincrust of desert soils. Ecology 29: 95-100.

Frasier, G.W. 1974. Water harvesting for livestock, wildlife and domestic use. Proc. Water Harvesting Symp., Phoenix, Ariz. March 26-28, 1974. ARS W-22, p. 40-49.

Frickel, D.G. 1972. Hydrology and effects of conservation structures, Willow Creek Basin, Valley County, Montana, 1954-68. U.S. Geol. Surv. Water-Supply Pap. 1532-G: 34 p.

Geddes, H.J. 1963. Water harvesting, *In:* Water Resource Use and Management. Mellbourne Univ. Press, Mellbourne, Australia.

Gifford, G.J. 1975. Beneficial and detrimental effects of range improvement practices on runoff and erosion. *In:* Watershed Management, Proc. Amer. Soc. Civil Eng. Symp., Utah State Univ., Logan, May: 216-248.

Gifford, G.F., and R.H. Hawkins, 1976. Grazing systems and watershed management—a look at the record. J. Soil & Water Conserv. 31: 281-283.

Gifford, G.F., and C.M. Skau. 1967. Influence of various rangeland cultural treatments on runoff and sediment production from the big sagebrush type, Eastgate Basin, p. 137-148. *In:* Nevada. Third Amer. Water Resour. Conf. (San Francisco), Proc., Nov. 8-10.

Glymph, L.M., and H.N. Holtan. 1969. Land treatment in agricultural watershed hydrology research. p. 44-68 *In:* Effects of watershed changes on streamflow, Water Resources Symp. No. 2, Univ. Texas Press. Austin.

Hanson, C.L., A.R. Kuhlman, C.J. Erickson, and J.K. Lewis. 1970. Grazing effects on runoff and vegetation on western South Dakota rangeland. J. Range Manage. 23: 418-420.

Hewlett, J.D. 1961. Soil moisture as a source of baseflow from steep mountain watersheds. USDA Forest Service Southeast. Forest Exp. Sta. Res. Pap. SE 132: 11 p.

Hewlett, J.D. 1974. Comments on letters relating to "Role of subsurface flow in generating surface runoff, 2, upstream source areas" by R. Allan Freeze. Water Resources Res. 10: 605-607.

Hewlett, J.D., and A.R. Hibbert. 1963. Moisture and energy conditions within a sloping soil mass during drainage. J. Geophys. Res. 68(4): 1081-1087.

Hewlett, J.D., and A.R. Hibbert. 1967. Factors affecting the response of small watersheds to precipitation in humid areas. *In:* Int. Symp. on Forest Hydrology, edited by W.E. Sopper and H.W. Lull, Pergamon, New York. p. 275-290.

Hewlett, J.D., and W.L. Nutter. 1970. The varying source area of streamflow from upland basins. *In:* Proc. of the Symp. on Interdisciplinary Aspects of Watershed Manage. Amer. Soc. Civil Eng., New York. p. 65-83.

Hibbert, A.R. 1971. Increases in streamflow after converting chaparral to grass. Water Resources Res. 7: 71-80.

Hibbert, A.R., E.A. Davis, and T.C. Brown. 1975. Managing chaparral for water and other resources in Arizona. p. 445-468. *In:* Watershed Management, Proc. Amer. Soc. Civil Eng. Symp., Logan, Utah, Aug. 11-13.

Hickey, W.C., Jr., and E.J. Dortignac. 1964. An evaluation of soil ripping and soil pitting on runoff and erosion in the semi-arid southwest. Int. Union Geodesy and Geophys. Int. Ass. Sci. Hydrol., Land Erosion. Precipitation Hydrometry. Soil Moisture, Comm. Pub. 65: 22-33.

Hodder, R.L., D.E. Ryerson, R. Mogen, and J. Buckholtz. 1970. Coal mine spoils reclamation research project located at Western Energy Company, Coalstrip, Montana. Anim. and Range Sci. Dept., Montana State Univ., Bozeman. 56 p.

Hoover, M.C. 1944. Effect of removal of forest vegetation upon water yields. Trans. Amer. Geophys. Union, Pt. 6: 969-975.

Horton, R.E. 1933. The role of infiltration in the hydrologic cycle. Trans. Am. Geophys. Union, 14: 446-460.

56 p.

Hoover, M.C. 1944. Effect of removal of forest vegetation upon water yields. Trans. Amer. Geophys. Union, Pt. 6: 969-975.

Horton, R.E. 1933. The role of infiltration in the hydrologic cycle. Trans. Amer. Geophys. Union 14: 446-460.

Hubbard, W.A., and S. Smoliak. 1953. Effects of contour dykes and furrows on short-grass prairie. J. Range Manage. 6: 55-62.

Hursch, C.R. 1936. Storm water and absorption. Eos Trans. Amer. Geophys. Union 17: 301-302.

Hursch, C.R., and E.F. Brater. 1941. Separating storm hydrographs from small drainage areas into surface and subsurface flow. Eos Trans. Amer. Geophys. Union 22: 863-871.

Hursh, C.R., and P.W. Fletcher. 1942. The soil profile as a natural reservoir. Soil Sci. Soc. Amer. Proc. 7: 480-486.

Iseri, K.T., and W.B. Langbein. 1974. Large rivers of the United States. U.S. Geol. Circ. 686. 10 p.

Jennings, M.E., R.O. Hawkinson, and D.P. Bauer. 1972. Application of a stream system model for evaluating effects of floodwater retarding structures on small basin flood flows, p. 166-170. *In:* Proc. Amer. Water Resources Ass. Symp. on "Watersheds in Transition," Ft. Collins, Colo., June 12-22.

Keppel, R.V., and K.G. Renard. 1962. Transmission losses in ephemeral stream beds. Amer. Soc. Civil Eng. Proc., J. Hydraul. Div. 88 (HY-3): 59-68.

Kincaid, D.R., and G. Williams. 1966. Rainfall effects on soil surface characteristics following range improvement treatments. J. Range Manage. 19: 346-351.

Kleiner, E.F. 1968. Comparative study of grasslands on Canyonlands National Park. Unpub. Diss., Univ. Utah 58 p.

Koberg, G.E., R.R. Cruse, and C.E. Shrewsburg. 1963. Evaporation control research, 1959-60. U.S. Geol. Surv., Water-Supply Pap. 1692. 53 p.

Lane, L.J. 1972. A proposed model for flood routing in abstracting ephemeral channels. Hydrol. and Water Resources in Arizona and the Southwest. Amer. Water Resources Ass., Arizona Section—Arizona Academy of Science, Hydrology Section, Proc. of May 5-6 meeting, Prescott, Arizona, II: 439-453.

Lane, L.J., M.H. Diskin, and K.G. Renard. 1970. Input-output relationships for an ephemeral stream channel system. J. Hydrology 13: 22-40.

Lauritzen, C.W. 1961. Collecting desert rainfall. Crops and Soils, Aug.-Sept. 2 p.

Lewis, D.C. 1966. Hydrologic change in a watershed with conversion from an oak woodland to annual grasses. Amer. Geophys. Union Symp. Pap. H8. 23 p.

Loope, W.L., and G.F. Gifford. 1972. Influence of a soil microfloral crust on select properties of soils under pinyon-juniper in southeastern Utah. J. Soil and Water Conserv. 27: 164-167.

Love, Dudley. 1960. Water yield for alpine, sup-alpine, and montane vegetation zones, p. 3-15. *In:* Symp. Water Yield in Relation to Environment in the Southwestern United States, Proc.

Lusby, G.C. 1970. Hydrologic and biotic effects of grazing vs. non-grazing near Grand Junction, Colorado. J. Range Manage. 23: 256-260.

Marston, R.B. 1952. Ground cover requirements for summer storm runoff control on aspen sites in northern Utah. J. Forest. 50: 303-307.

Martin, W.P., and L.R. Rich. 1948. Preliminary hydrologic results, 1935-1948, "Base Rock" undisturbed soil lysimeters in the grassland type, Arizona. Amer. Soil Sci. Soc., Proc. 13: 561-567.

McBride, M.W., and L.W. Shiflet. 1974. Water harvesting catchments in the Stafford District, southeastern Arizona, p. 115-121. *In:* Proc. Water Harvesting Symp., Phoenix, Ariz., U.S. Dept. Agr., Res. Serv. ARS W-22.

Mickelson, R.M. 1974. Performance and durability of sheet metal, butyl rubber, asphalt roofing, and bentonite for harvesting precipitation, p. 93-102. *In:* Proc. Water Harvesting Symp., Phoenix. Ariz. ARS W-22.

Moore, W.L. and C.W. Morgan (editors). 1969. Effects of watershed changes on streamflow. Center for Res. *In:* Water Resour., Univ. Texas, Austin, Water Resour. Symp. No. Two: 289 p.

Myers, L.E. 1964. Harvesting precipitation. Int. Ass. Sci. Hydrology, Berkeley, Calif. Pub. 65: 343-351.

Myrick, R.M. 1971. Cibecue Ridge juniper project. 15th Arizona Watershed Symp. (Phoenix) Proc., Arizona Water Comm. p. 35-39.

National Academy of Science. 1974. More water for arid lands: promising technologies and research opportunities. Nat. Acad. Sci. Washington, D.C. 153 p.

Neff, E.L. 1973. Water storage capacity of contour furrows in Montana. J. Range Manage. 26: 298-301.

Nichols, J.T. 1969. Range improvement practices on deteriorated dense clay wheatgrass range in western South Dakota. South Dakota Agr. Exp. Sta. Bull. 552: 23 p.

Noble, E.L. 1963. Sediment reduction thorough watershed rehabilitation, p. 114-123. *In:* USDA Misc. Pub. 970.

Nutter, W.L. 1973. The role of soil water in the hydrologic behavior of upland basins. Soil Sci. Soc. Amer. Spec. Pub. 5. p. 1-14.

Orr, H.K. 1970. Runoff and erosion control by seeded and native vegetation on a forest burn, Black Hills, South Dakota. USDA Forest Service Res. Pap. RM-60. 12 p.

Osborn, H.G., and L. Lane. 1969. Precipitation-runoff relation for very small semiarid rangeland watersheds. Water Resources Res. 5: 419-425.

Osborn, H.B., and K.G. Renard. 1970. Thunderstorm runoff of the Walnut Gulch Experimental Watershed, Arizona, USA, p. 455-464. *In:* Proc. IASH-UNESCO Symp. on Results of Res. on Representative and Experimental Basin, Victoria University of Willington, New Zealand, Int. Ass. Sci. Hydrol. 96: 455-464.

Pase, C.P., and P.A. Ingebo. 1965. Burned chaparral to grass: early effects on water and sediment yields from two granitic soil watersheds in Arizona, p. 8-11. *In:* 9th Arizona Watershed Symp. (Tempe), Proc. Sept. 22.

Pearson, H.A., D.C. Morrison, and W.K. Wolke. 1969. Trick tanks: water developments for range livestock. J. Range Manage. 22: 359-360.

Piercy, A. 1917. The structure and mode of life on a form of *Hormidium flaccidum,* A. Brann. Ann. Bot. 31: 513.

Rauzi, R. 1956. Water-infiltration studies on the Big Horn National Forest. Wyoming Agr. Exp. Sta. Circ. 62. p. 1-7.

Rauzi, R., M.L. Fairborn, and L. Landers. 1973. Water harvesting efficiencies of four soil surface treatments. J. Range Manage. 26: 399-403.

Rauzi. F., R.L. Lange, and C.F. Becker. 1962. Mechanical treatments on shortgrass rangeland. Wyoming Agr. Exp. Sta. Bull. 396. 16 p.

Renard, K.G. 1970. The hydrology of semiarid rangeland watersheds. USDA-ARS 41-162, 26 p.

Rich, L.R. 1959. Hydrologic research using lysimeters of undisturbed soil blocks, p. 139-145. *In:* Symp. of Hannoversch—Muden, Int. Sci. Hydrol. Ass., Pub. 49 (2).

Rich, L.R. 1961. Effect of vegetation changes on water and sediment yields. Abstr. Pacific Southwest Inter-Agency Committee Meeting Proc., Richfield, Utah.

Rich, L.R., and H.G. Reynolds. 1963. Grazing in relation to runoff and erosion on some chaparral watersheds in central Arizona. J. Range Manage. 16: 3-2-326.

Roessel, B.W.P. 1950. Hydrologic problems concerning the runoff in headwater regions. Eos Trans. AGU, 31: 431-442.

Rothacher, J. 1970. Increases in water yield following clear-cut logging in the Pacific Northwest. Water Resources Res. 6: 653-658.

Rowe, P.B. 1948. Influence of woodland chaparral on water and soil in central California. USDA and State of California Dep. Nat. Resources Div. Forest. Unnumbered Pub. 70 p.

Rowe, P.B., C.M. Countryman, and H.C. Storey. Hydrologic analyses used to determine effects of fire on peak discharges and erosion rates. U.S. Forest Service, California Forest and Range Exp. Sta. (mimeo). 49 p.

Rozanov, A.N. 1951. Serozemy Srednei Azii (Serozems of Central Asia). Issued in transl. by Israel Program for Sci. Transl., Jerusalem, 1961. 550 p. (A-so O.T.S. 60-21834).

Ryerson, D.E., J.E. Taylor, L.D. Baker, H.A. Houton, and D.W. Stroud. 1970. Clubmoss on Montana rangelands. Montana Agr. Exp. Sta. Bull. 645. 116 p.

Schreiber, H.A., and D.R. Kincaid. 1967. Regression models for predicting on-site runoff from short-duration convective storms. Water Resources Res. 3: 389-395.

Shanon, L., M. Evenari, and E.H. Tadmor. 1969. Ancient technology and modern science applied to desert agriculture. Endeavor 28: 68-72.

Shields, L.M., C. Mitchell, and F. Drouet. 1957. Alga- and lichen-stabilized surface crusts as soil nitrogen sources. Amer. J. Bot. 44: 489-498.

Shown, L.M. 1971. Sediment yield as related to vegetation in semiarid watersheds, p 347-354. *In:* Monke, E.J. (editor), Biological Effects in the Hydrologic Cycle, Proc., 3rd Int. Seminar for Hydrol. Prof., NSF Advanced Sci. Seminar, Purdue Univ., West Lafayette, Ind.

Slatyer, R.O., and J.A. Mabbutt. 1964. Hydrology of arid and semiarid regions. Ven TeChow, Handbook Appl. Hydrol McGraw-Hill, New York. p. 24-1 to 24-46.

Smeins, F.E. 1975. Effect of livestock grazing on runoff and erosion, p. 267-274. *In:* Watershed Management, Proc. Symp., Logan, Utah, Aug. 11-13: 267-274.

Smith, R.E. 1972. Border irrigation advance and ephemeral flood waves. Irrigation and Drainage Division, Proc. A.S.C.E 98 (IR2): 289-307.

Soiseth, R.J., M.R. Wight, and J.K. Aase. 1974. Improvement of panspot (Solonetzic) range sites by contour furrowing. J. Range Manage. 27: 107-110.

Sturges, D.L. 1975. Hydrologic relations on undisturbed and coverted big sagebrush lands: The status of our knowledge. USDA Forest Service Res. Pap. RM-140. 23 p.

Tadmor, N.H., and L. Shanan. 1969. Runoff inducement in an arid region by removal of vegetation, Soil Sci. Soc. Amer. Proc. 33: 790-794.

Tennessee Valley Authority. 1964. Bradshaw Creek-Elk River, a pilot study in area-stream factor correlation. Res. Paper No. 4, TVA, Office of Tributary Area Development.

Troendle, C.A., and J.W. Homeyer. 1971. Storm flow related to measured physical parameters on small forested watersheds in West Virginia (abstract). Eos Trans. Amer. Geophys. Union. 52: 204.

USDA Forest Service. 1961. Watershed management: Some ideas about storm runoff and base flow. Annual report. Southeast Forest. Exp. Sta. Asheville, NC. p. 61-66.

Vallentine, J.F. 1971. Range Development and Improvement, Brigham Young University Press, Provo, Utah. 516 p.

Wight, J.R., and F.H. Siddoway, 1972. Improving precipitation-use efficiency on rangeland by surface modification. J. Soil and Water Conserv. 27: 170-174.

Wilm, H.G. 1966. The Arizona watershed program as it enters into its second decade. 10th Annual Arizona Watershed Symp. (Phoenix) Proc., Arizona Water Commission. p. 9-11.

Chapter 6

Erosion, Sediment Yield, and Water Quality

In the United States, there has been a concern for sediment problems in channels and reservoirs for more than 100 years. Recently, however, we have begun to more fully appreciate the water quality implications of sediment. Sediment is now recognized as a pollutant in the same context as industrial waste, sewage treatment plant effluents and other point pollution sources. This new treatment of sediment as a pollutant requires a new level of technology for estimating the magnitude of sediment problems and for selecting cost-effective measures for sediment management or control in the interest of attaining water quality goals. The recent water quality legislation has imparted an urgency to our need to know more about sediment yields and properties as they relate to sources and how to control or regulate damaging effects.

Glymph (1975) defined sediment yield ". . .as the effluent from a soil processing system. The system is a diffuse natural process, which we know as soil erosion. The system is distributed in time and space and can be acceler-

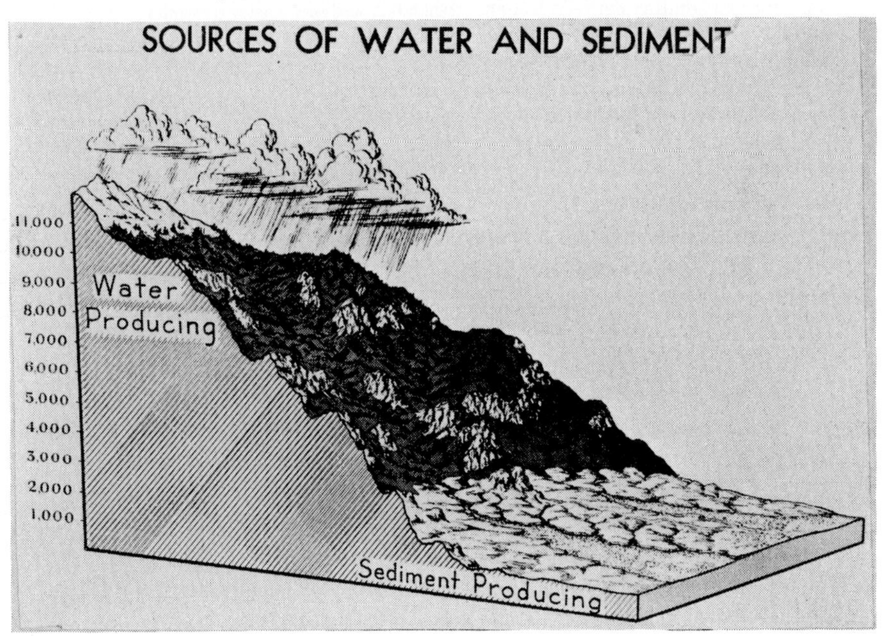

Fig. 6-1. Water and sediment sources as a function of elevation. (Divide elevation in ft. by 3.28 to obtain m.)

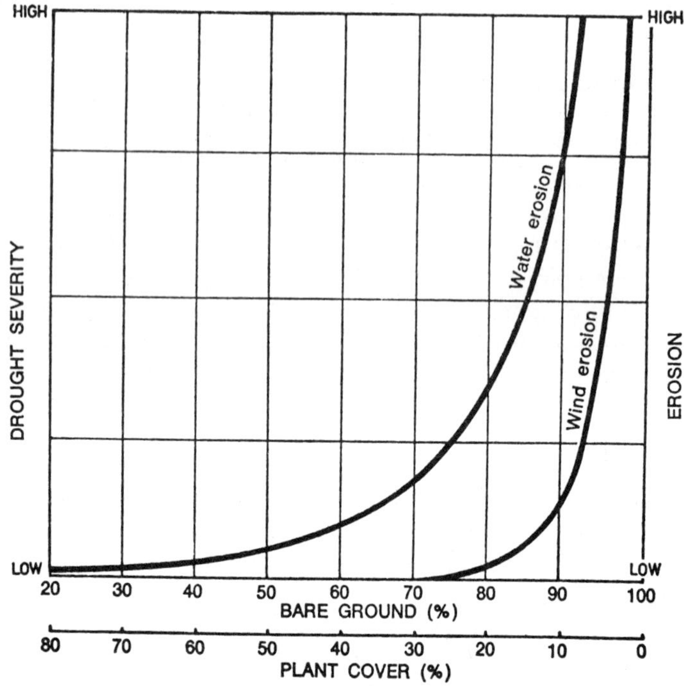

Fig. 6-2. Interrelationships between drought, plant cover and soil erosion by wind and by water. As drought severity increases, plant cover decreases. The markedly non-linear nature of the relationship of wind and of water erosion to plant cover is noteworthy (from Marshall 1973, *In:* the Environmental Economic and Social Significance of Drought, Edited by J.V. Lovett, Angus and Robertson, Publishers—used by permission).

ated or decelerated by a multitude of factors, including the activities of man. The system ranges from small to large, depending upon natural or man-induced constraints. The effluent from this soil-processing system is a heterogeneous mixture of mineral and organic matter in particle sizes ranging from small to large, with a variety of natural or acquired chemical properties. The effluent is transported and delivered from the system of flowing water."

Streamflow in rivers of the western United States originates mainly in the forested mountains, while sediment loads are derived mainly from rangelands (Figure 6-1). As an example of sediment yields from the drier part of the Rio Grande Basin, Dortignac (1956) states:

> The Rio Puerco, for example, which represents less than 20 percent of the Upper Rio Grande Basin, contributes almost half of the measured sediment but less than 8 percent of the water yield.

The relationships between sediment yield, erosion, precipitation, and vegetation are shown in Figures 6-2 and 6-3. From Figure 6-2 it is apparent that, for water erosion, sediment yield is comparatively slight down to perhaps 70% plant cover, but it increases rapidly thereafter for equal rainfall events. For wind erosion, there is little erosion until plant cover comes into the range of approximately 15 % or less, after which the increase in wind erosion is very rapid. With respect to Figure 6-3, Marshall (1973) states the following

> Consider first, water erosion as indicated by relative sediment yield. Curve A indicates that under conditions of natural vegetation cover, water erosion increases from a low value at the arid extreme of mean annual precipitation to a peak in rainfall corresponding to semi-arid conditions. There the amount of rain falling is not great enough to sustain a complete vegetation cover yearlong, but is sufficient to cause erosion of the bare soil. With further increases in precipitation the natural vegetation cover becomes more complete and water erosion decreases. Curve B indicates that which would be expected in the absence of natural vegetation. Here, the sediment yields continue to increase to values considerably in excess of normal erosion yields. What is more apparent from a general comparison of curves A and B over the range of precipitation from arid to humid is the variation in scope for containg erosion at rates below the potential maximum rates. At the arid and semi-arid end of the range there is relatively little scope for containing erosion rates at much below the maximum observed. However, with the substantial divergence of the two curves towards the more humid end of the spectrum, the possibility of containing erosion rates increases markedly in that direction.

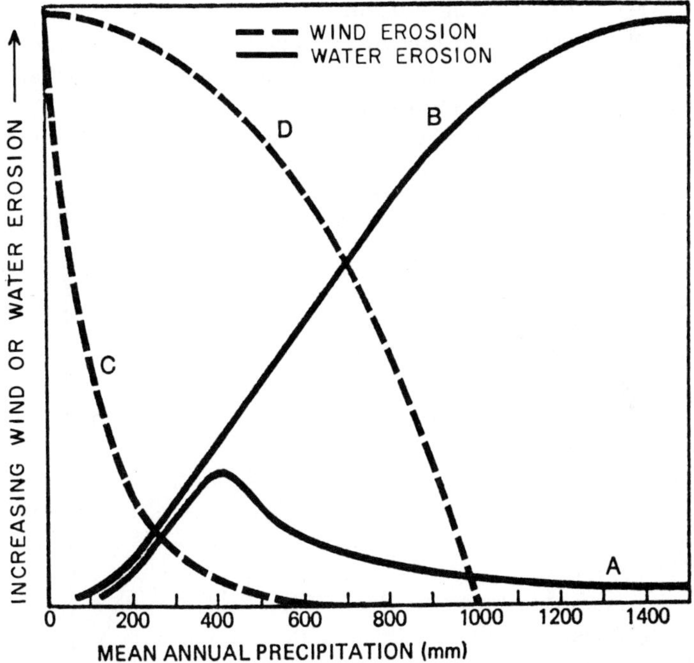

Fig. 6-3. Relationships of water erosion (continuous lines) and wind erosion (broken lines) with increasing mean annual precipitation. The curves for water erosion indicate the relationships with mean annual precipitation for (A) areas of natural vegetation cover and (B) bare ground (after Schumm 1969). The curves for wind erosion indicate the relationships with mean annual precipitation for (C) areas of natural vegetation cover and (D) bare ground. These curves are based on what would be expected from the relationship of wind erosion to vegetation cover and to moist soil (from Marshall 1973, *In:* The Environmental Economic and Social Significance of Drought, Edited by J.V. Lovett, Angus and Robertson, Publishers—used by permission).

> The reference curves for wind erosion, C and D, differ considerably from those for water erosion. Normal wind erosion, curve C, falls rapidly with increasing precipitation because the presence of a moderately sparse vegetation cover can considerably reduce the wind force at the ground surface. . . In the absence of vegetation, wind erosion remains at a relatively high level (curve D) until precipitation increases to the extent of keeping the surface soil moist for much of the time. The non-erodibility of moist soil then results in a reduction in wind erosion. . .The form of curve D depends not only on amount of rainfall but also on distribution. The curve falls more steeply if the increasing precipitation is distributed more evenly or if it coincides with the summer months, and less steeply if the precipitation is mainly confined to the winter months. In contrast to water erosion, the greatest scope for reduction of wind erosion (compare curves C and D) occurs in the semi-arid range of precipitation, the disparity between the two curves decreasing towards the extreme arid and the more humid climates.

Erosion (and therefore resultant sedimentation) has also been related to vegetation yields, and Figure 6-4 is an example.

Sediment yield is economically significant because of the following: (1) detrimental deposition on land and crops, (2) reduction of reservoir storage capacities, (3) aggradation of river channels causing increased flood hazards, (4) plugging of canals, ditches, and drains, (5) increased evapotranspirational losses by vegetation invading widening flood plains. Sediment yields and erosion are closely related phenomena. As a pollutant, sediment transported by runoff water greatly exceeds in volume the combined total of all other substances that pollute our surface water.

Erosion has been classified as geologic or normal for a natural environment

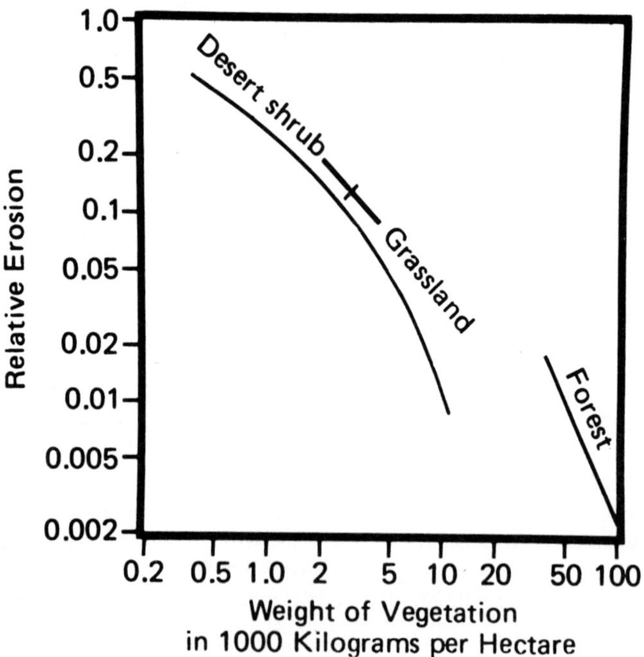

Fig. 6-4. Relative erosion compared with vegetation weight per unit of area (adapted from data of Langbein and Schumm 1958, by Branson 1975). (Kg/ha × 0.891 = lb/acre)

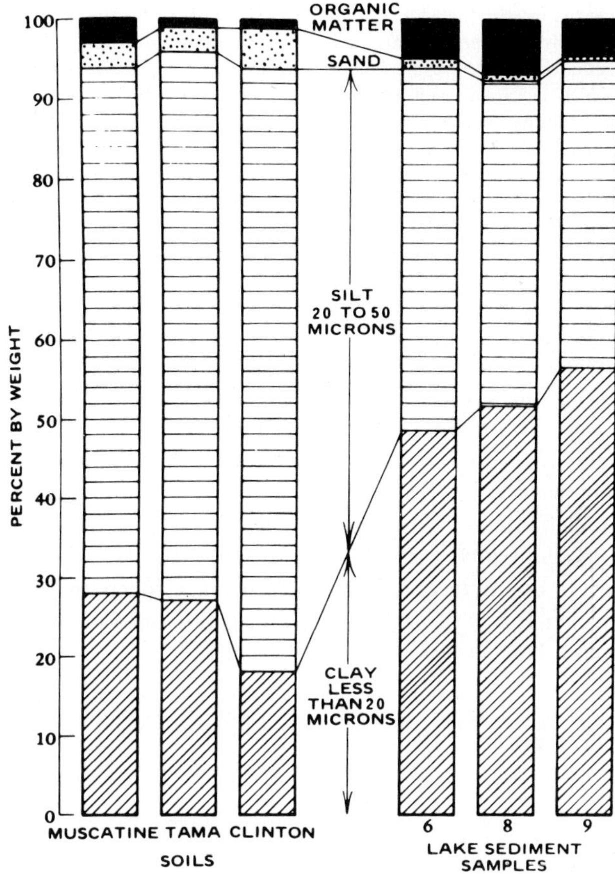

Fig. 6-5. Comparison of particle-size distribution for three soils typical for the watershed and three samples of lake sediments for Lake Braken in Illinois (from Stall 1972, used by permission).

undisturbed by man, and accelerated, or proceeding at a rate greater than normal reflecting man's activities at a site. Erosion rates, both geologic and accelerated, vary considerably from one place to another and actual rates are difficult to determine. Erosion may be divided into three related processes: soil particle detachment, sediment transport or entrainment, and sediment deposition or aggradation. The principal sources of sediment are: (1) sheet and rill erosion on agricultural, forest, and rangelands, (2) upland gullies, arroyos, stream channels, and valley trenches, (3) roads and rights-of-ways, and (4) various construction sites, including urban developments. Additional kinds of erosion and factors that shape land surfaces are discussed in Chapter 10, Geomorphology. Of major importance is whether or not accelerated erosion is taking place and, if it is, what can be done to reduce or prevent further damage from the process.

 Glymph (1975) stated:

> Sediment is the product of a selective process in which the finer and lighter particles are preferentially removed and carried away by runoff and streamflow (see example in Figure 6-5). Sediments, therefore, are generally higher in clay, silt, and organic matter than the soils from which they are derived. These particles and organic compounds have great capacity for absorption of pathogens, viruses, plant nutrients, pesticides, and other chemicals. Thus, the commitment for identification and control of pollutants from runoff sources requires improved understanding of the physical and chemical properties of sediment in respect to specific erosion sites. We can no longer be satisfied with knowing the volume of sediment involved; for many cases we now we must also know about the properties of the constituent sediments in relation to their source.

Glymph went on to report that persistent chlorinated hydrocarbons which have become a serious pollutant, have low solubility in water but are strongly absorbed on soil particles. Thus, although such a pesticide is not leached through the soil, it does move to water bodies in significant amounts when erosion occurs. Similar conditions exist with regard to phosphorus and its movements from soil erosion. Nitrogen fertilizers, although they are highly soluble in water, also leave the land absorbed on sediment particles. For example, Schuman et al. (1973) showed that 92% of the nitrogen discharged from some cultivated watersheds in southwestern Iowa was carried by the sediment. In this same experiment, sediment nitrogen losses were 86% of the total nitrogen losses from a level-terraced watershed in corn and 51% of the total from a bromegrass pasture watershed.

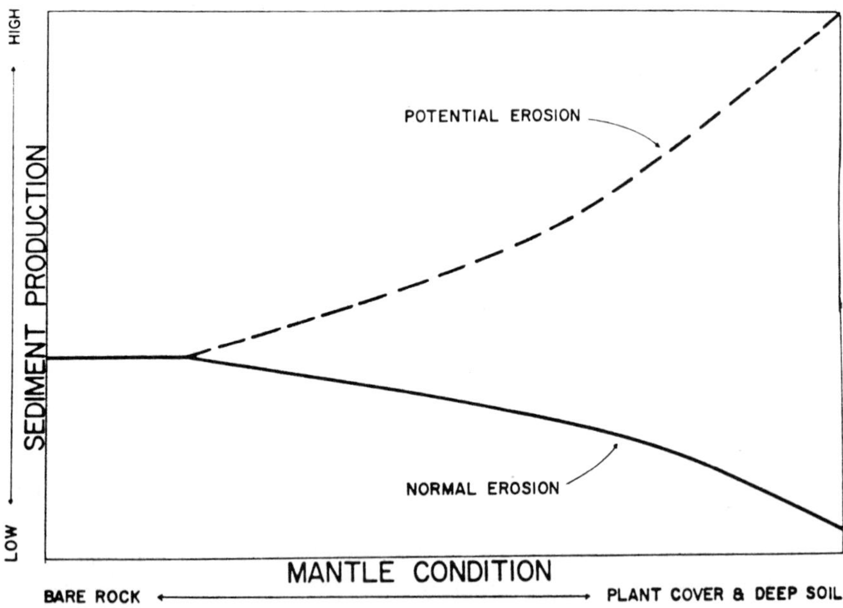

Fig. 6-6. The decrease in normal erosion and the increase in erosion potential that accompany development of a soil mantle under the protective influence of vegetation (from Bailey and Copeland 1961).

Erosion

Smith and Wischmeier (1962) emphasize that four factors and their interrelations have long been considered basic determinants of rate of rainfall erosion: (1) climate, largely rainfall and temperature; (2) soil, its inherent resistance to dispersion and its water intake and transmission rates; (3) topography, particularly steepness and length of slope; and (4) plant cover, either living or the residues of dead vegetation. The influence of a protective vegetal cover on a developing soil mantle as related to erosion is shown in Figure 6-6. Any one of the four factors can assume values which, alone, may create a rainfall erosion problem. Other reviews of erosion problems include Smith and Wischmeier (1957), Copeland (1965), and Roehl (1965). Baver (1965) showed in diagram form the factors affecting soil erosion by water (Figure 6-7).

Raindrops striking bare soil have two effects: impacts cause minor explosions" that dislodge and force soil particles upward (Figure 6-8) (quantities greater than 100 tons/acre, 225,000 kg/ha, have been measured), and raindrop impact helps seal the soil surface and reduces infiltration. Dislodged particles may remain in suspension and move downslope in overland flow. Efficiency of ground cover in controlling runoff and erosion under simulated rainfall on subalpine rangeland along the Wasatch Mountains front, Utah, is shown in Figure 6-9. The range manager can reduce splash erosion by practices that increase soil surface coverage by plant materials. Overland flow may be reduced by practices that reduce slope length and enhance infiltration, e.g., contour furrowing (see Chapter 4, Infiltration).

Erosion-Estimating Techniques

Although it is difficult to separate pure erosion-estimating techniques from watershed sediment yields, let's consider them separately, realizing their

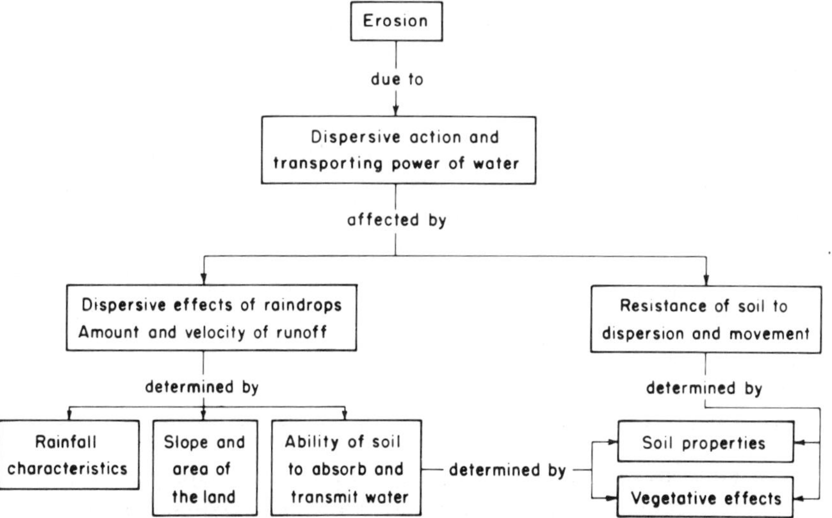

Fig. 6-7. Factors affecting soil erosion by water (Baver 1965).

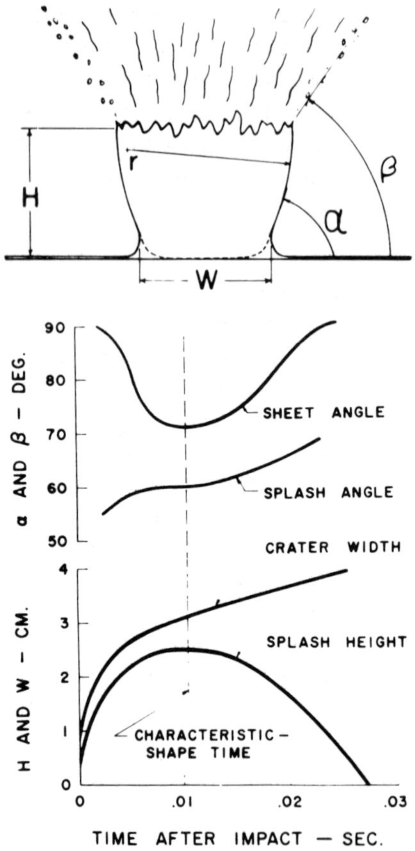

Fig. 6-8. Splash shape geometry and the change in the splash parameters with time after impact. The characteristic shape time was selected at the time for maximum splash height (from Mutchler and Young 1975).

interrelationships. Many methods have been developed for estimating erosion from small upstream areas (Pacific Southwest Interagency Committee 1974). Each of these methods involves a mathematical epxression for the erosion variables considered important for a particular area and application. Kilinc and Richardson (1973) stated that sediment discharge by means of overland flow is a function of

(1) hydraulic properties of flow,
(2) physical properties of soil, and
(3) surface characteristics.

The variables to be considered are sediment discharge, water velocity, inflow rate or rainfall excess, depth normal to slope, mean sediment particle size, distance downslope, kinematic viscosity of water, acceleration of gravity,

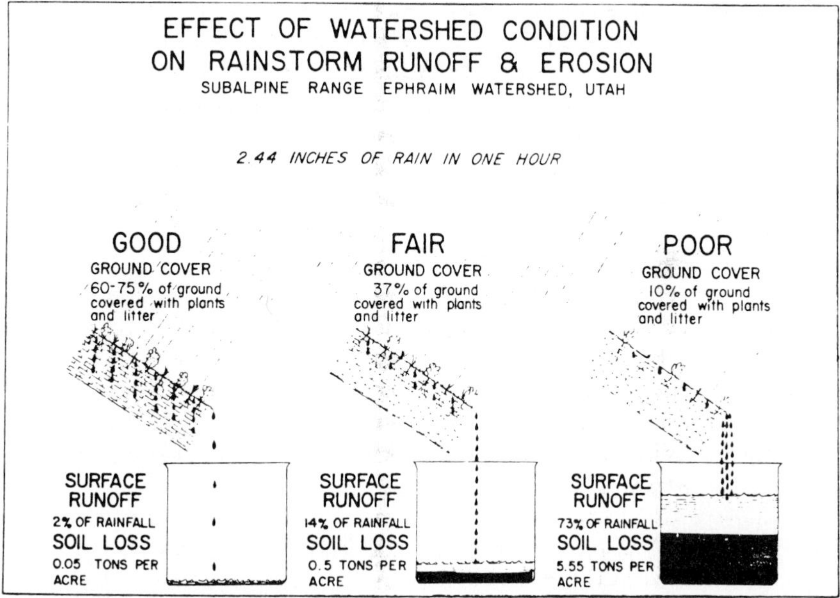

Fig. 6-9. Effects of various densities of ground cover in controlling overland flow and soil erosion (from Bailey and Copeland 1961). (Multiply t/acre by 2,250 to obtain kg/ha).

mass density of water, mass density of sediment, slope, and porosity of sediment.

The Pacific Southwest Interagency Committee Sediment Yield Estimate Method

Most of the erosion-estimating techniques were developed from data from cultivated agriculture (Pacific Southwest Interagency Committee 1974). An earlier task force of Pacific Southwest Interagency Committee (1968) developed an estimating procedure for conditions encountered in the western portion of the United States which considered the nine factors and assigned numerical values to each factor (Table 6-1, Figure 6-10), which, upon summing, gave an estimate of the erosion class that would be expected. The summed numerical value obtained from field evaluations of factors in Table 6-1 can then be used to obtain estimated sediment yields from the graph, Figure 6-10. High values are given to factors that would cause sediment yields to be large. Average sediment yields were grouped into erosion classes given by the following:

Classification		
1	> 3.0	acre-feet/square mile
2	1.0 – 3.0	acre-feet/square mile
3	0.5 - 1.0	acre-feet/square mile
4	0.2 - 0.5	acre-feet/square mile
5	<0.2	acre-feet/square mile

Table 6-1. Rating ranges for the factors used in the Pacific Southwest Inter-Agency Committee method for estimating sediment yields.[1]

Factor	Rating range	Main characteristics considered	
Surface geology	0–10	Rock type. Hardness	Weathering. Fracturing.
Soils	0–10	Texture. Texture. Shrink-swell. Rockiness.	Salinity. Caliche. Organic matter.
Climate	0–10	Storm frequency, intensity, and duration. Snow. Freeze-thaw.	
Runoff	0–10	Volume per unit area. Peak flow per unit area.	
Topography	0–20	Steepness of upland slopes. Relief. Fan and flood plain development.	
Ground cover	-10–10	Vegetation. Litter. Rocks. Understory development under trees.	
Land use	-10–10	Percentage cultivated. Grazing intensity. Logging. Roads.	
Upland erosion	0–25	Rills and gullies. Landslides. Wind deposits in channels.	
Channel erosion and sediment transport	0–25	Bank and bed erosion. Flow depths. Active headcuts. Channel vegetation.	

[1]From Pacific Southwest Inter-Agency Committee 1968.

(Multiply acre-ft/mi^2 by 476 to convert to m^3/km^2.) It is desirable to compare one's ability to assign realistic values for the nine factors in watersheds with known sediment yields before attempting to obtain estimates for unmeasured watersheds.

The classifications and guide material developed by the Task Force were intended for broad planning purposes only, rather than for specific projects where more intensive investigations of sediment yield would be required. They recommended that map delineations be for areas no smaller than 10 square miles (26 km^2). They also suggested that actual measurements of sediment yield be used to the fullest extent possible. This descriptive material and the related numerical evaluation system (Table 6-1) would best serve as a means of delineating boundaries between sediment yield areas and in extrapolation of existing data to areas where none is available.

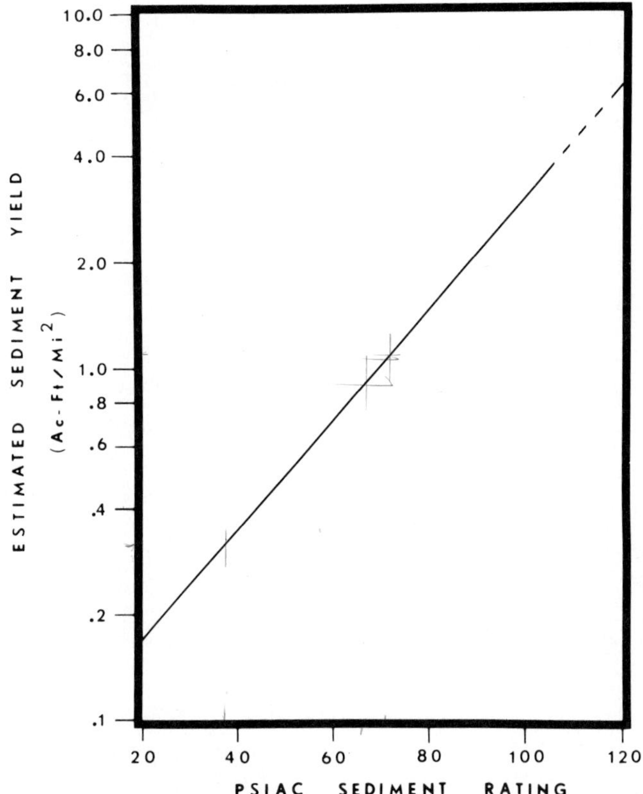

Fig. 6-10. Graph to be used to obtain sediment-yield estimates by the Pacific Southwest Inter-Agency Committee method (adapted from data by Shown 1970) (Multiply acre-ft by 476 to convert to m^3 per km^2.)

This Pacific Southwest Interagency Committee Task Force recommended different management and land treatment measures for reduction of erosion and sediment yield under various site conditions (Table 6-2). Their report also illustrated how the nine factors might change as a result of the land treatment and management measures and thus affect the sediment yield classification (see also Figures 6-11 and 6-12).

The Universal Soil Loss Equation

An erosion prediction method that has recently received wide attention is the Universal Soil Loss Equation (USLE) developed by Wischmeier and Smith (1965) for cultivated agricultural areas of the eastern United States. The USLE is given as:

$$A = RLKSCP \tag{6-1}$$

where: A = estimated soil loss (tons/acre/year)

R = rainfall factor

Table 6-2. Management and land treatment measures recommended for reduction of erosion and sediment yield under various site conditions (from Pacific Southwest Interagency Committee 1968).

	Climate environment[1]				Soils[2]			Upland slope topography		
Measures	Arid	Semi-arid	Sub-humid	Humid	Fine textured	Medium textured	Coarse textured	Steep	Moderate	Gentle
	A	B	C	D						
Forest and Range Plants										
Brush control	x	x			x	x	x		x	x
Contour furrowing and trenching		x	x		C	B		x	x	
Contour trenching		x	x			x	x	x	x	
Critical area planting		x	x	x	x	x		x	x	
Fire prevent and suppression		x	x	x	x	x	x	x		
Livestock exclusion	x	x	x	x	x	x	x	x	x	x
Proper grazing use—trespass control	x	x	x	x	x	x	x	x	x	x
Range seeding		x	x		C	B			x	x
Rotation-deferred grazing		x	x		x	x	x	x	x	x
Tree and shrub planting	x	x	x	x	x	x		x	x	x
Cultivated Land										
Chiseling and subsoiling			x	x		x			x	x
Contour farming		x	x	x	x	x	x		x	x
Contour terracing		x	x	x	C-D	x			x	x
Critical area planting		x	x	x	x	x		x	x	
Crop residue and mulching		x	x		x	x		x	x	x
Field diversion		x	x		x	x			x	x
Proper cropping and use		x	x	x	x	x	x		x	x
Strip cropping		x	x		x	x	x		x	x

[1]A climatic map based on Visher's (1954) classification can be used.
[2]Mechanical treatments are not recommended for shallow soils.

K = soil-erodibility factor
L = slope length factor
S = slope gradient factor
C = cropping-management factor, and
P = erosion control practice factor.

Numerical values of the seven factors have been determined from research conducted east of the Rocky Mountains on plots with both natural and artificial rain. Limited data for areas west of the Rocky Mountains restricts the usefulness of the USLE in that area, although additional information is now becoming available to permit use of the equation in these western U.S. rangeland areas.

The *rainfall factor* R is the average number of erosion-index (EI) units in a year's rainfall, or the EI units in an individual storm if the equation is used to predict individual storm erosion. The EI units are determined by multiplying the total kinetic energy of the storm times its maximum 30-minute intensity

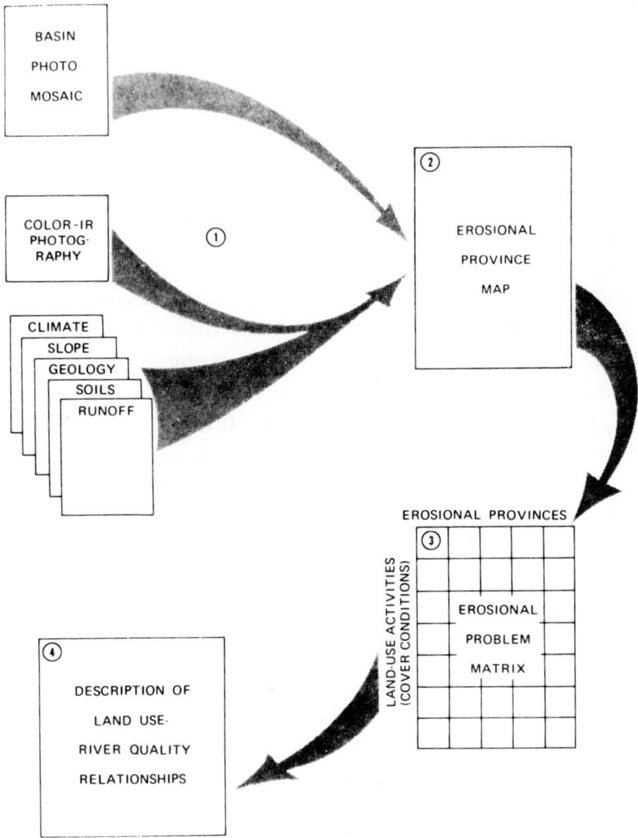

Fig. 6-11. Process of assessing the impacts of land-use activity on erosion and river quality (from Rickert and Hines 1975).

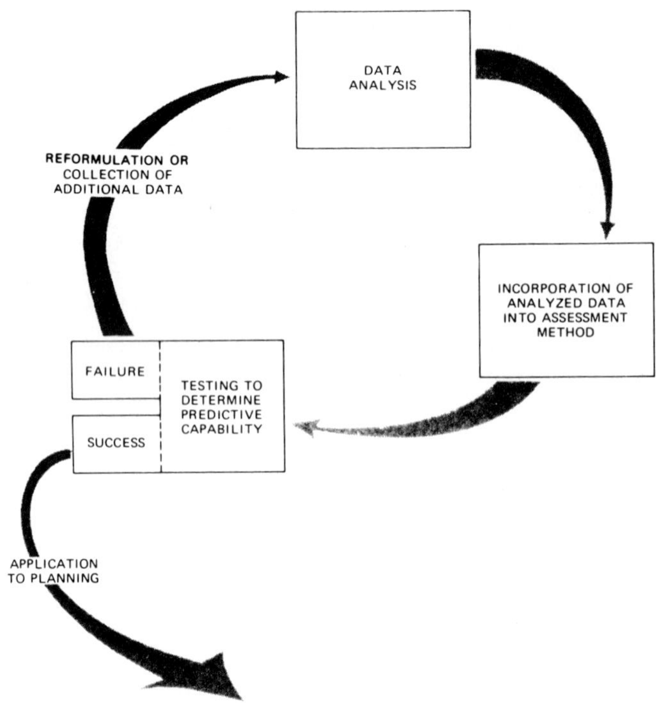

Fig. 6-12. Interrlation of data analysis, method formulation, and the testing of predictive capability (from Rickert and Hines 1975).

(Wischmeier and Smith 1958). Each storm's EI value is accumulated to obtain a yearly EI value or R factor. Isoerodent maps, lines of equal R factors, are available for portions of the United States east of the 104th meridian (Wischmeier and Smith 1958). To use the term elsewhere requires computing the EI value using precipitation data from recording raingages or estimating the EI value from an empirical method such as the equations developed by Ateshian (1974). Ateshian showed that the average annual EI (R factor) could be obtained from Figure 6-13 if the 2-year frequency, 6-hour rainfall depth is known and if the precipitation producing that depth was describable with one of two depth distributions. He showed that the Type I distribution had generally lower rainfall intensities such as might be encountered along the west coast of the United States (Woodward 1975) while Type II storms included higher intensities during the storm duration such as might be encountered with thunderstorms. Renard (1975) and Renard and Simanton (1975a) showed that thunderstorms such as experienced in the southwestern United States are not adequately defined by the Type II distribution and that the average annual EI values may vary appreciably even within a watershed (Figure 6-14). Thus, there might be severe problems of trying to use a raingage as an indication of what might be expected over a wide area (Renard and Simanton 1975b).

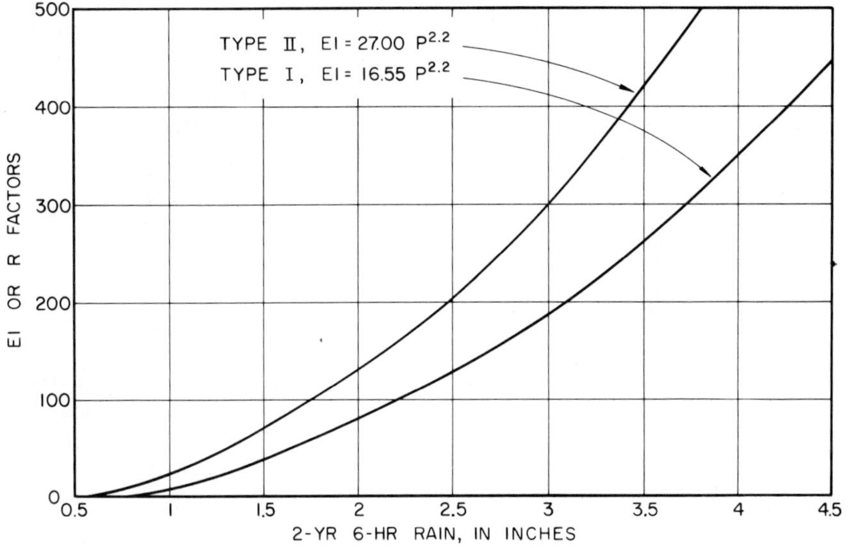

Fig. 6-13. Relation between annual average erosion index (EI on vertical axis) and 2-yr, 6-hr rainfall depth, rainfall types from (Ateshian 1974) p = precipitation.

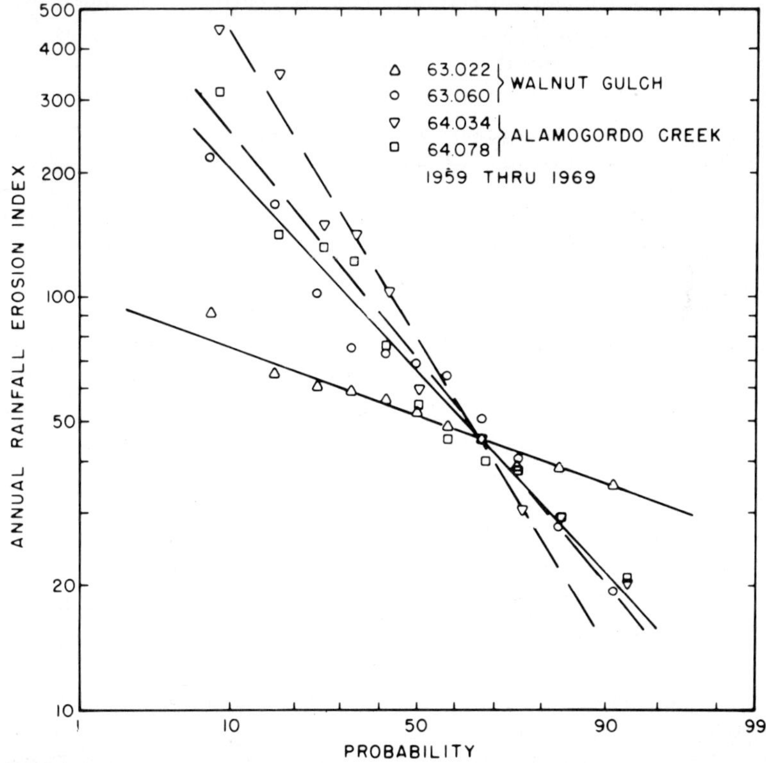

Fig. 6-14. Probability of annual rainfall energy in foot tons per acre for two raingages on the Alamogordo Creek and Walnut Gulch experiment watersheds from 10 years of records (from Renard and Simanton 1975a).

The *soil erodibility factor* K, a soil-erodibility index, is experimentally determined from "unit" plot data. A unit plot is 72.6 ft (22.1 m) long, with a uniform 9% slope, in continuous fallow, and tilled up and down the slope (Wischmeier and Smith 1965). The plots were 30 feet (8.6 m) wide but, in usual use of the equation, values selectd for K and a selected period for R usually compute A in terms of tons per acre per year, but other units can be selected. For a given soil, it is the erosion rate per unit of erosion index from "unit" plots on that soil. Values of K have been determined for many eastern soils but not for many western soils.

Wischmeier Johnson, and Cross (1971) presented a soil erodibility nomograph (Figure 6-15) which assists in determining K values for any soil. To use the nomograph, the percent silt and very fine sand (0.05–0.10 mm), percent sand (0.10–2.0 mm) percent organic matter, soil structure, and permeability must be known. These parameters can be readily obtained for most soils with standard laboratory analyses. For the Walnut Gulch Watershed soils (Rilliot-Nickel soil), the K value determined from the nomograph is 0.08.

Many western soils are quite poorly developed and may have erodibility characteristics quite different from that experienced in other parts of the country. For example, many western soils contain large amounts of carbonates (caliche) which may provide additional soil particle cohesion. Western soils are often dominated by very coarse materials, such as sand, gravel, and cobbles, which may affect the erodibility factor.

The *cropping-management factor* C is a ratio of soil loss from land cropped

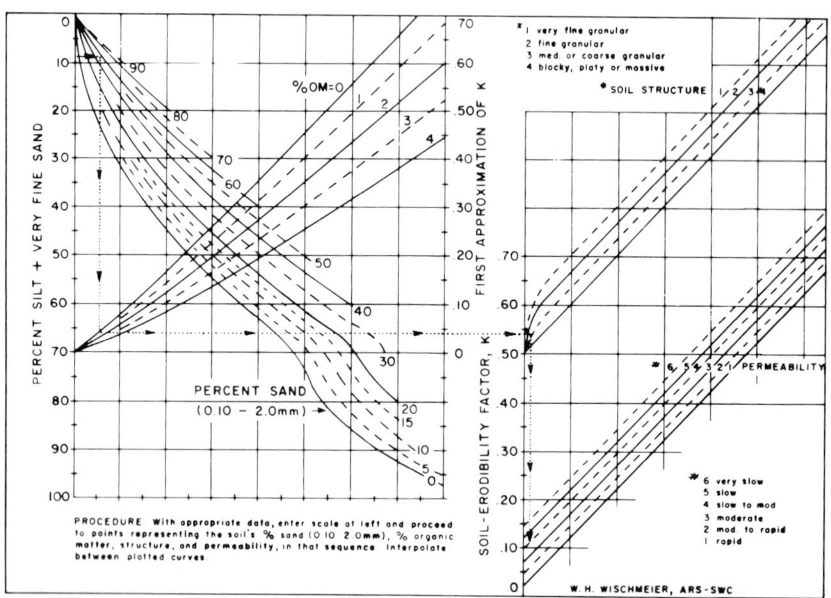

Fig. 6-15. Wischmeier (1971) soil erodibility nomograph. The dotted line illustrates the sequence for determining a K value using values for a Rillito-Nickel soil (8% silt and fine sand, 1% OM, 55% sand, medium granular soil structure, and moderate permeability) (from Renard, Simanton, and Osborn 1974).

under specific conditions to the corresponding soil loss from tilled, continuous fallow land. The term was developed primarily to handle conditions connected with crops and rotations. On rangeland areas, guidelines for determining a value of C are not generally available. The term also varied seasonally, reflecting crop growth stage (Wischmeier and Smith 1965). Such a seasonal variability should not be necessary for rangeland conditions, unless the brush encountered loses its foliage during the winter or for ranges with a large annual plant component. Of greater consequence is the relative density of plant cover. On the brush-grass areas in southeastern Arizona, the vegetative cover is generally less than 10% basal area and approximately 30% crown cover.

Wischmeier, working with range scientists of the U.S. Soil Conservation Service (1972), has proposed some C values for permanent pasture, rangeland, and idle land. A portion of this information from the Soil Conservation Service Technical Release No. 51 is reproduced in Table 6-3. The percent ground cover has a very dramatic effect on the C factor as would be expected. For example, on the first line of the table, if the ground cover is composed of a grass or grass-like plant, the erosion would be expected to be 150 times greater for bare soil than for one with nearly 100% ground cover.

The western rangeland areas are often dominated by erosion pavements, consisting of gravel-size material that are residual from normal erosion which has removed the finer materials from soil profiles. The erosion pavement role in the C factor can be very significant. Often, it may have an effect similar to vegetation in reducing erosion. This erosion pavement can withstand the energy of the falling raindrop and the erosive force of flowing water. The

Table 6-3. "C" Values for permanent pasture, rangeland, and idle land[1] (from U.S. Soil Conserv. Serv. 1972).

Vegetal canopy			Percent ground cover					
Type and height of raised canopy[2]	Canopy cover[3] (%)	Type[4]	0	20	40	60	80	95-100
No appreciable canopy		G	.45	.20	.10	.042	.013	.003
		W	.45	.24	.15	.090	.043	.011
Canopy of tall weeds or short brush (0.5 m fall height)		G	.36	.17	.09	.038	.012	.003
	25	.	.36	.20	.13	.082	.041	.011
		G	.26	.13	.07	.035	.012	.003
	50	W	.26	.16	d.11	.075	.039	.011
Appreciable brush or bushes (2 m fall height)	25	G	.40	.18	.09	.040	.013	.003
		W	.40	.22	.14	.085	.042	.011
	50	G	.34	.16	.085	.038	.012	.003
		W	.34	.19	.13	.081	.041	.011

[1]All values assume: (1) random distribution of mulch or vegetation, and (2) mulch of appreciable depth where it exists.

[2]Average fall height of waterdrops from canopy to soil surface.

[3]Portion of total-area surface that would be hidden from view by canopy in a vertical projection.

[4]G: Cover at surface is grass, grasslike plants, decaying compacted duff, or litter at least 2 inches deep.
W: Cover at surface is mostly broadleaf herbaceous plants (like forbs)

effects of erosion pavement on soil erosion seemingly is a very fruitful area for future research.

The *erosion-control practice factor* P is the ratio of soil loss from the supporting practice to the soil loss with up-and-down hill culture. This ratio should obviously be less than one, if the erosion-control practice is effective. Since there are generally no cultivation practices involved on rangelands, the P ratio should be 1.0. However, since rangeland rejuvenation is becoming increasingly common, P values may be needed to reflect rangeland treatment practices, such as pitting, subsoiling to break up caliche layers, and root plowing for brush removal. In most rangeland erosion predictions, the C and P terms can be combined.

In practice, slope length (L) and gradient (S) are often considered as one term. This factor is the ratio of soil loss per unit area on a field slope to the soil loss from the base "unit" plot (9% slope and 72.6 ft (22.1 m) long). The ratio for specific length/gradient combinations may be obtained from the slope-effect chart shown in Figure 6-16.

Application of the USLE to some small watersheds in Southeastern Arizona (Renard, Simanton and Osborn 1974) showed that an additional term may be required on even very small watersheds (<8 acres, <3.2 ha) when the alluvial sand streambed is encountered because in such instances the sediment delivery ratio may have a value larger than unity. It was also found that the erosion pavement probably should be included with the plant cover for determining the "C" term in the USLE.

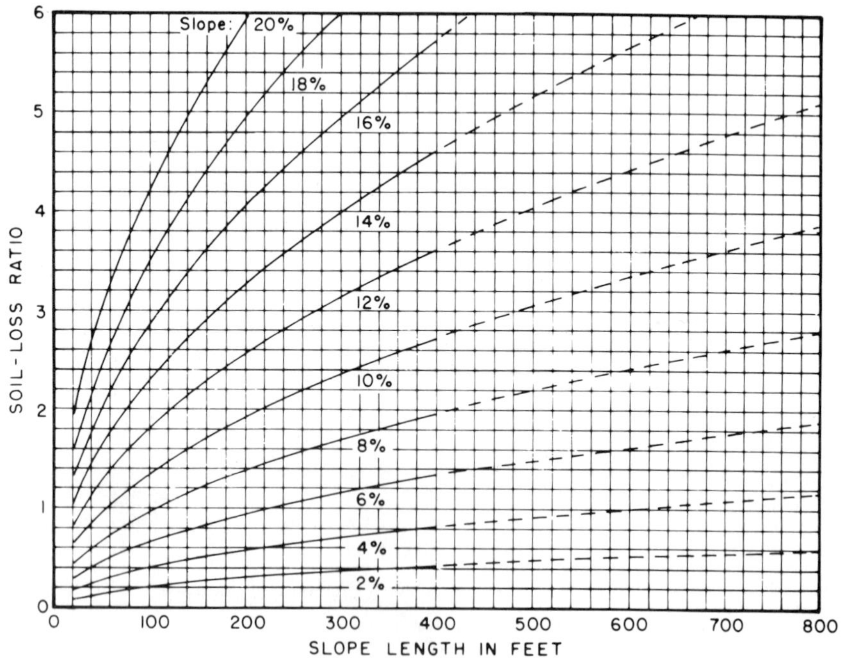

Fig. 6-16. Slope-effect chart from Wischmeier and Smith (1965).

In research application, the R term of the USLE has been replaced by a term E defined by Foster, Meyer and Onstad (1973) as in Equation 6-2.

$$E = (aR + bcQq_p{}^{1/3}) \qquad \textbf{6-2}$$

where E = the estimated soil loss,
$a,b,$ = weighting parameters (a+b=1),
c = equality coefficient,
R = rainfall factor from USLE,
Q = runoff volume, and
q_p = runoff rate.

A physically based modeling effort developed for chemical transport uses this modification of the USLE for estimating the watershed slope erosion. Figure 6-17 illustrates how a complex watershed shape was transformed into computational units for this hydrologic-sediment-chemical model (more description in Chapter 11). Each slope increment of a watershed was expressed by a series of planes (zones I, II, III, IV). The width and slope of each zone were representative of the prototype watershed. Conceptually, flow passes off the land in unit width computational units (designated as tubes A,B,C-,D,E,F,G in Figure 6-17) and cascades from zone to zone with infiltration persisting while surface water is available. Soil moisture movement occurs with an accounting scheme for the water moving in the various soil layers (layer 1,2,3 and 4 in Figure 6-17). Results of this work are encouraging and the

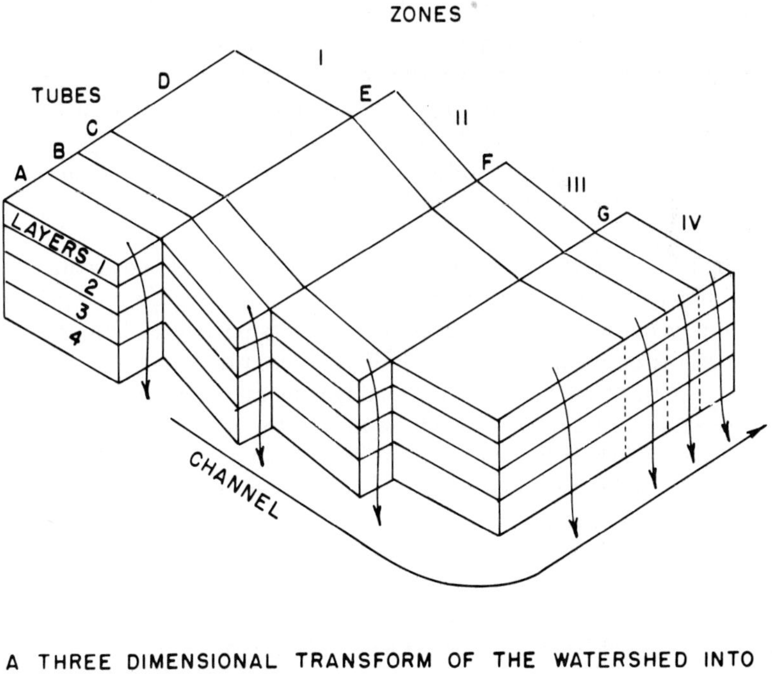

A THREE DIMENSIONAL TRANSFORM OF THE WATERSHED INTO COMPUTATIONAL UNITS

Fig. 6-17. A conceptual watershed transformation into analytic computational units (from Frere, Onstad, and Holtan 1975).

 equation is now being tested at other locations. The results indicate better agreement for predicted and measured data when the original rainfall factor is modified to account for the transport of eroded material by overland flow (the form of Equation 6-2).

There are numerous other methods available for estimating erosion and sediment yields. A summary of some of the factors considered in the available methods identified in an Interagency Task Report (PSIAC 1974) is given in Table 6-4. This report also gives a brief summary of each method as well as comparisons of some of the methods for locations where data was available to compare predicted with actual measurements.

Effects of Soils on Erodibility

On rangelands, the effects of geologic materials on erosion are often apparent. In general, soils derived from marine shales are the most erodible, soils from sandstone the least, and soils derived from mixed sandstones and shales are intermediate. Studies on six watersheds in Utah (ranging in size from 253 to 7,349 ha, 6,221 to 18,147 acres) showed that soil depth also affects erodibility; the most obvious relation observed between accelerated erosion and soils concerned the depth of the friable material (all soil materials that are readily permeable to water) over bedrock or tight subsoils (Olson 1949). These soils that have friable surfaces underlain by clay or bedrock at shallow depth occupied only one-fifth the total area studied, but 85% of the severe erosion occurred on them.

Erodibility is also related to soil chemistry. Wallis and Stevan (1961) reported soil erodibility indices that were related to amount and kind of cations present in the soil. Erodibility of 20 soils was indexed using Middleton's (1930) dispersion ratio and Anderson's (1951) surface-aggregation ratio. Of the four most plentiful soil cations (Ca, Mg, K, and Na), linear and curvilinear terms involving Ca and Mg were significant when used to predict an erosion index.

A similar study (Andre and Anderson 1961) utilized the above data, and in addition included the soil-forming factors of geology, vegetation type, elevation, and geographic zone to compute an erosion index. This technique improved the predictability of erodibility reported by Wallis and Stevan by 30% ($r^2 = .63$). Soil developed from acid igneous rock was about 2-1/2 times as erodible as soil developed from basalt. Of the three types of vegetation cover studied, erodibility was highest for soils under brush, next highest under trees, and least erodible under grass. No clear-cut relation of erodibility to elevation was found.

On many areas with steep slopes (45%) in north-coastal California watersheds where landslides are a problem, Anderson (1975) reported that brushland converted to grassland is not a stable vegetation type. In fact, conversion of 14.8% of steep watershed lands to grasslands multiplied sediment yield by 4.7 times (Anderson 1970). In general, however, on moderate slopes soils developed under brushland areas tend to be more erodible—particularly for geologic rock types of marine and nonmarine sediments of

Table 6–4. Summary table of on-site erosion and sediment yield methods (Pacific Southwest Interagency Committee 1974).

	METHOD	FACTOR		GEOLOGY AND SOILS				CLIMATE				RUN-OFF		TOPOGRAPHY							GROUND COVER					LAND USE			UP-LAND EROSION		CHANNEL[1/] EROSION AND SEDIMENT TRANSPORT		COST OF FIELD DATA			METHOD OF SOLUTION			LIMITATIONS AND APPLICABILITY
		PROPORTION NONERODIBLE	SOIL SIZE DISTRIBUTION	SOIL DENSITY	% ORGANIC	SOIL STRUCTURE	PERMEABILITY	ANNUAL PRECIP.	PRECIP. INTENSITY	RELATIVE RAIN AREA	ANNUAL TEMP.	RUNOFF	STREAM HYDROGRAPH	SLOPE	SLOPE LENGTH	AREA	DRAINAGE DENSITY	RELIEF RATIO	SHAPE OF BASIN	ANGLE OF JUNCTION	% BARE GROUND	EFFECT OF FIRE	VEGETATIVE COVER	ROADS, URBAN	PRACTICE FACTOR	GRAZING	LOGGING	TEST PLOTS	FARM PONDS	CHANNEL X-SECTION AND SLOPE	SUSPENDED LOAD	RESERVOIR SEDIMENTATION	LOW	MEDIUM	HIGH	ANALYTICAL	NOMOGRAPH-CHARTS	COMPUTER	
ON-SITE EROSION	BRYAN 1968		X O		X O				X O					X O														X											NO PREDICTIVE EQUATION. DATA RELATES SOIL ERODIBILITY TO WATER SOLUBLE AGGREGATES > 3mm
	FOSTER AND MEYER 1972		X O						X O					X O	X O													X						*		*			EQUATION UNIVERSALLY APPLICABLE IF COEFFICIENTS ARE DETERMINED.
	MEEUWIG 1971	X O	X O	X	X O				X O					X O									X O					X						*		*	*		EQUATIONS DETERMINE RELATIVE ERODIBILITY BUT AT A CONSTANT RATE OF PRECIPITATION.
	MEYER AND WISCHMEIR 1966		X O	X O					X O			X O		X O	X O	X O												X					*			*			EQUATION UNIVERSALLY APPLICABLE BUT COEFFICIENTS DIFFICULT TO DETERMINE.
	MUSGRAVE 1947		X O			X O	X O		X O					X O	X O										X O			X					*			*			EQUATION UNIVERSALLY APPLICABLE BUT LIMITED TO SLOPES LESS THAN 20 PERCENT.
	WISCHMEIR AND SMITH 1965		X O			X O	X O	X	X O					X O	X O										X O			X						*		*	*		EQUATION UNIVERSALLY APPLICABLE BUT LIMITED TO SLOPES LESS THAN 20 PERCENT.
SEDIMENT YIELD	ANDERSON 1970	X	X					X	X O	X O		X O	X	X O		X O			X			X O	X O	X O		X	X			X	X	X O			*	*		*	EQUATION APPLICABLE MAINLY IN LARGE WATERSHEDS BUT NEEDS SEDIMENT TRANSPORT DATA.
	BRANSON AND OWEN 1970		X									X		X			X O	X O		X	X O											X	*			*			EQUATION UNIVERSALLY APPLICABLE BUT CORRELATION COEFFICIENTS FOR EACH AREA MUST BE DETERMINED.
	FLAXMAN 1972		X O					X O			X O	X O		X O															X			X	*			*			EQUATION UNIVERSALLY APPLICABLE ESPECIALLY IN WESTERN UNITED STATES.
	NEGEV 1967		X						X				X																		X								EQUATION UNIVERSALLY APPLICABLE BUT NEEDS PRECIPITATION AND RUNOFF DATA.
	RENARD 1972											X O	X O	X O		X O														X O	X	X		*				*	EQUATION APPLICABLE UNIVERSALLY IF RUNOFF GENERATION METHOD AVAILABLE. 2/
	TATUM 1965								X O					X O			X O		X O			X O										X	*				*		EQUATION APPLICABLE TO SOUTHWESTERN UNITED STATES. NEEDS PRECIPITATION & BURN EFFECT DATA.

X DATA NEEDED TO FORMULATE METHOD
O DATA USED TO SOLVE DERIVED METHOD
* MISCELLANEOUS INFORMATION ON COST AND DATA OUTPUT
1/ CHANNEL SLOPES ARE NEEDED FOR RENARD'S METHOD
2/ USED WITH EPHEMERAL STREAM MODEL IN SOUTHWESTERN UNITED STATES

 Cenozoic Age and younger. Such soils have erodibilities 76% higher than the average of all other geologic soil types. Soils under brush occurring on granodiorite and quartz diorite are less erodible.

Effects of Grazing

As indicated in Chapter 4, several plot studies have shown lower infiltration rates for ranges in poor condition that for those in good or excellent condition. A decrease in infiltration is accompanied by an increase in overland flow, which results in more water available for sediment transport. Grazing caused about 45% increase in sediment yields in western Colorado (Figure 6-18) (Lusby 1970). In this 14-year hydrologic study, vegetation changed very little on the four ungrazed watersheds, but under grazing there was a slight increase in bare soil and rock and a decrease in ground cover; salt desert shrubs and erodible marine shale soils were present on these watersheds. On permeable, granitic soils in eastern Colorado (Dunford 1949), moderate grazing resulted in almost no increase in erosion but a marked increase in erosion occurred under heavy grazing (Fig. 6-19). These results in Colorado indicate that moderate grazing (4 acres per cow-month) on similar soils should not increase erosion and, as cited elsewhere, should result in

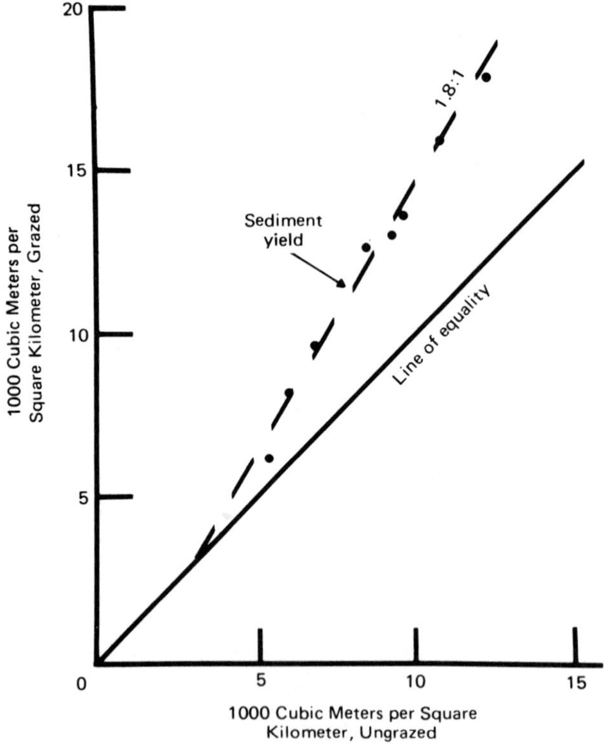

Fig. 6-18. Mass diagram of sediment yields from four grazed and four ungrazed watersheds in western Colorado. The data are for 14 years of measurement (adapted from Lusby 1970) (1,000 $m^3/km^2 \times 2.1$ = acre-ft/sq. mile).

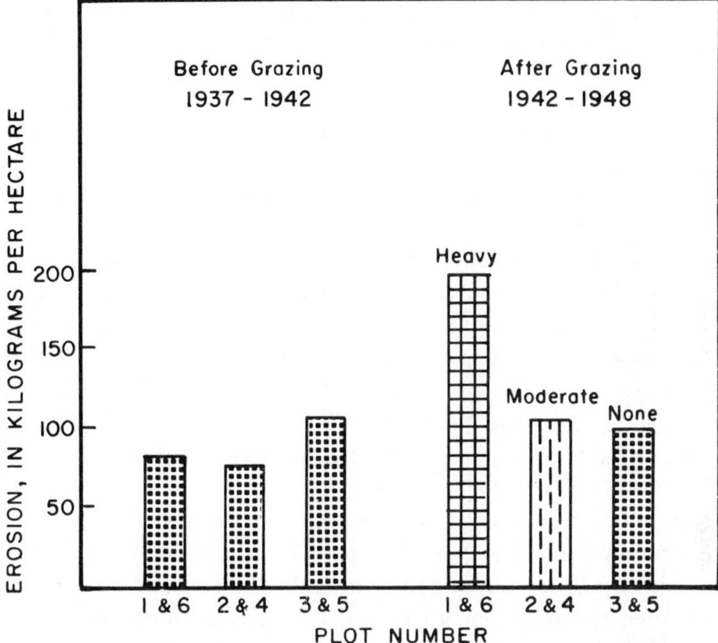

Fig. 6-19. Average erosion from pairs of plots, June to September, on pine-bunchgrass range subject to different intensities of grazing near Colorado Springs, Colorado (after Dunford 1949) (kg/ha ÷ 0.891 = lb/acre).

increased runoff.

The influence of several natural and introduced factors on erosion was determined on the Boise River watershed (Renner 1936); variance in erosion associated with each of seven factors, expressed in percent, appears in ascending order of importance as follows:

Gradient	2.9
Rodent infestation	4.9
Plant density	6.8
Aspect	10.6
Soil conditions	12.8
Plant type	13.4
Cumulative effect of past grazing as expressed by accessibility of range to livestock	15.3

This indicates that grazing, and more particularly overgrazing, and its attendant effects of depletion of plant cover and litter and trampling of the soil, was the most important factor contributing to erosion. To control the accelerated erosion on the watershed, the author recommended that plant cover be restored to at least 30%, that the rodent population be reduced, and that range and livestock management improvements be initiated which would relieve grazing pressures on areas particularly susceptible to erosion. On the Gallatin elk winter range in Montana (Packer 1963), ground cover densities of at least 70% and soil bulk densities no greater than 1.04 gm/cc were considered necessary for restoring and maintaining soil stability.

Improved grazing management (a change from yearlong to winter grazing and grazing controlled to attain 55% use of the key species) in New Mexico (Aldon 1964) resulted in improved watershed conditions over a 3-year period. Average ground cover for the three experimental watersheds doubled, bare ground decreased, and there were marked reductions in sediment yields and runoff.

At the Great Basin Experimental Station near Ephraim, Utah, studies of the effect of ground cover changes caused by domestic livestock on surface runoff and erosion were made on two 4-ha (9.9 acres) watersheds, A and B, during the period 1912-1958. Figure 6-20 shows the effect of plant cover changes on surface runoff and erosion during a 46-year period from the A and B experimental watersheds. The history of the experiment, as indicated in the figure, was divided into six periods. The relationship of plant and litter cover to both erosion and runoff is obvious.

The impact of grazing on sediment production appears to be accumulative. Though grazing eventually has an impact, as evidenced above, the impact may not come immediately. For example, Buckhouse and Gifford (1976a) found that one grazing period on a sandy loam site in southeastern Utah was sufficient to have some impact on infiltration rates but no impact was detected on potential sediment production rates.

Anderson (1975) reported that land use changes and its effect on sedimen-

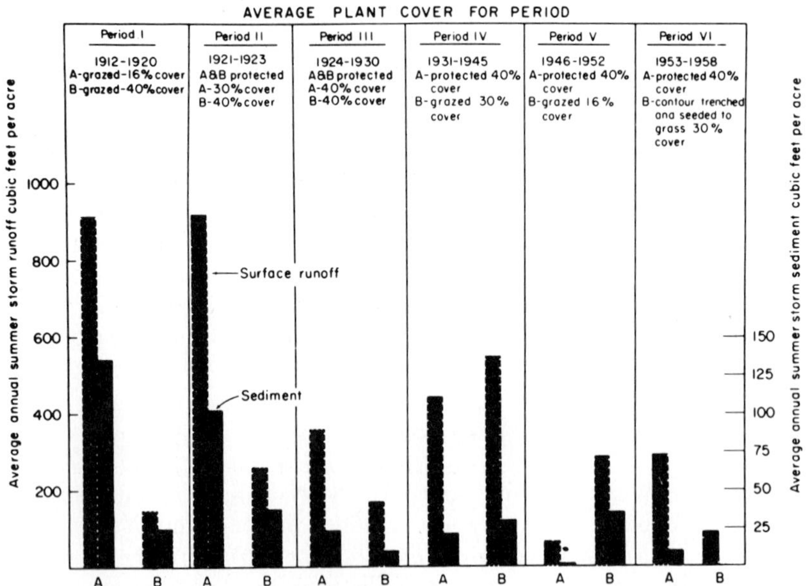

Fig. 6-20. The effect of plant cover changes on surface runoff and erosion during a 46-year period from A and B watersheds. Great Basin Experimental Area, Utah (from Croft and Bailey, 1964). (Divide ft³/acre by 14.3 to obtain m³/ha.)

tation must be treated with caution: "Highly variable responses in land use effects on sediment occur both in space and time." He reported (Table 6-5) that the mean effect for the land use changes may be estimated by multiplying the mean sediment discharge by a factor varying from 1.1 to 4.7 but that the extremes may be as much as 10 times these mean values. He also reported that major floods may establish new land and channel conditions in watersheds that augment post-flood sediment production. He reported that the effect of major floods or wildfires causes sediment production increases which last for varying durations with the recovery rate depending on many factors.

Table 6-5. Land uses effects on suspended sediment concentration and sediment discharge[a]. (Anderson 1975)

	Mean area considered, percent by	Mean effect, multiply by	Local extreme effect[b], multiply sediment by
Conversion steep forest to grassland (northern Calif.)	14.8	4.7	26[c]
Forest and brusland fires (10-year northern Calif.)	5.3	2.3	13[c]
Brushland fires (southern California)	2.0	1.4	5[d]
Poor logging (northern California)	1.4	1.26	19[c]
Unimproved roads (northern California)	0.6	1.24	9[c]
Roads (Oregon)	0.3	1.92	19[e]
Recent cutover (Oregon)	6.0	1.13	3.2[e]
Bare cultivation (Oregon)	4.0	1.29	8.2[e]
Eroding channel banks	8.0[f]	2.24	17[e]

[a]Mean sedimentation concentration 577 parts per million for northern California, and 62 parts per million for Oregon watersheds, mean Q were 1.67 and 2.97 cubic feet/second/mile, respectively; average sediment discharge 950 and 180 tons per square mile per year respectively.
[b]Local extreme is 100 percent in that land use condition, ecept roads, which is for 6 percent. These are extrapolated values.
[c]Wallis and Anderson (1965).
[d]Anderson (1949).
[e]Anderson (1954).
[f]Percent of either bank main channel length.

Impacts of Burning

As discussed elsewhere, burning may have an impact on both infiltration rates and runoff. Therefore, an impact on erosion and sedimentation would also be expected.

An analysis by Rowe, Countryman, and Storey (1954) was made to determine the effects of fire on erosion in southern California. Figure 6-21 represent the influence of time on sediment yields (see also Table 6-5). This same general type relationship—a recovery period—probably exists for most range sites throughout the country. Rich (1962) indicates this is true within the ponderosa pine type in Arizona, and Pase and Ingebo (1965) and Suhr (1967)

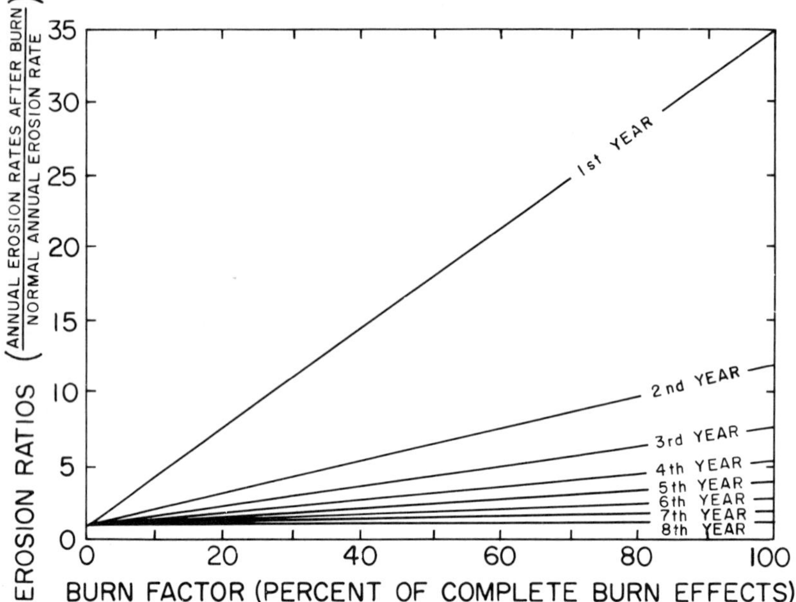

Fig. 6-21. Effect of partial burns on annual erosion rate (from Rowe, Countryman, and Storey 1954).

indicate the same within the chaparral type, also in Arizona.

Wright, Churchill, and Stevens (1974) looked at prescribed burning of ashe juniper (*Juniperus ashei*) on six paired micro-watersheds in central Texas. Erosion losses, runoff, and water quality were unaffected on level areas, but effects lasted for 9 to 15 months on moderate slopes (45-53%). Erosion losses stabilized within 18 months on all slopes when vegetative cover reached 60 to 70%. Figure 6-22 shows accumultive loss of sediment for two slope classes on burned watersheds.

Table 6-6 summarizes a number of studies as to the impact of fire on sediment production.

Impact of Chaining Pinyon-Juniper

As was the case in Chapter 4, summary statements pertinent to water quality *as related to chaining* will simply be extracted from the recent review by Gifford (1975):

1. Indications are that sediment discharges have not generally increased (or decreased) on pinyon-juniper sites due to vegetation manipulation practices. An exception to this is the increased quantity of sediment produced from debris-windrowed sites during high intensity thunderstorms in Utah.

2. Factors influencing sediment yields at given points on a pinyon-juniper sites are variable from site to site. Minimum sediment yields (equal to that from disturbed woodland) may be expected where surface soil disturbance is minimized (as with spraying a herbicide) or where

debris is left in place on a chaining project.

3. Chemical aspects of water from pinyon-juniper sites indicate water of quality suitable for irrigation, public water supply, and for aquatic life.

4. Given a runoff event, runoff during the first year from burned and debris-in-place sites may contain increased amounts of phosphorus and potassium, but not calcium, sodium, or nitrate-nitrogen.

Sediment Yield Estimates

The sediment transported past the outlet of a basin from all fluvial erosion sources is called the sediment yield. At numerous points in a watershed, the energy of the transporting water is insufficient to move all the sediment eroded and deposition occurs in reservoirs, valleys, or stream channels as well as the toes of long slopes where gradient changes may occur. Although we can estimate the amounts and location of some of this deposition with low precision, much more work is needed to eliminate such deposition and the adverse effects it may create. Because of deposition, the amount of erosion in a watershed is generally more than the amount of sediment leaving the watershed in the stream (sediment yield). Thus, sediment yield cannot be estimated from the erosion estimates within the watershed without additional

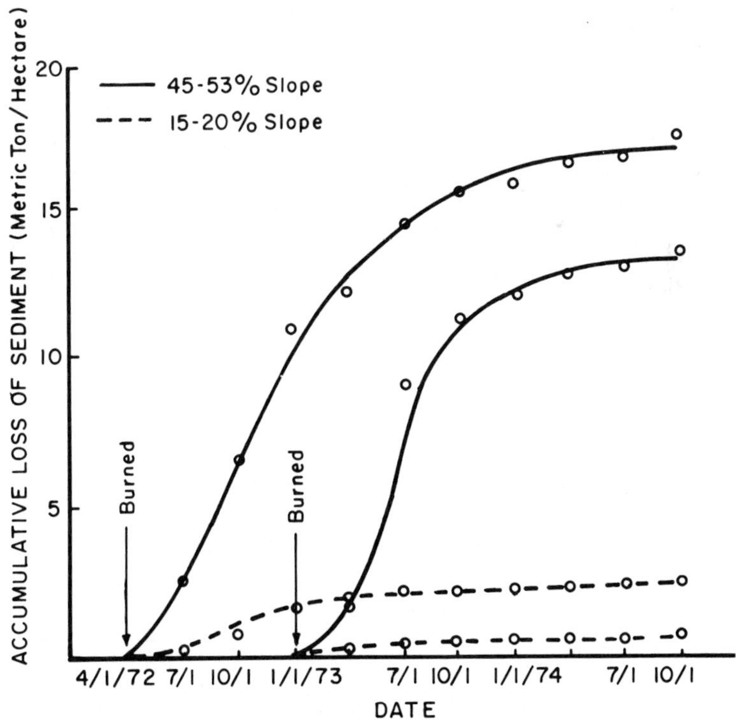

Fig. 6-22. Accumulative loss of sediment on burned watersheds for moderate and steep slopes in relation to date (from Wright, Churchill, and Stevens 1974). (Multiply metric tons/ha by 0.45 to obtain tons/acre.)

Table 6-6. Impact of fire on annual sediment yields.[1]

Author	Location	Type of burn	Soil	Vegetation type	Sediment increase (metric tons per hectare)
Connaughton (1935)	Idaho	Various intensities (wildfires)	Coarse, loose granitics	—	Increased where fire was intense or where slope was steep
Rowe (1941)	Nevada	Annual controlled	Sandy clay loam	Digger pine, California black oak	9.1 to 256.2
Shantz (1947)	California	Wildfire	—	Brush ranges	19.3
Biswell and Schultz (1957)	California	Light controlled burn	Gravelly loam	Ponderosa pine, black oak, manzanita	None
Cooper (1961)	Arizona	Light controlled	Basaltic materials	Young ponderosa pine	Significant increase
Rich (1962)	Arizona	Intense wildfire	Loam to clay loam	Ponderosa pine, Gambel oak, Douglas fir	79.4
Loveland	Nevada	Wildfire	—	**Wheatgrass-cheatgrass range to aspen-subalpine herbaceous range.**	6.8
Fredriksen (1970)	**Oregon**	Light slash burn	—	**Old growth Douglas fir**	**14 times the control**
Nobel and Lundeen (1971)	Idaho	Wildfire	Coarse-textured over granite	Steep river break lands	6.8
Ursic (1970)	Mississippi	Light controlled burn	Loess, silt loam, sandy loam	Upland hardwoods	0.5
Pase and Lindenmuth (1971)	Arizona	Various intensities (controlled)	Coarse, poorly	Chaparral	2.5

Brown and Krygier (1971)	Oregon	Intense slash burn	—	Douglas fir	2.0
DeByle and Packer (1972)	Montana	Light to intense, controlled	Silt loam	Western larch, Douglas fir, Engleman spruce	0.2
Helvey (1973)	Washington	Intense wildfire	fine sandy loam over pumice	Ponderosa pine, Douglas fir	509.6
Wright, Churchill and Stevens (1974)	Texas	Light controlled burn	limestone derived soils	Ashe juniper	2.3 15% slope 18.2 53% slope
Megahan and Moliter (1975)	Idaho	Intense wildfire	sandy loam, loamy sand	Douglas fir	2.0
Buckhouse and Gifford (1976c)	Utah intense to light	Controlled burn,	sandy loam	Chained pinyon-juniper	No increase (first

[1]Modified after Wassen (1976). Values given in this table should be regarded as short term impacts of burning on sediment yields. The data represent results from both small plots and also calibrated watershed. For more specific information, consult the original reference (tons ha × 0.446 = t/acre)

 data.

In the main, sediment yield is predicted by using one or more of four methods or approaches:

1. The sediment-rating curve/flow-duration method requires concurrent field measurement of streamflow and sediment discharge to establish an average relationship between parameters of streamflow and sediment quantity. The method is often used for establishing the amount of sediment expected to reach large reservoirs or other points on principal tributaries and main rivers. The method, however, is generally not applicable in estimating sediment yields at problem sites in upstream watersheds because of the difficulty and expense of obtaining the required data for the large number of sites potentially of concern. Problems are also encountered in using the sediment-rating curve-flow duration method for estimating the effects of treatment measures upon sediment yield because of imponderables and uncertainties in arriving at a sediment-rating curve for conditions with the treatments in effect.

2. The reservoir sediment-deposition survey method involves measurements by field survey of the volume of sediment accumulated in a pond or a reservoir. The measured volumes are converted into weights, adjusted for reservoir trap efficiency, and expressed as rates of accumulation according to the age of the reservoir or the time interval between surveys. Deposition surveys on a number of reservoirs in a land resource area, watershed, or river basin are often compiled and summarized to show relationship between sediment yields and size of drainage area. This approach gives useful general information of the magnitude and variation of sediment yield in the region of interest, but has limited value for forecasting sediment yield from an individual watershed where no measurements have been obtained. The method also has limitations in estimating the effects of program measures on sediment yield.

3. The sediment-delivery ratio method requires a factor expressing the percentage relationship between sediment yield from a watershed and gross erosion in the watershed in the same time period. Sediment-delivery ratios are developed from the sediment yields obtained by reservoir surveys or measurements at suspended-load stations in comparison with erosion in the watershed. The erosion quantities for sloping uplands are computed by erosion prediction equations and are estimated by various procedures for gullies, stream channels, and other sources. The erosion quantities for sloping uplands are used with a sediment-delivery ratio to give estimates of sediment yield for a watershed. Since the method deals with calculated and estimated quantities of erosion on the watershed with and without treatments, it has a rationale for estimating program effects upon sediment yield, under the assumption that the same sediment-delivery ratio applies for both the "before" and "after" conditions. Sediment-delivery ratios have been developed for much of the eastern half of the United States, but scant rainfall data and other complicating factors have precluded the use of the method in the West.

Table 6-7. Conditions indicative of high sediment-yield potential that can usually be identified by observation (after Stewart et al. 1975).

Cropland
1. Long slopes farmed without terraces or runoff diversions
2. Rows up and down moderate or steep slopes
3. No crop residues on surface after new crop seeding
4. No cover between harvest and establishment of new cropy canopy
5. Intensively farmed land adjacent to stream without intervening strip of vegetation
6. Runoff from upslope pasture or rangeland flowing across cropland
7. Poor stands or poor quality of vegetation
Other Sources
1. Gullies
2. Residential or commercial construction
3. Highway construction
4. Poorly managed range, idle, or wood areas
5. Unstabilized streambanks
6. Surface mining areas
7. Unstabilized roadbanks
8. Bare areas of noncropland
9. Logging roads poorly constructed
10. Recently burned areas

4. The bedload function methods make use of mathematical equations developed for calculating the rate and quantity of movement of materials constituting the bed of alluvial channels. Applications of these equations requires information on sediment particle sizes, channel gradients and cross sections, and a flow-duration curve. The equations often give widely differing results for the same conditions.

In using these methods for estimating sediment yield, observations within the watershed may also be used to modify the computed yield. Table 6-7 lists some likely intensive sediment sources. Presence of such sources within a watershed may indicate that the predicted sediment yield should be increased. Because erosion hazards (such as in Table 6-7) are so extremely local in nature, soil loss estimates and control guides are often subjective or, at best, accurate only on a site basis.

To the methods of predicting sediment yield might be added another class that includes predictive equations based on climatic and physiographic characteristics of the watershed for which the sediment yield value is required. One example of such a method is the Flaxman (1972) model (revised March 1974, Pacific Southwest Interagency Committee 1974) which uses multiple regression techniques (see Chapter 11, Rangeland Hydrologic Models, Theory to Practice for a discussion for such a model) and data from numerous watersheds in the western United States. The prediction equation developed is:

$$\log(Y + 100) = 524.37231 - 270.56525 \log(X_1+100) + 6.41730 \log(X_2+100) - 1.70177 \log(X_3+100) + 4.03317 \log(X_4+100) + 0.99248 \log(X_5+100) \quad \textbf{(6-3)}$$

(logs are to the base 10)

where: Y = Average annual sediment yield, tons per sq. mile,
X_1 = The ratio of average annual precipitation in inches to average annual temperature, in ° Farenheit,
X_2 = Watershed slope,
X^3 = Soil particles, greater than 1.0 mm in
X_4 = Soil aggregation or dispersion, less than 0.002 mm, in %, and
X_5 = 50% mean annual flood in cms, which has a recurrence interval of 2.33

This model has been widely used by the Soil Conservation Service, United States Department of Agriculture in the western U.S. and has correlated quite well with existing data. When the method was compared with data from Walnut Gulch, it fit the data quite well (Renard and Simanton 1973). The independent variables used in this model have physical significance, but the equation certainly does not describe the individual processes involved nor is this its intent. Some experience is probably required to adjust variable X_1 for orographic variations in basin and range topography.

Anderson (1975) developed a regression model very similar to the one developed by Flaxman with the exception that he used many more variables. The equation is presented in tabular form in Table 6-8. It is interesting to note that the coefficients of many of the variables are quite low, thus indicating that their contribution to the predicted sediment yield is quite low except when the parameter itself is large. Detailed discussion of the parameters is given in the original paper.

Weber, Fogel, and Duckstein (1976) discuss, on a limited basis, the use of multiple regression models in predicting sediment yield.

Geologists in the Soil Conservation Service have made wide use of sediment delivery ratios for routing sediment from small areas to the outlet of larger watersheds (Boyce 1975, Renfro 1975) to predict sediment yield. Experience has shown however that the delivery ratios are not location invariant. For example, five differnt delivery ratio relationships are presented in the proceedings of the 1972 Sediment Yield Workshop, (Figure 6-23). Unfortunatley there are no known delivery ratio curves for rangeland condition such as might be applicable for areas considered in this report. The Soil Conservation Service curve, based on widely scattered data is undoubtedly the best to use in the absence of site specific data.

In ephemeral streams, which are common in most rangeland areas, there is a decrease in annual runoff with increases in drainage area (Hadley and Schumm 1961, Renard 1970). Thus, the mean annual rate of reservoir sediment accumulation (sediment yield assuming very little sediment moves through the principal or emergency spillway) should reflect a similar decrease

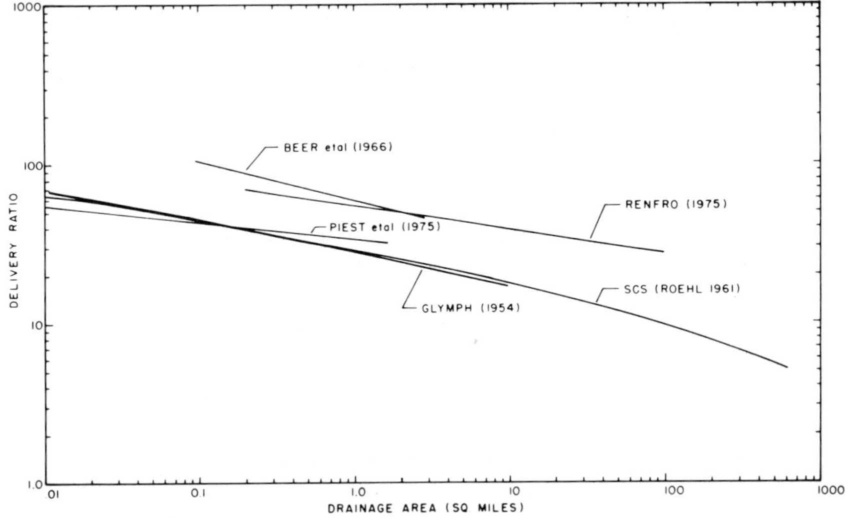

Fig. 6-23. Sediment delivery ratio variations with drainage area as reported in the 1972 Sediment Yield Workshop (U.S. Agr. Res. Serv. S-40, 1975). (Multiply mi² by 2.59 to obtain km².)

in sediment yield with increasing watershed area. Hadley and Schumm (1961) stated "The marked decrease in rate of sediment movement with increasing drainage area size... can be attributed, it is believed, to absorption of water in channels together with a trend toward gentler slopes and wider flood plains in a downstream direction." The results of the Hadley and Schumm (1961) investigations of the Cheyenne River in Wyoming are shown in Figure 6-24 along with the sediment yield-area results from other investigators. As might be expected, the results are extremely varied.

Renard (1972) and Renard and Laursen (1975) presented a method for estimating sediment yield from watersheds containing ephemeral streams where the runoff results from limited areal extent thunderstorms. The computer model used in this effort incorporated a stochastic runoff model (see Chapter 11 for a description of the model) and a deterministic sediment transport relation (Laursen, 1958).

Summation of the sediment yield from each runoff event led to the sediment yield relationship with drainage area shown in Figure 6-24, which was obtained by calibrating the model with data from the Walnut Gulch Experimental Watershed in southeastern Arizona. The stochastic runoff model used to produce the synthetic flow events incorporates the transmission losses observed in ephemeral streams.

The sediment yield-drainage area relationship of Strand in Figure 6-24 has a slope very similar to that of Renard (1972) as might be expected because both sample a similar climatic-physiographic set of conditions. Strand's data covered drainage areas in the Southwest from California to New Mexico. The Strand data, however, indicate a higher yield for all watershed sizes.

Table 6-8. Reservoir deposition regression model, coefficients, units, and means of variables (from Anderson, 1975).

Symbol	Definition[1]
Log SED = +0.166	Regression constant, for deposition, cubic meters/hectare/year, mean log SED + 0.53.
+0.5589 Log Y1	Year of first reservoir capacity measurement 1850 = 0, log mean 1.8861.
+0.1548 Log EF	Effect of forest fires, annual percent of area burned, linearly depleting to zero effect after 26th year, %, mean 2.35.
+0.4470 Log EBR	Effect of high elevation large brush fields, percentage of catchment area in brush times percent of area above 1500 meters elevation, %/10, mean 0.1058.
+0.7831 Log CV	Coefficients of variation of basin flowpath lengths,[2] with path lengths as suggested by Busby and Benson[3], unitless, mean 0.1635.
+0.3087 Log SL	Slope of streams of 800-meter mesh length[4], m/Km, mean 1.9909.
+0.0043 USED	Area of consolidated sedimentary rock types (Cenozoic and younger), %, mean 31.1.
-0.0003 PCEN	Area of pre-Cenozoic marine sediments and metamorphic rocks, %, mean 46.
-0.0012 GR	Area of granitic rock type, %, mean 21.1.
-0.0024 BA	Area of basalt rock, %, mean 4.5.
+0.0965 R12	Length of highways on ridges, km/km^2 of watershed area, mean 0.035.
+0.1216 SL12	Length of highways on slopes, km/km^2 of watershed area, mean 0.093.
+0.0426 ST12	Length of highway near streams, km/km^2 of watershed area, mean 0.049.
+0.1977 R3	Length of improved (surfaced) roads on ridges, km/km^2 of watershed area, mean 0.080.
+0.4013 SL3	Length of improved (surfaced) roads on slopes, km/km^2 of watershed area, mean 0.151.
+1.3496 ST3	Length of improved (surfaced) roads on streams, km/km^2 of watershed area, mean 0.072.
-0.1064 R4	Length of unimproved (dirt) roads on ridges, km/km^2 of watershed area, mean 0.161.
+0.4236 SL4	Length of unimproved (dirt) roads on slopes, km/km^2 of watershed area, mean 0.320.
+0.3020 ST4	Length of unimproved roads near streams, km/km^2 of watershed area, mean 0.101.
+0.1210 URB	Length of urban roads, km/km^2 watershed area, mean 0.037.
+0.011 RSP	Average of yearly mean flow time maximum daily flow during period of sediment accumulation divided by long term, 1881–1965, average, mean 1.32.
-0.0128 SA	Mean "long term" snow frequency (1-Relative Rain Area)[5], percent, mean 15.5.
+0.0778 Log CW	Capacity of reservoir at year Y1 divided by drainage area, m^3/ha, mean 2.7420.

+0.0137 Log CWSQ	Log CW squared, mean 7.5186.
+0.1397 Log MAP	Mean annual precipitation[6], mm, mean 2.9745.
+0.0255 Log PSQ	Square of log MAP, mm^2, mean 8.8476.
-1.9665 DEN	Average density of reservoir sediment, calculated from geology soil-texture of Wallis and Willen[7] and texture-density-age relation of Koelzer and Lara[8], g/cm, mean 1.44.
+0.0096 CLAJ	Percentage of clay in surface soil, calculated from geology soil-texture relation and adjusted for climate[7], mean 15.3.
-0.0011 LI	Percentage of area in landslide class 1[9], mean 7.0.
-0.0014 L2	Percentage of area in landslide class 2, mean 31.0.
+0.0021 L3	Percentage of area in landslide class 3, mean 18.3.
+0.0012 L4	Percentage of area in landslide class 4, mean 25.5.
-0.008 L5	Percentage of area in landslide class 5, mean 10.6.
-0.004 L6	Percentage of area in landslide class 6, mean 7.6.
+0.0591 FAUL	Extent of geologic fault zones per unit area of watershed, State of California 1966, m/10ha, mean 0.93.

[1]Multiply m^3/ha/yr by 0.21 to obtain acre-ft/sq. mile/yr. Multiply m/km by 5.28 to obtain ft/mile. Multiply km/km^2 by 1.6 to obtain miles/sq. mile. Multiply g/cm by 0.067 to obtain lb/ft. Multiply m/10 ha by 0.13 to obtain ft/acre.
[2]Wallis and Anderson (1965)
[3]Busby and Benson (1960)
[4]Anderson (1956)
[5]Anderson and Wallis (1965)
[6]U.S. Geological Survey (1969)
[7]Wallis and Willen (1963)
[8]Koelzer and Lara (1958)
[9]Radbruch and Crowther (1973)

The Livesey (1975) data for the area in the Upper Mississippi Basin is surprisingly close to the data from southeastern Arizona. Livesey showed in his paper that other land resource areas in the Upper Mississippi Basin have sediment yield-drainage area relationships that parallel the one shown in Figure 6-24 and would constitute a family of curves covering the range of yields shown on the ordinate of the figure.

The Piest Kramer, and Heinman (1975) data in Figure 6-24 are based on data from cultivated areas (corn) in southwestern Iowa. Each of these sediment yield-drainage area relationships represents a sediment delivery ratio. Assuming the sediment yield for the smallest area represents a delivery ratio of 100%, subsequent larger areas are then a smaller delivery ratio. As Renfro (1975) points out, the use of a sediment-delivery curve must be tempered with experienced judgement of the characteristics of the drainage basin such as the texture of soils, drainage density, relief and opportunities for deposition within the basin being studied.

Sediment Transport

Civil engineers have been concerned with sediment transport problems for a long time. This concern has led to numerous laboratory flume studies of the factors controlling the movement of sediment in streams. Major portions of recent books by Henderson (1966), Blench (1966), Graf (1971) and an American Society of Civil Engineers manual edited by Vanoni (1975) describe the sediment transport process in detail.

Sediment transport is to a large extent a function of velocity of flow. The largest particle of a given specific gravity that a stream can transport (competency) varies approximately as the fifth power of velocity. The maximum quantity of sediment that a stream can transport (capacity) varies approximately as the 3.2 to fourth power of stream velocity (Twenhofel 1939). Several factors determine stream velocity. A widely accepted empirical formula that related stream velocity to channel roughness, shape, and slope is the Manning equation—

$$V = \frac{1.5R^{2/3}S^{1/2}}{n} \qquad \textbf{(6-4)}$$

where V is mean stream velocity in ft/sec, n is a roughness coefficient which ranges from 0.02 for unlined canals to 0.10 for streams obstructed by vegetation and debris (Barnes 1967), R is the hydraulic radius which is defined as the cross-sectional area of a stream divided by the wetted perimeter, and S is the slope of the energy gradient which closely approximates the slope of the streambed. Reduction of n will increase V and increase sediment transport; reduction in R or S will cause a decrease in sediment transport. The factor having the greatest effect and one most easily changed is the roughness

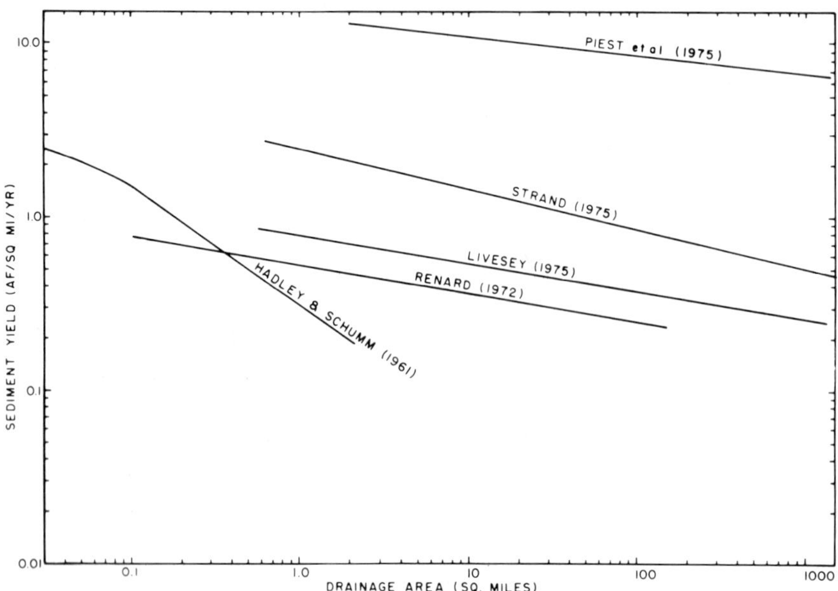

Fig. 6-24. Examples of the relationship of sediment yield for various sizes of watershed. Data reported as tons was converted to volume assuming 90 lbs/cu ft (acre-ft/yr = 0.00123 hm³/yr).

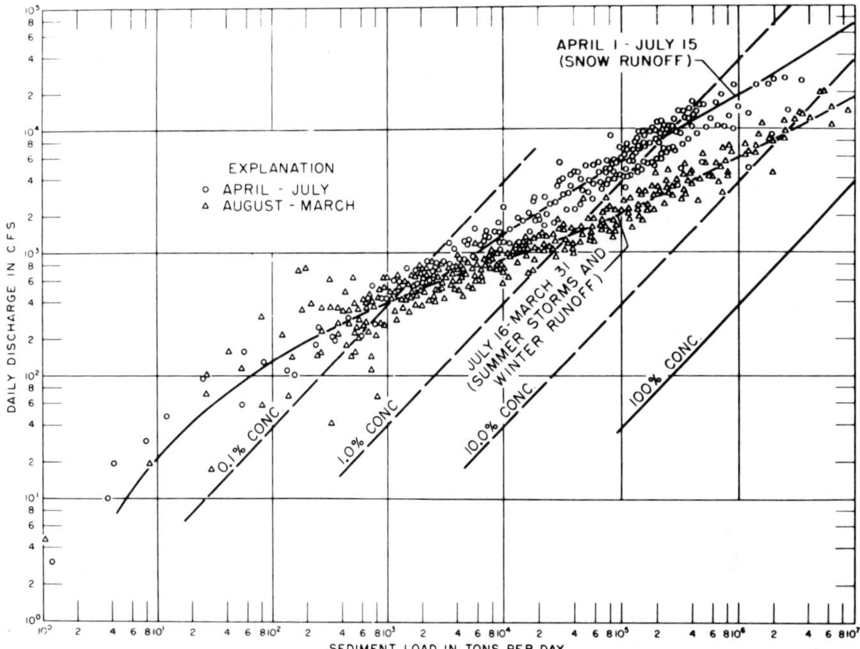

Fig. 6-25. Sediment-discharge rating curves for San Juan River at Bluff, Utah (Strand 1975) (ft³/sec = 28.32 liters/sec).

coefficient, n. Mechanical structures can be used but vegetation may offer a more practical means of modifying roughness.

Sediment Yield with Flow-duration Curves

The flow-duration, sediment-rating curve method is widely used for computing sediment yield, especially for larger rivers. The use of the technique is illustrated by using Figures 6-25 and 6-26 for the San Juan River at Bluff, Utah with Table 6-9 and 6-10 showing typical computations. Interestingly, the sediment concentration-discharge data for this stream is quite seasonally dependent and the computations such as those in Table 6-11 must be made for both seasons using the corresponding values for the seasons as shown in Figures 6-25 and 6-26. The computed sediment yield for the thunderstorm periods at this station is 58% of the yearly total, although the runoff during that period is only 36% of the average annual value.

Corrections are also made in the sediment yield computation for the unmeasured portion of the sediment load, i.e., that portion of the material in transport not sampled with conventional sampling equipment. If sufficient data are not available to make total-load computations (using modified Einstein Procedure [Colby and Hambree 1955] or other bedload formulas), the U.S. Bureau of Reclamation makes adjustments to the suspended-sediment loads using the data in Table 6-10. The percentages of anticipated bedload shown in this table indicate the need for collecting data to compute

 Table 6-9. Bedload correction table (from Strand 1975).

Suspended- load concentration (p/m)	Streambed material	Texture of suspended material	Percent bed-load in terms of measured suspended load
<1,000	Sand	Similar to bed of stream .	25-150
<1,000	Compacted clay, gravel, cobbles and boulders.[1]	Small amount of sand ...	5-12
1,000-7,500	Sand	Similar to bed of streams.	10-35
1,000-7,500	Compacted clay, gravel, cobbles and boulders.[1]	25 pct sand or less	5-12
>7,500	Sand	Similar to bed of stream .	2-8
>7,500	Compacted, clay, gravel cobbles and boulders.[1]	25 pect sand or less	5-15
Any Silt and clay . concentration	Clay and silt, <2 unconsolidated		

[1]The bed material of the stream may contain any one or all of these very fine or coarse-sized sediments.

the quantity of unmeasured load.

Stochastic Computer Models for Sediment Yield

Chapter 11 discusses some of the modeling techniques which are available for water resource problems. Stochastic models for sediment yield estimating offer many possibilities although the progress in some of these has not carried them sufficiently far to make them ready for operational problem solving.

Woolhiser and Todorovic (1971) presented a stochastic model for predicting the sediment yield from an ephemeral stream where the yield was produced by a series of events of relatively short durations. They suggested that sediment yield from a watershed with ephemeral streams may be treated as the sum of a random number of random variables. In this model, the distribution of sediment yield per event must be obtained emphirically, but the sediment-yield counting process can be related to the precipitation counting process by a rainfall-runoff model. They compared their model to data from two small watersheds near Hastings, Nebraska. They also applied the stochastic model of sediment yield to the problem of estimating mean and variance of annual sediment yield when only short concurrent records of sediment yield and runoff are available, which can be supplemented by longer records of precipitation and runoff.

Sakhan, Riley, and Renard (1972) developed a simulation model to describe the dynamics of a channel in terms of: (1) two one-dimensionsal streamflow equations, (2) a one-dimensional sediment transport equation (for the stream bed), and (3) a stochastic sediment transfer at the stream bed which also included the bed load. This complex model was then solved using a hybrid (combination digital and analog) computer. The model thus postulated illustrates what often happens, i.e., the discriptions of most systems incorporates aspects of various types of modeling schemes.

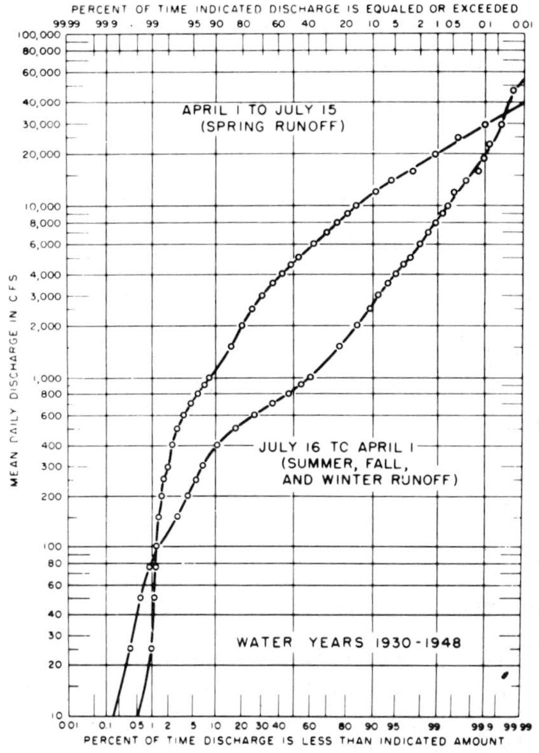

Fig. 6-26. Seasonal flow-duration curves for San Juan River at Bluff, Utah (Strand 1975) (ft^3/sec = 28.32 liters/sec).

Erosion Control Structures and Mechanical Land Treatments

Numerous attempts have been made to control erosion on rangelands but failures have been surprisingly frequent. In reviewing the effectiveness of erosion abatements practices on semiarid rangelands in the western United States, Peterson and Hadley (1960) include conditions of 191 common erosion control structures—ranging from 6 to 10 years in age—located in the upper Gila River Basin (Table 6-11). The authors state that the almost complete absence of any appreciable benefits to vegetation from structures of this type plus the excessive costs of maintenance do not warrant the further use of such units in areas where climate and soil conditions are similar to those in southern Arizona. In general, structures like these have dropped from all conservation programs on rangelands in the southwestern United States.

In a related study in the upper Gila River and Mimbres River basins in New Mexico (Peterson and Branson 1962), appraisals were made to determine effectiveness of various land treatments measures constructed by the Civilian

Table 6-10. Average annual sediment load from the San Juan River snowmelt season (from Strand 1975).[1]

Percent of time (1)	Interval (%) (2)	Cumulative (%) (3)	Qw[2] (4)	Qs[3] (5)	Columns 2X4 (6)	Columns 2X5 (7)
00.00 - 0.02	0.02	0.01	40,000	3,500,000	8.0	700
0.02 - 0.1	0.08	0.06	32,000	2,400,000	25.6	1,920
0.1 - 0.5	0.4	0.3	25,000	1,550,000	101.2	6,200
0.5 - 1.5	1.0	1.0	20,7000	1,130,000	207.0	11,300
1.5 - 5.0	3.5	3.25	16,300	74,000	570.5	20,450
5 - 15	10	10	11,700	500,000	1,170.00	40,000
15 - 25	10	20	8,900	245,000	890.0	24,500
25 - 35	10	30	7,000	157,000	700.0	15,700
35 - 45	10	40	5,650	107,000	565.0	10,700
45 - 55	10	50	4,720	79,000	472.0	7,900
55 - 65	10	60	3,750	52,000	375.0	5,200
65 - 75	10	70	2,940	33,500	294.0	3,350
75 - 85	10	80	1,950	17,000	195.0	1,700
85 - 95	10	90	1,100	7,000	110.0	700
95 - 98.5	3.5	96.75	540	2,100	18.9	74
98.5 - 99.5	1.0	99.00	25	s12	0.25	0.12
99.5 - 99.9	0.4	99.7	7.5	4.5	0.03	0.02
99.9 - 99.98	0.08	99.94	2.5	2.0	0.02	0.0
99.98 - 100	0.02	99.99	0	0	0	0
Total					5,702.5	150,394

[1]Computations from table data:
Runoff = 5,702.5 × 106 × 1,9835 = 1,200,000 acre ft/Yr
Sediment = 150,394 × 106 = 15,950,000 Tons/Yr
Unmeasured correct (20%) = 3,200,000 Tons/Yr
Total Sediment = 19,150,000 Tons/Yr

$$\text{Yield} = \frac{19{,}150{,}000 \text{ Tons/Yr}}{(1351 \text{ Tons/acre feet})(23{,}000 \text{ mi}^2)} = 0.612 \text{ acre feet/sq. mile/yr.}$$

[2]Qw = water discharge in cubic feet per second
[3]Qs = total sediment load in tons

Conservation Corps in the mid-1930's with respect to vegetation improvement, longevity of structures, and quantities of sediment retained. Treatments included earth fill dams, earth dike spreaders, loose rock spreaders, hand placed rock spreaders, brush spreaders, "cement worm" spreaders,

Table 6-11. Effectiveness of land treatment structures in the Upper Gila River Basin in Arizona.

Type of structure	Number examined	Number breached	Effect on vegetation
Small earth diversion dams	37	14	Considerable sediment deposited above dams but vegetation negligible.
Hand-placed rock spreader	65	33	Sediment deposited above spreader but no perennial vegetation established.
Loose rock spreader	27	16	Little sediment deposition, vegetative recovery negligible.
Brush spreader	30	30	Results negative
Wire and brush spreader	32	16	Fair sediment deposition but no appreciable amount of vegetation established.

Data from Peterson and Hadley (1960).

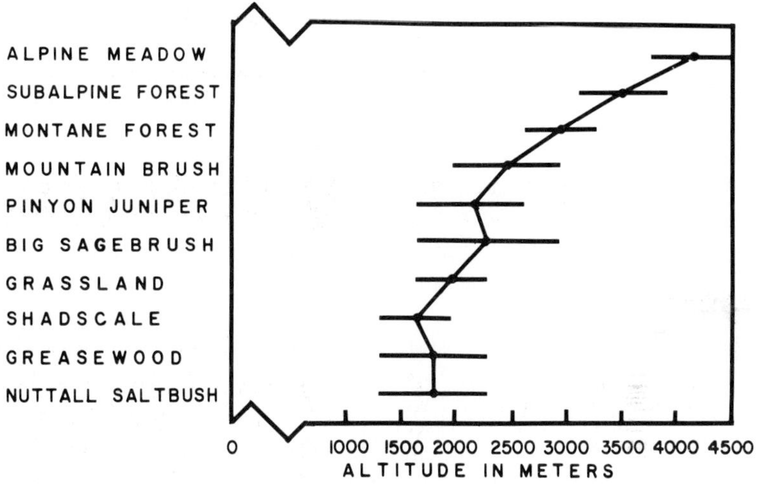

Fig. 6-27. Approximate altitudinal ranges of important vegetation types found in the Upper Colorado River Basin (from Branson 1975) (m × 3.218 =ft).

cable and wire spreaders, and rock rubble gully control structures. More than half of the structures were breached within a few years after construction, with the highest percentage of failure occurring in the loose rock spreaders, brush and rock spreaders, and rock rubble structures. Where earth dikes were not breached and water reached the spreader system, there was improvement in vegetation even in the driest areas at lower altitudes.

Gully-control structures installed on National Forests in Colorado 20 to 25 years ago were evaluated to show the role of check dams in gully control (Heede 1960). Based on an analysis of four gully streams, the following principles and recommendations were derived to guide future gully-control measures:

1) Determine the type of system, i.e., *continuous* (the main active [critical] location is the lowest segment near the gully mouth), or *discontinuous* (gully tends to expand upstream and downstream. The two different types probably represent only stages in development, the discontinuous representing a more juvenile stage (Heede 1970). Heede (1975b) further discusses the two types as follows:

> Continuous gullies begin their downstream course with many small rills, while continuous gullies start with an abrupt head cut. The headcut of the discontinuous gully may be located at any position on the slope of a hillside, while a continuous gully always starts high up on the mountainside and continues its course down to the main valley floor. The discontinuous gully may intersect the surface of the slope at any point, and thus be terminated.
>
> After coalescence of many fingerlike rills, the continuous gully soon attains relatively great depth. It maintains approximately the same depth until the lowest reach above the gully mouth is approached, where depth decreases quickly along a concave profile that terminates

at the gully mouth. A discontinuous gully rapidly decreases in depth downstream and thus deveiops a gully bottom gradient much gentler than that of the original valley floor. Where the gradients intersect, a sediment fan is built. There, a new gully may begin with a headcut. Discontinuous gullies. . .generally occur in series along the length of the drainageway.

Discontinuous gullies develop into continuous gullies.

2) Determine the critical locations in the gully system.

3) Estimate the highest expected storm runoff for the period of structural treatment.

4) Estimate the highest expected peak flow at all structural sites.

5) Design structures to accommodate this flow (see Heede and Mufich 1974).

6) In discontinuous gully systems, control headcuts and lower channel segments near the gully mouth by structures, and install water spreader below the system if the lowest gully mouth is not on the main flood plain.

7 In continuous gully systems, start structural treatment at the gully mouth; place strongest check dam there.

8) Install structures below headcuts located in the channel bottom.

9) Treat critical locations, such as sharp curves in meandering gullies or cuttie banks, by revetments; install these devices parallel to the bank.

10) Do no deflect flow toward channel banks.

11) As a general principle, do not plant woody vegetation within the high-water channel where obstruction of flow would cause the stream to go out of banks or undercut gully walls.

12) Remove boulders, trees, bushes, and other flow restrictions that would result in bank cutting or overtopping of the channel.

13) To reduce the amount of earth movement and surface soil disturbance required, defer bank sloping until channel bottom is raised.

14 After treatment, inspect gully periodically to determine needs for maintenance and further control measures.

Another study (Heede 1968) tested the effectiveness of four carefully engineered and revegetated waterways for treatment of steep, gullied hillsides of a watershed on the White River National Forest near Silt, Colorado. Three years after treatment, the waterways had lost only 9% as much soil as comparable untreated gullies. Design criteria for the waterways were as follows:

1) Place waterway on undisturbed soils away from the filler placed in the existing gully.

2) Increase the length of waterway over that of the original gully and thus decrease the gradient.

3) Design broad, flat cross-sections that allow waterflows to spread.

4) establish grass or other low-growing vegetation on waterway and all other disturbed areas as quickly as possible.

Heede (1975a) indicates that a few criteria will demonstrate if a watershed is within or outside of dynamic equilibrium (adjustment of various components of stream flow to accommodate the conveyance of flow and load). These are as follows:

A. Indicators of dynamic equilibrium:
Missing channel nickpoints and heatcuts; concave longitudinal bed profile, relatively small sediment loads; relatively similar behavior of streamflow from station to station.

B. Indicators of accelerated landform development:
Occurrence of channel nickpoints and headcuts; linear to convex longitudinal bed profile; relatively large sediment loads; relatively large differences in the behavior of streamflow from station to station.

Heede also points out some of the implications for watershed management:

What are the inferences for the land manager? Where dynamic equilibrium does not exist, it is obvious that the forces exerted by streamflow lead to sediment production and deposition—in short, to active landform development. If man plans to utilize the streams of such a watershed, extensive and expensive channel stabilization structures will be required. Maintenance may a long-lasting task, since under such conditions man works against nature rather than with it.

In contrast, where watersheds in dynamic equilibrium are to be managed for the development of the water resource, it will only be necessary to establish the upper limits of use in order to avoid misuse. Aspects of equilibrium will be of great importance where streams are to be used for the conveyance of additional water, or where sources of water pollutants must be determined. Man may not have moved one finger, but water quality may be poor because the watershed is in an active stage for landform development rather than equilibrium.

Some very interesting conclusions regarding channel stabilization measures in the mountain and foothill area of the Angeles National Forest in California can be found in a report prepared by Rudy (1974). Because information of this type is not readily available, most of the channel stabilization conclusions section follows.

1) Check dam systems tend to promote the growth of channel vegetation where none existed before, such as alder and sycamore trees and other riparian species.

2) Check dam systems may appear to be a highly efficient debris control treatment until a large flood occurs to cause an unload of the excess debris from the debris cone.

3) The debris cone needs to be designed wide enough to permit side slope storage to come to rest and still not constrict the flood channel of a major flood.

4) Both the check dams and their debris cones need periodic maintenance and especially after a major flood.

5) Check dam systems are valuable watershed treatment in a few selected areas, but as a uniform prescription for the Los Angeles Watershed they are marginal to negative in the return on the investment.

6) The permanent decrease in debris yield to a basin is probably something less than 10% of the normal, as a result of installing a crib dam system.

7) Construction-induced debris accounts for 25% to 35% of the design storage capacity of a check dam system in the Los Angeles Watershed.

8) There are many other than check dam watershed stabilization treatments that need to be installed on this watershed. Each must be a custom-designed structure to treat a specific problem for each subwatershed rather than a massive, uniform type of treatment.

9) Check dam systems are effective watershed treatment for:: (a) stopping channel degradation, (b) stabilizing a residual bedload, (c) raising and widen-

 ing a channel, (d) directing flood flow down a defined channel, (e) storing debris up to a level line from spillway, and temporary storage above the level line for as much as 15 years, (f) stabilizing side slope debris deposits on the wide part of the check dam debris cone, (g) aiding landscape improvement work, (h) the support of road crossings, bridge abutments, fill slopes, and numerous road construction needs, (i) stabilization of land fill areas, and (j) areas that need a small cleanout basin designed to withstand overtopping. 10) Check dam systems are of marginal or questionable value for: (a) decreasing the debris yield to a debris basin, (b) water yield improvement work, (c) stabilization in channel beds where end-cutting may be a problem, (d) channels that parallel roadfills that can be undercut by meandering flows, (e) stabilizing the bedload in channels that have degraded to bedrock, (f) reducing the velocity of flood flows, except where the system is designed such that the toe of an upper dam is level with the spillway of the lower dam, (g) channels that carry large boulders regularly, (h) post-fire treatment, except as storage structure, (i) channels that have no debris basin or control structure downstream of the check dam system, and (j) areas where post-construction access for check dam maintenance is not possible.

The influence of low dams and barriers on particle-size distribution of material deposited, slope deposit, volume, and extent of deposit and original channel gradient were determined by Lusby and Hadley (1967). Results indicate that the slope of deposition is dependent somewhat on the particle-size distribution of transported sediment. The rate of filling of steep-sided gullies is dependent on the availability for transport of material approaching the size of the original bed material.

Deposits behind low permeable barriers had steeper surface gradients than the original stream channels, and deposits behind low dams and surface gradients less steep than the original channels. A probable reason for over-steepening behind permeable barriers was backwater causing enough reduction in velocity to deposit coarse sediment at a point upstream, while sufficient velocity was maintained near the barriers to carry the fine sediment through.

Impact of Contour Furrows and Gully Plugs

The effects of gully plugs and contour furrows on the soil moisture regime, erosion and sedimentation rates, and vegetation response in treated areas on Mancos Shale near Cisco, Utah (average annual precipitation 7.12 inches, (183 mm) are discussed by Coltharp (1967) and Thomas (1975). They found that contour furrows constructed with a Holt trencher on shallow shale-derived soils were effective in holding runoff and sediment on site. Life expectancy of the furrows range from 7 to 12 years. Gully plugs (small earth-fill structures constructed in small gullies) were effective in holding runoff and sediment. Life expectancy of the treatments were from 14 to 33 years, depending on site location. Wein and West (1972, 1973) indicate, however, that because of harsh microclimates and lack of improvement in either vegetation or soils, continuance of these treatments cannot be recommended on sites similar to the Cisco area. From an economic standpoint, a rather unfavorable benefit-cost ratio of 0.17 to 1 was derived based on results of treating the Cisco area with both gully plugs and contour furrows (Workman and Keith 1971). Soil moisture data showed that gully plugs and contour furrows significantly increased moisture storage immediately beneath the treatment depression but not laterally around them. Soiseth, Wight, and Aase (1974) found that contour furrowing increased infiltration rates 0.25 to 3.11 cm/hr (0.1 to 1.2 inches/hr) and forage yields 498 to 770 kg/ha (443 to 685 lb/acre). Data on erosion were not presented by increased water retention on the solonetzic soils studied probably reduced sediment movement significantly. King (1966) reports that a small basin in south central Utah which was treated with gully plugs in 1954 significantly reduced runoff and sediment production during the following 10-year period. Little storage capacity remained in the structures after 10 years. Peterson and Branson (1962) found that earthfill dams (gully plugs) built in the Southwest in the 1930's soon failed by breaching, piping, or washing out of inadequate or poorly protected spillways. These failures were probably due in part to low construction standards which required little or no moisture control or compaction.

Waterspeading

Waterspreading is a technique that involves diversion of water from arroyos or gullies onto the surrounding landscape through use of a system of dikes. Design criteria for waterspreading practices have been suggested by Monson and Quesenberry (1940), Hubbell and Gardner (1950), Hadley and McQueen (1961), and Miller et al. (1969). These studies indicate that spreading areas should be of such size that most of the sediment will settle on the upper 20%, and they must be large enough to accommodate exceptionally large flows if a severe erosion hazard at the lower end is to be avoided. Slope of the spreading area should not be steeper than two to five percent. Spreading of runoff from badlands and other sparsely vegetated areas is not advisable except as a means of storing sediment and thus protecting areas and installations at lower levels. Floodwater spreader construction should be restricted to sites whose soils are deep enough to store at least 10 cm (4 inches) of water and, to ensure forage increases, sites should receive at least one flooding per year. Ponding and poor drainage situations should be avoided where possible. Miller et al. (1969) have presented a summary of vegetation responses from various water spreader investigations, and this information is given in Table 6-12.

Water Quality

Chemical Quality of Surface Water

A detailed treatment of the chemical quality of natural water is beyond the scope and purpose of this discussion on rangeland hydrology; however, comprehensive presentation of the many aspects of the subject is found in Hem (1970). The chemical quality of surface water derived from arid and semiarid lands is determined mainly by the kinds of rocks that underlie each watershed and the soils that developed from them.

The effects of activities of man on the chemical quality of water are often subtle and difficult to detect. This is not necessarily true in humid lands where the ecology of whole drainage basins may be profoundly altered by bringing forested lands into cultivation.

There is a dearth of specific information on the effects of range management practices on dissolved solid content of runoff water but information of a general nature is available. A summary statement of the extent and nature of the field of water quality as stated by Hem (1970) is:

> The chemical composition of natural water is derived from many different sources of solutes, including gases and aerosols from the atmosphere, weathering and erosion of rocks and soil, solution of precipitation, reactions occuring below the land surface, and cultural effects resulting from the activities of man.

Altitudinal zonation is a striking feature of vegetation in the western United States. Increases a precipitation and decreases in temperature with increases in altitude are the probable causes of vegetation zones. At lower altitudes, however, zonation is a useless concept because plant distribution is determined by factors such as available soil moisture and salinity. Vegetation zones and lower altitude vegetation types for the Upper Colorado River basin are shown in Figure 6-27. Dissolved solids in water derived from the vegetation zones and vegetation types show a trend similar to that for sediment yields, i.e., generally more dissolved solids come from the zones with lower

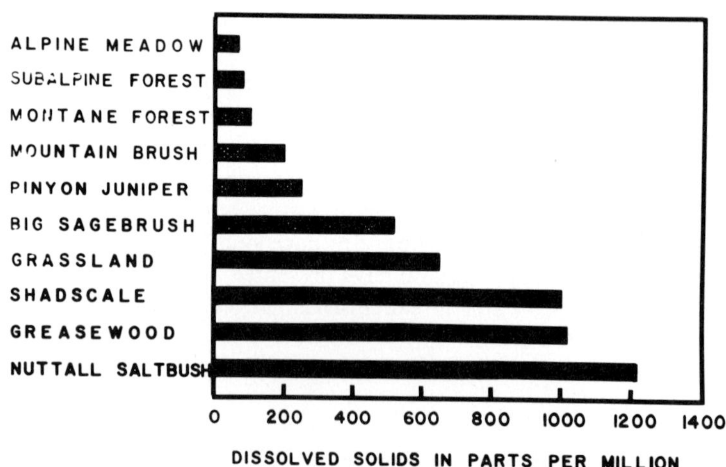

Fig. 6-28. Approximate weighted average concentration of dissolved solids in streams draining differing prominent vegetation types in the Upper Colorado River Basin (adapted from data of Irons et al. 1965 by Branson 1975).

precipitation (Figure 6-28). As stated by Irons, Hembree, and Oakland (1965) "... most of the water comes from the mountains and high plateaus, but most of the dissolved solid content comes from lower parts of the basin." In addition to greater precipitation, with its dilution effect at higher altitudes, rocks exposed in the mountains are generally much more resistant to the solvent action of water than are the rocks (often marine shales) that underlie large parts of the lowlands. Figure 6-29 shows in a general way the processes by which diffuse salt movement takes place on rangelands.

The effects of vegetation on quality of runoff water are not well understood but a few clues are available. Sodium and other soluble salts accumulate in surface soils under the canopy of plants such as greasewood (*Sarcobatus vermiculatus*) and shadscale (*Atriplex confertifolia*) but in adjacent soils under sagebrush (*Artemisia tridentata*) soluble salts decrease (Fireman and Hayward (1952). Salts at the soil surface can be readily transported by overland flow. However, Malekuti and Gifford (1978) have recently found for marine shale sites in the Price River Drainage near Price, Utah that plants were not a significant source of diffuse salt within the Colorado River Basin. In this study, plants were found to contribute between .01 and .02% or less of the total annual salt load to the Price River, this based on two range sites that would be expected to be high salt producers in the first place. Hawkins et al. (1976) report, in addition, that the magnitude of the salt mass added to the Price River annually by *surface runoff* (salt contributions from gully and channel flow excluded) occurring over Mancos (marine) Shale lands was

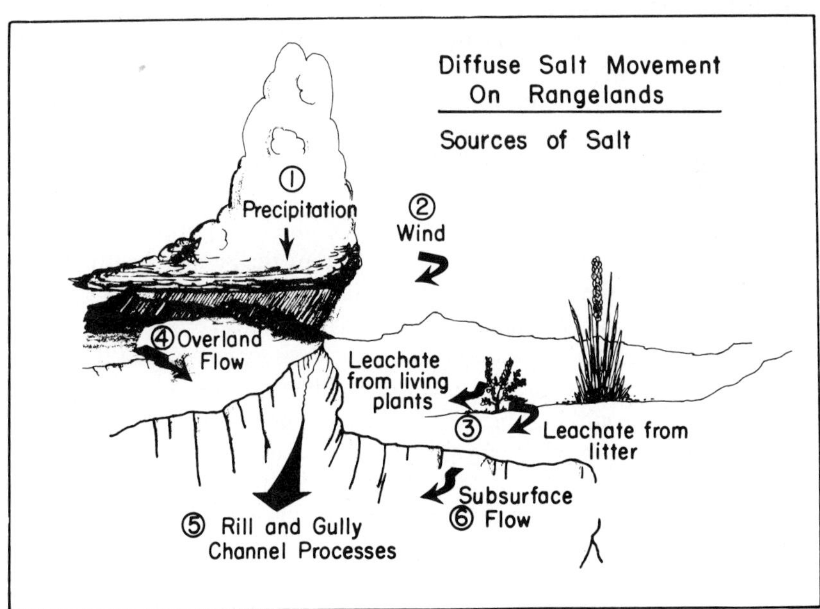

Fig. 6-29. Processes resulting in diffuse salt production from rangelands (from Malekuti and Gifford 1978).

Table 6-12. Data from previous waterspreader investigations (Date from Miller, McQueen, Branson, Shown, and Buller (1969).

Investigators	Date	Mean annual precipitation (millimeters)	Soil	Grasses	Yield of Grass
Valentine	1947	221	Coarse	Black grama	Slight increase
			Fine	Tobosa	Large increase
Hubbell and Gardner	1950	287	Fine	Alkali sacaton	2.72 tons/ha
				Western wheatgrass	Large increase
				Galleta	Large increase
				Vine mesquite	Large increase
Hubbard and Smoliak	1953	284	Medium	Western wheatgrass	4.25 tons/ha
Branson	1956	226	Fine	Western wheatgrass	1.40 tons/ha
Houston	1960	129	Fine	Western wheatgrass	3.80 tons/ha
			Medium	Western wheatgrass	8.64 tons/ha
Hadley and McQueen	1961	137	Fine	Western wheatgrass	Large increase

(Multiply tons/ha by 0.405 to obtain tons/ac.)

calculated to be less than 1.0 % of the total annual salt mass transported by the river.

Vegetation effects on water quality may be impressive. Data of Cleaves, Godfrey, and Bricker (1970) from a small watershed in Maryland show that dissolved solids load in tree stemflow water exceeded the combined values for dissolved solids in precipitation, base flow, flood flow, groundwater, and interflow. The quantity of stemflow water appearing as runoff was not determined. Their detailed input (dissolved materials in precipitation plus dissolved materials contributed by chemical weathering)—output (dissolved materials in runoff plus dissolved materials in water temporarily stored in the watershed plus material temporarily taken up by the biomass) shows chemical weathering for the humid climate of Maryland to be nearly five times as effective in removing watershed materials as mechanical erosion. Their long-term weathering model indicated that approximately one-half of the erosion in the watershed is caused by chemical weathering. We can speculate that different results would be obtained from arid and semiarid rangelands but quantitative data on such lands are presently unavailable.

The concentration (ppm or other units) of sediment generally increases with an increases in stream discharge. Differences in suspended sediment concentration between locations are a function of availability of sediment for transport as affected by rock and soil types, vegetation cover, and man-induced impacts on erosion rates. For areas with soils largely derived from igneous or metamorphic rock, annual discharge of dissolved solids may exceed that of suspended sediment. In Idaho, Emmett (1975) found that annual discharge of dissolved solids for the Salmon River to be 75 % greater than discharge of suspended sediment. Ion concentrations were found to decrease as discharge increased.

Land treatment practices may have undesirable effects on water quality (Borman and Likens 1969, Likens et al. 1970). Tree cutting plus use of herbicides to restrict plant growth increased water yields (39% the first year) but also caused nitrate levels to exceed health levels recommended for drinking water. In contrast to the results of Likens et al. (1970) where nitrate levels remained high for three years following clearcutting and herbicide application, in the Pacific northwest neither clearcutting nor patchcutting caused nitrate levels to exceed federal water quality standards (Fredriksen 1970). Fredriksen's results show that only for 12 days following broadcast burning

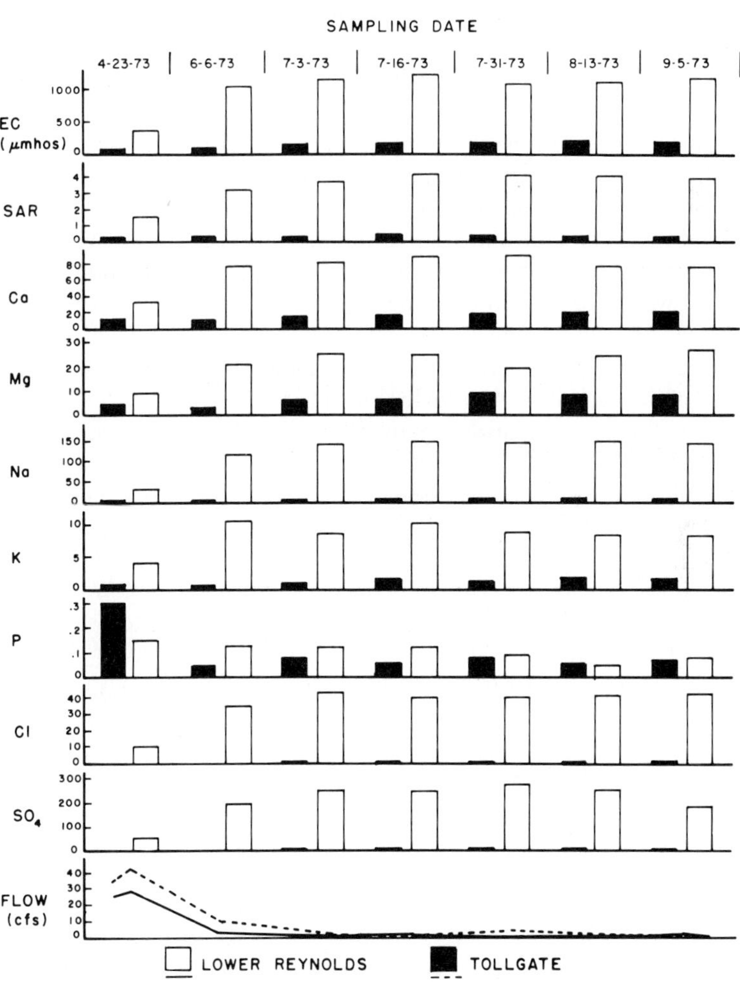

Fig. 6-30. Major chemical constituents (mg/liter) sodium absorption ratio (SAR), electrical conductivity (EC), and channel flow at Tollgate and Lower Reynolds sites (written communication, G.R. Stephenson, U.S. Dep. Agr., Agr. Res. Serv. Boise, Idaho).

 did ammonia levels exceed federal water quality standards. Nitrate levels did not approach the permissible limit following application of the various treatments.

An attempt to quantify the impact of extreme hydrologic events on loss of particulate organic materials from three semiarid watersheds in Utah has been made by Gifford and Busby (1973). Although the results have not been published, some information on water quality (chemicals) has been determined by the staff at two Agricultural Research Service Watershed Research Centers, namely the Boise, Idaho and Tucson, Arizona groups. At the Reynolds Creek Experimental Watershed 50 mi (80 km) southwest of Boise, significant water quality variations have been observed to occur temporarily and spatially. Data are presented for the Lower Reynolds sampling site (90 square miles 234 km^2) which characterizes waters receiving irrigation return flows, and the Tollgate sampling site (21 square miles, 54.4 km^2) which characterizes streamflow from open rangeland, upstream from an irrigation return flow.

Comparing the dominant chemical constituents at the two sites (Figure 6-30) shows a marked increase in concentration levels at the Lower Reynolds site during the irrigation season. Concentration levels at the Tollgate site remained relatively constant. The sodium adsorption ratio (SAR), an alkali hazard index computed using sodium, calcium, and magnesium concentrations, shows that the water at Lower Reynolds is not dominated by sodium and has a low alkali hazard.

Total phosphate (P), often used as an indicator of eutrophic conditions because of the solubility of the orthophosphates from applied fertilizers, does not follow the same variation trends in Figure 6-30 as do the other constituents. Phosphate ions characteristically are adsorbed to the suspended solids in streamflow and to channel sediments. This condition appears to be true in Reynolds Creek because the higher the percent of suspended solids, the higher the concentration of total phosphate. Because no commerical fertilizers have been applied to range or croplands on the Reynolds Creek Watershed, the total phosphates are derived from natural sources and very little soluble orthophosphates are present.

From the above analyses, based upon a relatively dry year, the increase in chemical concentration progressing downstream on Reynolds Creek is caused by irrigation and very little change is related to grazing conditions on open range. Based on data from infiltrometer plots, Buckhouse and Gifford (1976b) also found little impact of grazing livestock on chemical water quality parameters from pinyon-juniper sites in southeastern Utah.

Coliform analyses were also made of the water samples collected but conclusive results are not available because of the wide variability encountered. [1]Stephenson, G.R., USDA Agr. Res. Serv., Boise, Idaho.(1975, written communication) reports that the average fecal coliform counts are higher at sites where there are concentrations of cattle. The analysis of the coliform data showed an increase in both total and fecal counts occurs with a corresponding increase in flow. The average fecal coliform count at the Lower

Reynolds site was 104 whereas the Tollgate site with a lower concentration of cattle had 56 counts during the 1973 sampling.

An invertebrate sampling program was also conducted in 1973 at the Reynolds Creek Experimental Watershed. Stephenson[1] reported:

> All aquatic samples were dominated by immature insecta with a scattering of leaches and snails, especially at the lower elevations. Total populations ranged from a high of over 1300 individuals per square foot at the outlet weir (3600 ft.) in October to 90 per square foot at the Tollgate Wier (4600 ft.) in July. Average sample composition was: Trichoptera 47%, Ephemeropter 23%, Diptera 15%, Plecoptera 7%, and Coleoptera 6%. Invertebrate forms other than the insecta accounted for 2% of the average sample composition.
>
> Although the data are quite limited, tentative inferences as to the spatial and temporal variations in total population can be made at this point. The results of a temporal and spatial analysis of the data, using elevation as the space defining variable and month of the year as the time defining variable, are given in Figure 6-31. Time and elevation trends are evident. Generally, total populations are highest at all stations in October and lowest in June. Since August totals are higher than in June, additional sampling should define the peak of this apparent cyclical change. Channel relationships being equal, the elevation to population relationship indicates that less dense populations will be found at increasingly higher elevations. Quantification of this relationship will also require additional sampling. An examination of Figure 6-31 suggests the possibility that the combination of a log transform of the space parameter and an autoregressive model, using the time parameter, could define and possibly predict the population response surface.

[1]Stephenson, G.R., U.S. Dep. Agr., Agr. Res. Serv., Boise, Idaho.

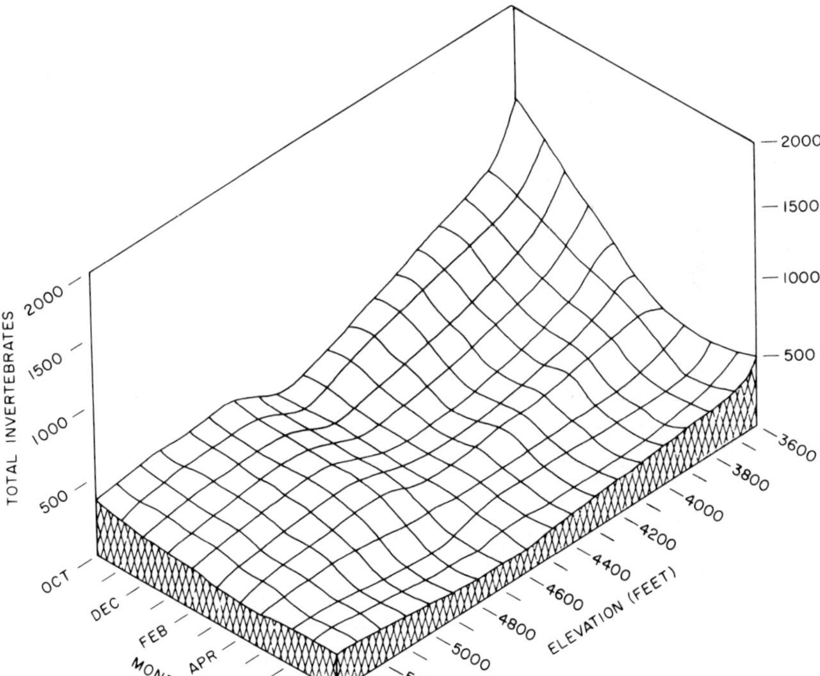

Fig. 6-31. Elevation and month versus total invertebrates per square foot at Reynolds Creek (Stephenson, 1976, written communication).

The Walnut Gulch Watershed has non-saline soils throughout; return flow or base flow does not occur and there is no cultivated land nor is irrigation practiced. The higher elevation lands (1670-2000 m) covered with grass have soils with unsaturated base exchange in thin A horizons usually overlying petrocalcic C horizons. Lower lying lands (1330-1670 m) are brush covered, with soils calcareous throughout the profile. Runoff-producing rainfall occurs as localized storms in late summer and early autumn that usually cover only a small portion of the source areas on the same day. The runoff water qualities described in the following paragraphs resulted from rains occurring from mid-July to late October, 1974.

The water quality from three unit source areas (areas of nearly uniform soil and vegetation) were examined for the attributes shown in Table 6-13 (Schreiber 1976). K2 is located near the upper end of the watershed, on a heavily grazed, grass-covered watershed. Sediment erosion is not excessive here. The second area, H121, is recently urbanized (current housing density is about 1 family dwelling per ha), and has brush covered, calcareous soil. The third area, LH 1,3, and 4, is a similar brush covered, calcareous soil area where grazing and other man-related activities have been excluded for 15 years. Sediment transport in runoff from parts of the latter two areas is appreciable.

Walnut Gulch water is comparable to lower Reynolds Creek in electrical conductivity, K, and Na concentration. Magnesium concentrations are usually lower and calcium concentration higher in southeastern Arizona because of the presence of calcium carbonate. This material also influences orthophosphate solubility, causing about a magnitude decrease in P concentrations (relative to Idaho) in waters originating in calcareous parts of Arizona watersheds.

Within Walnut Gulch, the presence or absence of calcareousness in a unit source area allows a separation to be made on the basis of pH; water in calcareous areas invaraibly has a pH greater than 7.4. Overland flows sampled at K2 never showed concentrations of Ca greater than 14 ppm, while overland flow in all other areas generally had greater Ca concentrations. K concentrations were lower on the calcareous areas although, interestingly, the urbanized watershed appeared to resemble the non-calcareous areas in K and P; the NO_3-N from the urbanized area was also higher than from the other areas.

Some rise in concentration of individual ions as well as total concentration occurred with surface drainage of water from the unit source areas to Flume 1, the outlet of the entire 58 square miles (150 km^2) watershed. NO_3-N and Mg were exceptional in being consistently higher in the flume outlet.

Screening studies determining total coliform bacteria and fecal coliform bacteria indicated the grazed watershed, K2, had 1 to 2×10^5 total coliform per 100 cc and from 4 to 7×10^4 fecal coliform early in the season. A higher proportion of fecal coliform seemed to be originating from the urbanized

Table 6-13. Ranges of attributes in runoff-water from various areas of Walnut Gulch, 1974.[1]

	PH	EC umhos/cm	SUM CATIONS	NO_3-N ppm	PO_4-P ppm	HCO_3 ppm	Na ppm	K ppm	Ca ppm	Mg ppm
H 121[2]	7.8 8.7	85 167	0.72 1.80	.03 1.09	.007 .498	53 119	.65 3.05	1.60 7.15	10.9 29.2	0.52 1.27
K 2[3]	7.0 7.4	60 159	0.60 1.02	.20 .67	.05 .70	24 45	.92 2.00	4.0 11.5	7.1 13.9	0.48 1.04
LH[4]	7.4 8.9	80 196	0.74 1.94	.08 .57	.001 .25	.001 .25	.56 3.65	1.6 5.4	12.0 28.0	0.49 2.38
F1.1[5]	7.4 8.5	77 216	1.17 2.10	.01 .72	.001 .164	53 109	1.20 2.40	2.8 5.6	17.0 34.0	0.93 2.05

[1]Schreiber (1976)
[2]Recently urbanized (1 family dwelling per hectare)
[3]Heavily grazed, grass-covered watershed
[4]Grazing and other man-related activities excluded for 15 years.
[5]Outlet of the entire watershed

 H121 watershed a little later in the season. LH4 at the end of July had about 4 to 5 = 10^3 total coliform/100 cc.

Other Water Quality Concerns

Wadleigh (1968), in a discussion of "Wastes in Relation to Agriculture and Forestry," described the mounting concern in the United States for the deterioration of environmental quality. The categories selected were radioactive substances, chemical air pollutants, sediments, plant nutrients, inorganic salts, organic wastes, infectious agents and allergens, agricultural and industrial chemicals, and heat. He offered a few comments on each category about the importance of the problems, the extent to which agriculture and forestry are involved, research contributions that have been made to ameliorate the problem, and an indication of the need for new or better information and technology towards meeting pressing problems.

Several excellent reviews of water quality models are available to the planner. Harper (1971) assessed mathematical models simulating dissolved oxygen and biochemical oxygen demand in streams and estuaries. Roesner (1969) reviewed models of heat transfer through the air-water interface of a flowing stream. Orlob (1972) and Ward and Espey (1971) evaluated mathematical modeling techniques applies to estuarine systems. Lombardo's critique (1973) covers six of the more comprehensive and flexible models. These, plus a more recent model by Waddell et al. (1973) have been analyzed by Bovet (1974).

Radioactive Substances

Radioactive substances may be produced from several sources such as from mining and refining radioactive minerals, accidental or unauthorized releases from power reactors or from research laboratories (radioactive tracers have had some applications in water resources), and from atmospheric fallout from nuclear testing. Agricultural, forested lands, and rangelands contribute very little to either actual or potential contamination from radioactivity—some phosphate rocks used for fertilizers have very low levels of radioactive uranium and thorium. Natural radioactivity, which itself varies many orders of magnitude, may be greater than all the sources combined.

Sediment

The magnitude of the sediment problem that Wadleigh (1968) describes has already been discussed. It is interesting to note that of the 4 billion tons of sediment washed into streams in the United States each year, only about one-fourth is transported to the sea. Five to 10% of the silt delivered is estimated to come from forests and rangelands. Perhaps most important is the loss of nutrients with the sediments. For example, if the sediment has an average analysis of 0.15% nitrogen, 0.15% P_2O_5 and 1.50 and percent K_2O, more than 50 million tons of primary nutrients are lost from our lands annually with the sediment delivery.

Plant Nutrients

Wadleigh (1968) defined plant nutrients as the inorganic chemicals essential for the mineral nutrition of plants. In surface water, some of these nutrients may become contaminants because they provide for unwanted aquatic plant growth. Nitrogen, phosphorus and potassium are the principal nutrients discussed because of their role in eutrophication. Besides industrial sources of such nutrients, barnyards, feedlots, and chemical fertilizers applied to the land are sources of the nutrients, but there are also losses from rangelands where little or no fertilization is done. Phosphates have been extensively investigated as a pollutant. For example, each 1000 tons of suspended sediment can be expected to contain about 1000 pounds of phosphorus, most of which is in the adsorbed or fixed state and therefore unavailable for plant growth. Likewise, sediment derived from subsoils low in phosphate may actually deactivate much of the soluble phosphorus in a stream. Wadleigh reported:

> Far more study is needed on methods of determining phosphorus content of water in terms of biological significance. Total phosphorus values are meaningless. We ought to be able to distinguish four different states of phosphorus in water samples: (a) That which is in true solution as orthophosphate; (b) that which is in solution in a nonpolar form as an organic polyelectrolyte; (c) that which is adsorbed in suspensois; and (d) that which is a component part of the mineral of suspensoids.

Because of its toxic characteristics, nitrate problems have been most common in groundwater. Evidence from a valley under intensive study in California indicates that sewage effluent and nitrification processes in the semiarid soil may be the main sources of the nitrate found in the groundwater. The diversity of observations and conclusions with respect to nitrate movement into groundwater indicates the need for better information.

Inorganic Salts and Minerals

This category includes neutral inorganic salts, alkaline substances, mineral acids, and fine suspended metal, and metal compounds. The substances enter streams or into the soil from the effluent of various smelting, metallurgical, and chemical industries; from mine drainage; and from natural sources. Undoubtedly, increased salinity is the major problem in this category although additions of toxic substances by increased strip mining or other industry may create problems in rangeland areas. Most of the salinity problems in the western United States are associated with irrigated crops, although saline seeps have become an increasing problem in some dryland farming areas in the northwestern United States.

Organic Wastes

This category includes material such as sewage, animal wastes, crop residues, forest trash, and food and fiber processing wastes. When such substances reach streams, they produce a high biochemical oxygen demand. When they occur on the land, some are combustible, produce odors, and attract insects and vermin.

Animal wastes can be a pollution problem of sizeable portions on range-

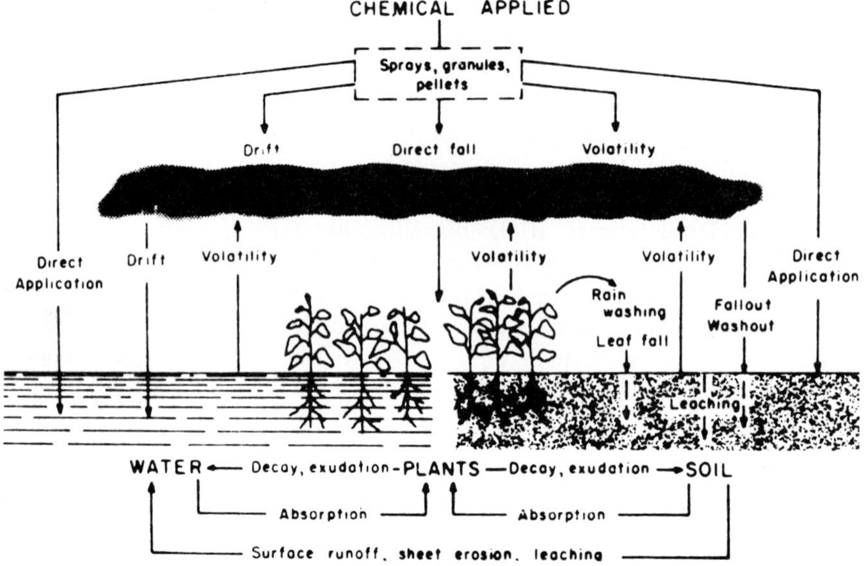

Fig. 6-32. The distribution and fate of chemicals in the environment (from Norris and Moore 1971. Originally from Foy and Bingham 1969).

lands. Wadleigh (1968) reported that domestic animals produce over 1 billion tons of fecal wastes a year with over 400 million tons of liquid wastes (Table 6-14). In fact, waste production by domestic animals in the United States is equivalent ot that of a human population of 1.9 billion. As much as 50% of this waste production is produced in concentrated locations and as such represents a potentially severe pollution problem. However, scattered animal wastes on rangelands are usually considered beneficial. Uniform terminology for rural waste management is published by American Association of Agricultural Engineers (1975).

Table 6-15 shows the population equivalent of the fecal production by various kinds of livestock in terms of biochemical oxygen demand (BOD). For example, a feedlot carrying 10,000 head of cattle has about the same sewage disposal problem as a city of 164,000 people. The city will be using 8,200,000 gallons of water a day to carry off the sewage. Such amounts of water are never used and are seldom even available at the feedlot.

BOD is an important measurement in determining the quality of water. BOD indicates the amount of oxygen required for the oxidation of the organic matter present in a sample of water. It is expressed by the amount of oxygen the organics in the water will consume when it is incubated for a 5-day period at 68° F.

Wastes carried in runoff from barnyards and feedlots may vary in BOD depending on dilution and degree of deterioration of wastes. Public Health authorities object to runoff entering a stream if it exceeds 20 ppm in BOD, so the problem is obvious. Use of lagoons for disposal of wastes has not been fully satisfactory. Additional research will undoubtedly solve some of the problem but the economics will be restrictive in many instances.

Plant residue wastes are generally not a problem on rangeland areas nor are municipal sewage wastes. Irrigation with sewage effluent on rangelands offers some potential for increasing forage production and ground-water recharge.

Oxygen-demanding wastes from processing of agricultural and forestry products include runoff or effluent from sawmills; pulp, paper, and fiber-board manufacturing; fruit and vegetable canning; cleaning dairies; slaughtering and processing of meat animals; tanning; manufacturing cornstarch and soy protein; sugar refining; molting, and fermenting, and distilling; scouring wool; and wet processing in textile mills. Although these sources are not widely encountered in rangeland areas, when present they increase the stream oxygen and make the streams unsightly, unpalatable, and malodorous.

Infectious Agents

Wadleigh (1968) suggested:

> The presence of coliform bacterial in water has been used as an indicator of bacterial contamination for about 75 years. Although the coliform bacteria are not pathological, their presence has been taken as an indication that infectious bacteria might also be present. Recent evidence suggests that more definitive information on bacterial contamination of water may be gained by making counts of both fecal coliform and fecal strepotococcal bacteria. Empirical observations indicate that if the bacterial contamination of water is coming from animals, the f. coliform/f. streptococci ratio is less than 0.5. If the bacterial contamination is from human sources, this ratio usually exceeds 4. On this basis, a water bacteriologist may use the following as a general guide: If this ratio is less than 1.0, the pollution is coming from nonhuman sources; if the ratio exceeds 2.5 the bacterial pollution is most probably coming from human sources. Ratios between 1.0 and 2.5 suggest that the pollution is a mixture of human and nonhuman sources.

Much needs to be learned about the complete reliability of the f. coliform/f. streptococci ratio as an indicator of the source of infectious agents in river water.

Diesch (1970) has reported that numerous diseases can be transmitted from one warm blooded organism to another via water. Commonly observed diseases related to contamination of water supplies from cattle feedlots are salmonellosis and leptospirosis. A leptospirosis outbreak among several young people in Iowa was traced to their swimming hole on the Cedar River. Leptospirosis-infected cattle had access to the river upstream from the swimming hole (Willrich 1967). Bryon (undated) has also reported a leptospirosis outbreak associated with a swimming hole near Columbus, Georgia. Cattle, swine, and dogs found to have *Leptospirosis canicola* infection all had access to the stream which fed the swimming hole.

Geldreich (1970) has concluded that the frequent occurrence of pathogens in domestic animals and wildlife supports concern about fecal contamination from all warm blooded animal sources.

Kunkle (1970) related bacterial densities to patterns of land use in Vermont. He found an increase in fecal coliform counts when cattle were adjacent to the stream as opposed to further away. However, he determined that only a "minor fraction of the total available live bovine fecal material ever washed into the stream." He concluded, therefore, that only the area immediately adjacent to the stream, rather than the entire watershed, is of major importance in terms of introducing this sort of pollution into the stream.

Working with a wildland stream in Colorado, Morrison and Fair (1966) attempted to determine several environmental effects on microbial dynamics. They did note, as a peripheral observation, that cattle grazing adjacent to the stream caused an increase in coliform counts.

Kunkle and Meiman (1967) studied the Little South Fork of the Cache la Poudre River in Colorado in order to assess several water quality characteristics. They found that fecal coliforms proved to be a better indicator of water quality deterioration than turbidity and suspended sediment.

The same team also investigated the water quality impact of grazing on a mountain meadow. They found that the indicator bacteria produced evidence of more pollution in a grazed watershed than in a natural ungrazed catchment (Kunkle and Meiman 1967).

Slawson and Everett (1973) found that the water quality of the main stream Colorado River was relatively constant in terms of chemical and biological parameters. However, the tributary streams showed extreme temporal variability in these parameters as a result of summer rain and flood patterns. They concluded that the side streams pose a definite health hazard to unwary travelers.

Kunkle and Meiman (1968) also investigated the variation in bacterial numbers with time. They noted a daily variation with afternoon lows and evening highs. This cycle followed the cyclic diurnal stream stage, with the high bacterial counts correlating to high streamflow. They also noted an extreme bacterial die-off contributed to the low bacterial counts observed during the late afternoons.

Mack[2] demonstrated, however, that coliform bacteria can persist and even multiply in natural waters. He also noted that multiplication was greater at 35 35° C than at lower temperatures. Hendricks and Morrison (1967) have also reported the enteric bacteria's ability to multiply and grow in cold mountain streams. They suggest that the river's self-purification mechanism plays a role in suppressing the unrestrained growth of these organisms, however.

Walter and Bottman (1967) studied a pair of watersheds near Bozeman, Montana. They noted that one of the watersheds which had been protected from public use for over 40 years consistently showed higher bacterial counts than the adjacent watershed, which was open to the public. Stuart et al. (1971) observed the same watershed and theorized that the higher bacterial counts could be attributed to higher wildlife numbers. Apparently the wildlife favored this watershed due to its protected or "refuge" status. Bissonette et al. (1970), employing a serological technique, attempted to isolate the source of bacterial contamination on the same closed or refuge watershed. They concluded that the sources of contamination were indeed wild animals.

---

[2]Mack, W.W. 1974. Investigations into occurence of coliform orgamisms from pristine streams. Dep. Microbial and Public Health, Mich. State Univ., East Lansing, Mich. 18 p. (Unpublished data).

Skinner, Adams, and Richard (1972), observing a mountain watershed in Wyoming, noted a seasonal fluctuation in bacterial numbers. The authors found increasing concentrations of bacteria as the snowmelt hydrograph advanced towards its peak. They further speculated that increased bacterial numbers were correlated with grazing periods of domestic livestock and wildlife numbers.

In a recent wildland study in northern Utah, Darling and Coltharp (1973) found significant increases in bacterial counts during cattle and sheep grazing periods at stream locations immediately downstream from the grazing activity. Bacterial counts in streams draining the grazed watersheds reached seasonal maximum values during the grazing period, while counts from a nearby ungrazed watershed remained relatively low and constant. In evaluating the health hazard implications of livestock grazing on public wildlands they recommended that direct access of livestock to important water supply streams be prevented . Fencing of stream channels would provide a buffer zone between the stream and the livestock. Livestock watering places should be developed away from the live stream. They further recommend that, rather

Table 6-14. Production of wastes by livestock in the United States (Wadleigh 1968).

Livestock	U.S. population 1975	Solid wastes	Total production of solid waste	Liquid wastes	Total production of liquid wastes
	Millions	G./cap./day[2]	Million tons/yr.	G./cap./day	Million tons/yr.
Cattle	107	23,600	1,004.0	9,000	390.0
Horses[1]	3	16,100	17.5	3,600	4.4
Hogs	53	2,700	57.3	1,600	33.9
Sheep	26	1,130	11.8	680	7.1
Chickens	375	182	27.4		
Turkeys	104	448	19.0		
Ducks	11	336	1.6		
Total			1,138.6		435.4

[1]Horses and mules on farms as work stock.
[2]G./cap./day =grams/capita/day.

Table 6-15. Population equivalent of the fecal production by animals, in terms of biochemical oxygen demand (BOD) (Wadleigh 1968).

Biotype	Fecal	Relative BOD per unit of waste	Population equivalent
	G./cap./day	Lb.	
Man	150	1.0	1.0
Horse	16,100	0.105	11.3
Cow	23,600	.105	16.4
Sheep	1,130	.325	2.45
Hog	2,700	.105	1.90
Hen	182	.115	.14

than suggesting a curtailment of livestock grazing on public lands, the recreating public be advised of the possible dangers involved in drinking untreated stream water.

Buckhouse and Gifford (1976c) studied water quality implications of cattle grazing on a semiarid, chained and seeded, pinyon-juniper site in southeastern Utah. The area was treated in 1967 and protected from grazing until 1974. The 1974 livestock grazing was introduced and investigations continued to determine if any deleterious land use effects were present from fecal contamination by cattle. Small plot data indicated no significant changes in fecal and total coliform production (fecal pollution bacterial indicators) from grazing use. There is an element of risk involved whenever data generated from a small area are projected to larger land areas. However, it appeared that livestock grazing (2 ha/AUM) did not constitute a public health hazard in terms of fecal pollution indicators on the semiarid watershed.

Agricultural and Industrial Chemicals

The benefits of the use of synthetic organic chemicals are well recognized. The discharge of some of these chemicals into the environment has, however, induced some problems (Figure 6-32). These chemical include such substances as household detergents, insecticides, herbicides, fungicides, and nematocides.

Chemical control by herbicides having high physiological specificity to plant species that produce allergenic pollen offers a real possibility for allergen control. Before application of the herbicide, we must be certain that the drift from aerial application is controlled and that the herbicide will not cause offsite deleterious consequences if it gets into a stream or other water body.

Although the rangeland areas of western United States do not presently receive large inputs of agricultural chemicals, it seems likely that future demands upon this resource might include greater inputs of agricultural chemicals to improve the range productivity. Thus, attention will be required to develop management systems which minimize the water pollution.

Detergents have been used in agriculture to disperse colloidal particles and to disperse soil colloids. The dispersed soil has lower hydraulic conductivity (saturated or unsaturated states) which in connection with septic systems can cause malfunctioning. There is no apparent hazard from the presence of detergents surface water but the presence of the phosphates in detergents can promote the development of obnoxious algal blooms.

The potential contamination of lakes and streams by pesticides, particularly the chlorinated hydrocarbons, has been a matter of wide discussion. The use of chemical insecticides and selection herbicides have become important tools for range improvement practices (Valentine 1971). As in most chemical control work, poor application (e.g. spray drift) has led to damge of sensitive vegetation in non-target areas and residuals have been carried by runoff waters and underground perculation into streams. Many pesticides do not injure fauna in the waterways, but some may persist in the food chains and become concentrated. Answers to the questions about the results of pesticide residuals in streams and lakes might help minimize or avoid such problems of water quality.

Heat

Heat acts as a water pollutant because the amount of oxygen water can hold in solution decreases with increasing temperatures. Thus, increasing temperature of the water is analogous to introducing additional oxygen-demanding wastes. Heat increases may also adversely affect fish and other aquatic life. In rangeland areas, increased heat to streams may be a problem following wildfires and removal of riparian vegetation by animal browsing.

Pollution Control Manual

Although directed primarily toward croplands, Stewart et al. (1975) presented a manual for evaluating the nonpoint pollution contributions to streamflow. Their treatment of pollution includes sediment as well as common fertilizers like nitrogen, phosphorus, and potassium, animal wastes, and pesticides. The manual discusses extensively the use of pesticides in agriculture, and lists the many herbicides in common use, their primary mode of movement (e.g. primarily with the sediment, the water, or both), their toxicity (for test animals and fish), and their approximate persistence in soils (days). Also listed are the transport mode and toxicity for most agricultural insecticides and miticides. The manual discusses control practices intended to keep erosion rates within tolerances compatible with good water quality, wholesome environment, and preservation of the productive capacity of the land.

The manual has lists of the principal types of erosion-control practices and some of the favorable and unfavorable features. Under many conditions, it will be necessary to apply combinations of the practices. Modifications of specific practices within these general types affect their adaptability and their effectiveness. Detailed discussions of these erosion control practices with reference to their relation for pollution control are also found.

Literature Cited

Ackerman, W.D., and Corinth, R.L. 1962. An empirical equation for reservoir sedimentation. Comm. on Land Erosion, Inc. Ass. Sci. Hydrology, Pub. 59: 359-366.

Aldon, E.F. 1964. Ground-cover changes in relation to runoff and erosion in west-central New Mexico. U.S. Forest Serv. Res. Note RM-34, 4 p.

Anderson, H.W. 1949. Flood frequencies and sedimentation from forest watersheds. Trans. Amer. Geophys. Union 39(4): 576-584.

Anderson, H.W. 1954. Suspended-sediment discharge as related to streamflow, topography, soil and land use. Trans. Amer. Geophys. Union 2(35): 268-281.

Anderson, A.G. 1962. Sediment transport mechanics, erosion of sediment. J. Hydraulics Div., Proc. Amer. Soc. Civil Eng. 88(HY 84): 109-127.

Anderson, H.W. 1970. Relative contributions of sediment from source areas and transport processes. *In:* Proc., A Symposium, Forest Land Uses and Stream Environment, Oct. 19-21, Oregon State Univ., Corvallis: 55-63.

Anderson, H.W. 1951. Physical characteristics of soils related to erosion. J. Soil and Water Conserv. 6: 129-133.

Anderson, H.W. 1975. Sedimentation and turbidity hazards in wildlands. *In:* Watershed Management, Proc. Symp., Irrig. Drain. Div., Amer. Soc. Civil Engr., New York, p. 347-376.

Anderson, H.W., and J.R. Wallis. 1965. Some interpretations of sediment sources and causes, Pacific Coast Basins in Oregon and California. USDA Misc. Pub. No. 970, p. 22-30.

Andre, J.E., and H.W. Anderson. 1961. Variation of soil erodibility with geology, geographic zone, elevation and vegettion type in northern California wildlands. J. Geophys. Res. 66(10): 3351-3357.

Amer. Soc. Agr. Engr. 1975. Agricultural Engineers Handbook. St. Joseph, MI, p. 515-517.

Ateshian, J.K.H. 1974. Estimation of rainfall erosion index. J. Irrig. & Drain Div., Amer. Soc. Civil Engr. 100 (IR3): 293-307.

Bailey, R.W., and O.L. Copeland, Jr. 1961. Low flow discharges and plant cover relations on two mountain watersheds in Utah. Intern. Ass. Sci. Hydrol. Pub. 51: 267-278.

Barnes, H.H., Jr. 1967. Roughness characteristics of natural channels. U.S. Geol. Surv. Water-Supply Paper 1849. 213 p.

Baver, L.D. 1965. Soil Physics. John Wiley & Sons, Inc., New York, N.Y.

Beer, C.E., and H.P. Johnson. 1965. Factors related to gully growth in the deep loess area of western Iowa. *In:* Proc. Federal Inter-Agency Sedimentation Conf. 1963. USDA Misc. Pub. Pub. No. 970: 37-43.

Beer, C.E., C.W. Farnham, and H.G. Heinemann. 1966. Evaluating sedimentation prediction techniques in western Iowa. Trans. Amer. Soc. Agr. Engr. 9(6): 828-831, 833.

Bissonnette, G.K., D.G. Stuart, T.D. Goodrich, and W.G. Walter. 1970. Preliminary studies of serological types of emerobacteria occurring in a closed mountain watershed. Proc. Montana Acad. Sci. 30: 66-76.

Biswell, H.H., and A.J. Schultz. 1957. Surface runoff and erosion as related to prescribed burning. J. Forestry 55: 372-374.

Blench, T. 1966. Mobile bed fluviology. Dep. Tech. Services, Univ. Alberta, Canada.

Borman, F.H., and G.E. Likens. 1969. The watershed-ecosystem concept and studies of nutrient cycles. p. 309-334. *In:* The Ecosystem Concep in Natural Resource Management. Edited by G.M. Van Dyne. Academic Press, New York.

Bovet, E.D. 1974. Evaluation of quality parameters in water resource planning. U.S. Army Engr. Instit. for Water Resour., IWR Contract Rep. 74-13.

Boyce, R.C. 1975. Sediment routing with sediment-delivery ratios. p. 61-65. *In:* Present and Prospective Technology for Predicting Sediment Yields and Sources. USDA , Agr., Res. Serv., ARS-S-40, 1975.

Branson, F.A. 1956. Range forage production changes on a waterspreader in southeastern Montana. J. Range Manage. 9: 187-191.

Branson, F.A. 1975. Natural and modified plant communities as related to runoff and sediment yields. p. 157-172. *In:* A.D. Hasler, ed., Ecological Studies 10, Coupling of Land and Water Systems. Springer-Verlag, Inc., New York.

Branson, F.A., and J.R. Owen. 1970. Plant cover, runoff, and sediment yield relationships on Mancos Shale in western Colorado. Water Res. Res. 6(3): 783-790.

Brown, C.B. 1950. Sediment transportation. *In:* H. Rouse (ed.), Engineering Hydraulics, p. 769-857. Proc. 4th Hydraulics Conf., Iowa Ins. Hydraulic Res., June, 1949, John Wiley and Son, New York.

Brown, G.W., and J.T. Krygier. 1971. Clearcut logging and sediment production in the Oregon Coast Range. Water Resour. Res. 7: 1189-1199.

Bryan, R.B. 1968. The development, use and efficiency of indices of soil erodibility. Geoderma 2(1): 5-26.

Bryon, F.L. (undated) Water: Its association with health and disease. U.S. Dept. Health, Educ., and Welfare. Public Health Serv., Commun. Dis. Center, Community Serv. Training Sec. 20 p. (unpublished).

Buckhouse, J.C., and G.F. Gifford. 1976a. Sediment production and infiltration rates as affected by grazing and debris burning on chained and seeded pinyon-juniper. J. Range Manage. 29: 83-85.

Buckhouse, J.C., and G.F. Gifford. 1976b. Water quality implications of cattle grazing on a semiarid watershed in southeastern Utah. J. Range Manage. 29: 109-113.

Buckhouse, J.C., and G.F. Gifford. 1976c. Grazing and debris burning on pinyon-juniper sites—Some chemical water quality implications. J. Range Manage.: (in print).

Busby, M.W., and M.A. Benson. 1960. Grid method of determining mean flow distance in a drainage basin. Bull. Int. Ass. Hyd. 20: 32-36.

Cleaves, E.T., A.E. Godfrey, and O.P. Bricker. 1970. Geochemical balance of a small watershed and its geomorphic implications. Geol. Soc. Amer. Bull. 81: 3015-3032.

Colby, B.R. 1964. Discharge of sands and mean velocity relationships in sand bedstreams. U.S. Geol. Survey Prof. Paper 462-A. 47 p.

Colby, B.R., and C.H. Hambree. 1955. Computations of total sediment discharge. U.S. Geol. Surv., Water-Supply Pap. 1357. 187 p.

Coltharp, G.B. 1967. How effective are gully plugs and contour furrows on frail lands in Utah? Soil Conserv. Soc. Amer. Proc., 22nd Annual Meeting. p. 82-89.

Connaughton, C.A. 1935. Forest fires and accelerated erosion. J. Forestry 33: 751-752.

Cooper, C.F. 1961. Controlled burning and watershed condition in the White Mountains of Arizona. J. Forestry 59: 438-443.

Copeland, O.L. 1965. Land use and ecological factors in relation to sediment yields. p. 72-84. *In:* Proc. Federal Interagency Sediment Conf. 1963. USDA Misc. Publ. No. 970.

Croft, A.R., and R.W. Bailey. 1964. Mountain water. U.S. Forest Service, Intermountain Forest and Range Exp. Sta. Unnumbered Publ.: 64 p.

Darling, L.A., and G.B. Coltharp. 1973. Effects of livestock grazing on the water quality of mountain streams. *In:* Proc. Symp. Water-Animal Relat. Southern Idaho State Colleg,e Twin Falls, Idaho. June 1-8, 1973.

DeByle, N.V., and P.E. Packer. 1972. Plant nutrient and soil losses in overland flow from burned forest clearcuts. p. 296-307. *In:* Proc. AWRA National Symp. on Watersheds in Transition. Colorado State Univ., Fort Collins.

Diesch, S.L. 1970. Disease transmission of waterborne organisms of animal origins. p. 265-285. *In:* T.L. Willrich and G.E. Smith (eds). Agricultural Practices and Water Quality, Iowa State Univ. Press, Ames, Iowa.

Dortignac, E.J. 1956. Watershed resources and problems of the Upper Rio Grande Basin. U.S. Forest Serv. Rocky Mtn. Forest and Range Exp. Sta. Pap. 107 p.

Dragoun, F.J. 1962. Rainfall energy related to sediment yield. J. Geophys. Res. 67: 1-4.

DuBoys, 1879. LeRhone et les rivieres at lit affoullable. Ann. Des. ponts et chauses 18: 141-195.

Dunford, E.G. 1949. Relation of grazing to runoff and erosion on bunchgrass ranges. U.S. Forest Serv. Rocky Mtn. Forest & Range Exp. Sta. Res. Note 7, 2 p.

Einstein, H.A. 1950. The bedload function for sediment transportation in open channels. USDA, Tech. Bull. 1026.

Ellison, W.D. 1945. Some effects of raindrops and surface flow on soil erosion and infiltration. Trans. Amer. Geophys. Union 26: 415-429.

Emmett, W.W. 1975. The channels and waters of the upper Salmon River area, Idaho. U.S. Geol. Survey, Open-file Rep., 446 p.

Farnham, C.W., C.E. Beer, and H.G. Heinemann. 1966. Evaluation of factors affecting reservoir sediment deposition. Symposium of Garda, Int. Ass. Sci. Hydrology, Pub. 71: 747-758.

Fireman, M., and H.E. Hayward. 1952. Indicator significance of some shrubs in the Essalents Desert, Utah. Bot. Gaz. 114: 143-155.

Flaxman, E.M. 1972. Predicting sediment yield in western United States. Am. Soc. Civil Engr., Jour. Hyd. Div. 98(HY12): 2073-2085.

Fleming, G. 1969. Design curves for suspended load estimation. Proc. Institute of Civil Engr. 43: 1-9.

Fleming, G. 1975. Sediment erosion-transport-deposition simulation: state of the art. p. d274-285. *In:* Present and Prospective Technology for Predicting Sediment Yield and Sources. USDA Agr. Res. Serv., ARS-S-40.

Foster, G.R., and L.D. Meyer. 1972. A closed-form soil erosion equation for upland areas. p. 121-129. *In:* Shen, H.W. (ed.). Sedimentation: Symposium to Honor Professor H.A. Einstein, Fort Collins, Colo.

Foster, G.R., L.D. Meyer, and C.A. Onstad. 1973. Erosion equation derived from modeling principles. Paper 73-2550. Presented at 1973 Winter Meetings, Amer. Soc. Agr. Engr., Dec. 11-14, 1973, Chicago, Ill.

Foy, C.L., and S.W. Bingham. 1969. Some research approaches toward minimizing herbicidal residues in the environment. Residue Rev. 29: 105-135.

Fredriksen, R.L. 1970. Comparative water quality-natural and disturbed streams. p. 125-137. *In:* Proc. of a Symposium, Forest land uses and stream environment, Oregon State Univ., Corvallis.

Frere, M.H., C.A. Onstad, and H.N. Holtan. 1975. ACTMO, an agricultural chemical transport model. USDA Agr., Res. Serv., ARS-H-3.

Geldreich, E.E. 1970. Applying bacteriological parameters to recreational water quality. J. Amer. Water. Works Ass. 62: 113-120.

Gifford, G.F. 1975. Impacts of pinyon-juniper manipulation on watershed values. p. 127-141. *In:* Proc., The Pinyon-Juniper Ecosystem—A Symp., Utah State Univ., Logan, May.

Gifford, G.F., and F.E. Busby. 1973. Loss of particulate organic materials from semiarid watersheds as a result of extreme hydrologic events. Water Resour. Res. 9: 1443-1449.

Glymph, L.M. 1954. Studies of sediment yields from small watersheds. Int. Union of Geodesy and Geophys. Pub. 36: 178-191.

Glymph, L.M. 1975. Evolving emphasis in sediment-yield predictions in present and prospective technology for predicting sediment yields and sources. USDAd, Agr. Res. Serv., ARS-S-40: 1-9.

Graf, W.H. 1971. Hydraulics of Sediment Transport. McGraw-Hill Book Co., Inc., New York, N.Y.

Hadley, R.F., and I.S. McQueen. 1961. Hydrologic effects of water spreading in Box Creek Basin, Wyoming. U.S. Geol. Surv. Water-Supply Pap. 1532A, 48 p.

Hadley, R.F., and S.A. Schumm. 1961. Sediment sources and drainage basin characteristics in upper Cheyenne River basin. U.S. Geol. Survey Water-Suypply Pap. 1531-B, 163 p.

Harper, M.E. 1971. Assessment of mathematical models used in analysis of water quality in streams and estuaries. Washington State Res. Center.

Hawkins, R.H., J.P. Riley, J.J. Jurinak, S.L. Ponce, and G.F. Gifford. 1976. Diffuse sources of salinity in a desert watershed. Paper presented ASCE meeting, San Diego, April 6.

Heede, B.H. 1960. A study of early gully-control structures in the Colorado front range. U.S. Forest Service Rocky Mtn. Forest & Range Exp. Sta. Pap. 55: 42 p.

Heede, B.H. 1968. Conversion of gullies to vegetation-lined waterways on mountain slopes. U.S. Forest Service Res. Pap. RM-40, 11 p.

Heede, B.H. 1970. Morphology of gullies in the Colorado Rocky Mountains. Int. Ass. Sci. Hydrol. Bull. 15: 79-89.

Heede, B.H. 1975a. Watershed indicators of landform development. p. 43-46. *In:* Vol. 5, Hydrology and Water Resources in Arizona and the Southwest. Proc. 1975 meeting, Ariz. Sec. AWRA and Hydrol. Sec., Ariz. Acad. Sci. Tempe, April.

Heede, B.H. 1975b. Stages of development of gullies in the west. p. 155-161. *In:* Present and prospective technology for predicting sediment yields and sources. USDA, Agr. Res. Serv. Rep. ARS-S-40.

Heede, B.H., and J.G. Mufich. 1974. Field and computer procedures for gully control by check dams. J. Environ. Manage. 2: 1-49.

Helvey, J.D. 1973. Watershed behavior after forest fire in Washington p. 403-422. *In:* Proc. Amer. Soc. Civil Engr., Irrig. and Drain. Div. Specialty Conference, Fort Collins. Colo.

Hem, J.D. 1970. Study and interpretation of the chemical characteristics of natural water. U.S. Geol. Surv. Water-Supply Pap. 1473. 363 p.

Henderson, F.M. 1966. Open Channel Flow. MacMillan Co., New York, N.Y. 522 p.

Hendricks, C.W., and S.M. Morrison. 1967. Multiplication and growth of selected enteric bacterial in clear mountain streamwater. *In:* Water Research, Pergamon Press, London, England, 8 p.

Horton, R.E. 1945. Erosional development of streams and their drainage basins. Hydro-physical approach to quantitative morphology. Geol. Soc. Amer. Bull. 56: 275-370.

Houston, W.R. 1960. Effects of waterspreading on range vegetation in eastern Montana. J. Range Manage. 13: 289-293.

Hubbard, W.A., and S. Smoliak. 1953. Effects of contour dykes and furrows on short-grass prairie. J. Range Manage. 6: 55-62.

Hubbell, D.S., and J.L. Gardner. 1950. Effects of diverting sediment-laden runoff from arroyos to range and croplands. USDA Tech. Bull. 1012. 83 p.

Hydrocomp International. 1969. Operations Manual. 2nd Ed. Palo Alto, California.

Inglis, C.C. 1968. Discussion of systematic evaluation of river regime. J. of Waterways and Harbor Div., Amer. Soc. Civil Engr. 54(HY1): 1-36.

Irons, W.V., C.H. Hembree, and G.L. Oakland. 1965. Water resources of the Upper Colorado River Basin Technical Report. U.S. Geol. Surv. Prof. Paper 441. 370 p.

Kilinc, M., and E.V. Richardson. 1973. Mechanics of soil erosion from overland flow generated by simulated rainfall. Colorado State Univ. Hydrology Paper #63, Fort Collins, Colo.

King, N.J. 1966. Erosion characteristics of salt desert shrub areas. p. 69-87. *In:* Salt Desert Shrub Symposium, Cedar City, Utah. USDI, Bureau of Land Management.

Koelzer, V.A., and J.M. Lara. 1958. Densities and compaction rates of deposited sediment. Amer. Soc. Civ. Eng., J. Hyd. Div. HY2-1603: 1-15.

Kunkle, S.H. 1970. Sources and transport of bacterial indicator in rural streams. Presented at a symposium on the Interdisciplinary Aspects of Watershed Management, Montana State Univ., Bozeman, Montana. August 306, 1970. 31 p.

Kunkle, S.H., and J.R. Meiman. 1967. Water quality of mountain watersheds. Colorado State Univ. Hydrology Pap. No. 21, Fort Collins, Colo., 53 p.

Kunkle, S.H., and J.R. Meiman. 1968. Sampling bacteria in a mountain stream. Colorado State Univ. Hydrology Pap. No. 28, Fort Collins, Colorado. 27 p.

Langbein, W.B., and S.A. Schumm. 1958. Yield of sediment in relation to mean annual precipitation. Trans. Amer. Geophys. Union 39: 1076-1084.

Laursen, E.M. 1958. The total sediment load of streams. J. of the Hydraulics Div., Proc. Amer. Soc. Civil Engr. 84(HYl): 1530-1-1530.36.

Likens, G.E., F.H. borman, N.M. Johnson, D.W. Fisher, and R.S. Pierce. 1970. Effects of forest cutting and herbicide treatments on nutrient budgets in Hubbard Book watershed-ecosystem. Ecol. Monogr. 40: 23-47.

Livesey, R.H. 1975. Corps of engineers methods for predicting sediment yields. p. 16-32. *In:* Present and prospective technology for predicting sediment yields and sources. USDA., Agr. Res. Serv. ARS-S-40.

Lombardo, P.S. 1973. Critical review of currently available water quality models. Hydrocomp, Inc., Palo Alto, Calif.: 91 p.

Lusby, G.C. 1970. Hydrologic and biotic effects of grazing vs. nongrazing near Grand Junction, Colorado. J. Range Manage. 23: 256-260.

Lusby, G.C. and R.F. Hadley. 1967. Deposition behind low dams and barriers in the southwestern United States. J. Hydrol. (N.Z.) 6: 89-105.

Mack, W.N. 1974. Investigations into occurrence of coliform organisms from pristine streams. Dept. Microbiology and Public Health, Michigan State Univ., East Lansing, Michigan. 18 p. (Unpublished data).

Malekuti, A., and G.F. Gifford. 197 . Plants as a source of diffuse salt within the Colorado River Basin. (Manuscript in preparation).

Marshall, J.K. 1973. *In:* Lovett, J.V. (ed.). The Environmental, Economic, and Social Significance of Drought. Angus and Robertson, Publishers.

Meeuwig, R.O. 1971. Soil stability on high elevation rangeland in the intermountain areas. USDA, Forest Serv. Res. Paper INT-94, 10 p.

Megahan, W.F., and D.C. Molitor. 1975. Erosional effects of wildfire and logging in Idaho. p. 423-444. *In:* Watershed Management Symposium, Amer. Soc. Civil Engr., Irrig. and Drain. Div., Aug. 11-13, Utah State Univ., Logan.

Meyer, L.D., and W.H. Wischmeier. 1966. Mathematical simulation of the process of soil erosion by water. Trans. Amer. Soc. Agr. Engr. 12(6): 754-758.

Miller, R.F., I.S. McQueen, F.A. Branson, L.M. Shown, and W. Buller. 1969. An avaluation of range floodwater spreaders. J. Range Manage. 22: 256-247.

Monson, O.W., and J.R. Quesenberry. 1940. Range improvement through conservation of floodwaters. Montana Agr. Exp. Sta. Bull. 380: 20 p.

Morrison, S.M., and J.F. Fair. 1966. Influence of environment on stream microbialdynamics. Colorado State Univ. Hydrology Pap. No. 13, Fort Collins, Colo. 21 p.

Musgrave, G.W. 1947. The quantitative evaluation of factors in water erosion—A first approximation. J. Soil & Water Conserv. 2: 133-138.

Mutchler, C.K., and R.A. Young. 1975. Soil detachment by raindrops. p. 113-117. *In:* Present and Prospective Technology for Predicting Sediment Yields and Sources. USDA Res. Serv., ARS-S-40.

Negev, Moshe. A sediment model on a digital computer. Dep. Civil Engr., Stanford Univ., Tech. Rep. No. 76. 107 p.

Norris, L.A., and D.G. Moore. 1971. The entry and fate of forest chemicals in streams. p. 138-158. *In:* Symp. Proc., Forest Land Uses and Stream Environment, Oregon State Univ., Corvalls, Oct. 19-21.

Olson, O.C. 1949. Relations between soil depth and accelerated erosion on the Wasatch Mountains. Sci. 67: 447-451.

Orlob, G.T. 1972. Mathematical modelling of estuarial systems. p. 401-410. *In:* Intern. Symp. on Mathematical Modelling Tech. in Water Resourc. Systems, Environ. Canada, Ottawa, May.

Packer, P.E. 1963. Soil stability requirements for the Gallatin elk winter range. J. Wildlife Manage. 27: 401-410.

Pase, C.P., and P.A. Ingebo. 1965. Burned chaparral to grass: early effects on water and sediment yields from two granitic soil watersheds in Arizona. Proc. Annu. Ariz. Watershed Symp. 9: 8-11.

Pase, C.P., and A.W. Lindenmuth, Jr. 1971. Effects of prescribed fire on vegetation and sediment in oak-mountain mahogany chaparral. J. Forest. 69: 800-805.

Peterson, H.V., and F.A. Branson. 1962. Effects of land treatments on erosion and vegetation on rangelands in parts of Arizona and New Mexico. J. Range Manage. 15: 220-226.

Peterson, H.V., and R.F. Hadley. 1960. Effectiveness of erosion abatement practices on semiarid rangelands in western United States. Int. Ass. Sci. Hydrol. Pub. 53: 182-191.

Peterson, H.V., and R.F. Hadley. 1960. Effectiveness of erosion abatement practices on semiarid rangelands in western United States. Int. Ass. Sci. Hydrol. Pub. 53: 182-191.

Piest, R.F., L.A. Kramer, and H.G Heineman. 1975. Sediment movement from loessial watersheds. p. 130-141. *In:* Present and prospective technology for predicting sediment yields and sources. USDA Agr. Res. Serv., ARS-S-40.

PSIAC (Pacific Southwest Interagency Commmittee). 1968. Factors affecting sediment yield and measures for the reduction of erosion and sediment yield. Mimeo. Rep., USDA Soil Conserv. Serv., Portland, Oregon.

PSIAC (Pacific Southwest Interagency Committee). 1974. Erosion and sediment yield methods. Mimeo. Rep. USDA, Soil Conserv. Serv., Portland, Ore.

Radbruch, D.H., and K.C. Crowther. 1973. Map showing estimated relative amounts of landslides in California. U.S. Geol. Survey Misc. Investigations, Map I-747.

Renard, K.G. 1970. The hydrology of semiarid rangeland watersheds. USDA Agr. Res. Serv., 41-162: 26.

Renard, K.G. 1972. Sediment problems in the arid and semiarid Southwest. p. 225-232. Soil Conserv. Soc. Amer., Proc. 27th Annu. Meeting, Aug. 6-9, Portland, Ore.

Renard, K.G. 1975. Discussion of: "Estimation of rainfall erosion index" by J.K.H. Ateshian. (Am. Soc. Civil Engr. 100 (IR3):293-307, 1974). J. Irrig. Drain. Div., Amer. Soc. Civil Engr. 101 (IR3): 240-241.

Renard, K.G., and E.M. Laursen. 1975. A dynamic model of an ephemeral stream. J. Hydr. Div. Proc. Amer. Soc. Civil Engr. 101(HY5): 511-528.

Renard, K.G., and J.R. Simanton. 1973. Discussion of: "Predicting sediment yield in Western United States," by E.M. Flaxman. (Am. Soc. Civil Engr. 98(HY12): 2073-2085, 1972). J. Hydr. Div. Am. Soc. Civil Engr. 99(HY9): 1647-1649.

Renard, K.G., and J.R. Simanton. 1975a. A discussion of: "Estimation of rainfall erosion index" by J.K.H. Ateshian (Amer. Soc. Civil Engr. 100(IR3): 293-307, 1974). J. Irrig. & Drain. Div., Am. Soc. Civil Engr. 101(IR3): 242-247.

Renard, K.G., and J.R. Simanton. 1975b. Thunderstorm precipitation effects on the rainfall-erosion index of the USLE. J. Hydrology and Water Resources in Arizona and the Southwest. Am. Water Resources Assoc., Ariz. Section—Ariz. Acad. of Sci., Hydrology Section, Proc. of April 11-12 meeting, Tempe, Ariz., V: 47-56.

Renard, K.G., J.R. Simanton, and H.B. Osborn. 1974. Applicability of the universal soil loss equation to semiarid rangeland conditions. Hydrology and Water Resources in Ariz. and the Southwest, Amer. Water Resources Ass., Ariz. Section, Ariz. Acad of Sci., Hydrology Section, Proc. of April 19-20 meeting, Flagstaff, Ariz., IV: 18-32.

Renfro, G.W. 1975. Use of erosion equations and sediment-delivery ratios for predicting sediment yield. p. 33-45. *In:* Present and Prospective Technology for Predicting Sediment and Sources. USDA Agr. Res. Serv., ARS-S-40.

Renner, F.G. 1936. Conditions influencing erosion on the Boise River watershed. USDA Tech. Bull. 528:32.

Rich, L.R. 1962. Erosion and sediment movement following a wildfire in a ponderosa pine forest of central Arizona. USDA Forest Service, Rocky Mt. Forest and Range Exp. Sta. Res. Note No. 76. 12 p.

Rickert, D.A., and W.G. Hines. 1975. A practical framework for river-quality asessment. USDA Geol. Survey Circ. 715-A. 17 p.

Ritchie, J.C., and J.R. McHenry. 1975. Fallout CS-137: A tool in conservation research. J. Soil & Water Conserv. 30: 283-286.

Ritchie, J.C., J.R. McHenry, A.C. Gill, and P.H. Hawks. 1970. The use of fallout cesium-137 as a tracer of sediment movement and deposition. Proc. Mississippi Water Resources Conf., p. 149-162.

Roehl, J.W. 1965. Sediment source area, delivery ratios, and influencing morphological factors. Int. Ass. Sci. Hydrology Pub. No. 59, p. 202-213.

Roesner, L.A. 1969. Mathematical models of heat transfer through the air-water interface of a flowing stream. Ph.D. Thesis, Univ. of Washington, Seattle.

Rowe, P.B. 1941. Some factors of the hydrology of the Sierra Nevada foothills. Trans. Amer. Geophys. Union 22: 90-100.

Rowe, P.B., C.M. Countryman, and H.C. Storey. 1954. Hydrologic analysis used to determine effects of fire on peak discharges and erosion rates. USDA Forest Serv., Calif. Forest and Range Exp. Sta. (mimeo).

Ruby, E.C. 1974. Review report. Los Angeles river flood prevention project, mountain and foothill area. USDA Forest Serv., Angeles National Forest, Unnumbered Publ. (California Region).

Sakhan, K.S., J.P. Riley, and K.G. Renard. 1972. Hybrid computer simulation of sediment transport with stochastic transfer of the stream bed. Sedimentation Symp. to honor Professor H.A. Einstein, Colorado State Univ., Fort Collins.

Schreiber, H.A. 1976. Written Communication. USDA Agr. Res. Serv., Tucson, Ariz.

Schuman, G.E., R.E. Burwell, R.F. Piest, and R.G. Spomer. 1973. Nitrogen losses in surface runoff from agricultural watersheds on Missouri Valley loss. J. Env. Quality 2(2): 299-302.

Shantz, H.L. 1947. The use of fire as a tool in the management of the brush ranges of California. Calif. Div. of Forestry Rep.

Shown, L.M. 1970. Evaluation of a method for estimating sediment yield. USDA Geol. Survey Prof. Paper 700B, p. B245-249.

Shulits, S., and R.D. Hill. 1968. Bedload formulas, Part A: A selection of bedload formulas. Report PB 194, 950. Penn State Univ.

Skinner, Q.D., J.C. Adams, and P.R. Richard. 1972. Influence of summer use of a mountain watershed on bacterial water quality. p. 57-78. *In:* Age of Changing Priorities for Land and Water, Irrig. and Drain. Div. Specialty Conference, Spokane, Wash. Sept. 26-28, 1972. Published by Amer. Soc. Civil Engr.

Slawson, G.C., Jr., and L.G. Everett. 1973. Chemical and biological problems in the Grand Canyon. p. 63-72. *In:* Hydrology and Water Resources in Arizona and the Southwest. Proc. Ariz. Section, Amer. Water Resources Assoc. and the Hydrology Sec., Ariz. Acad. of Sci., Tucson, Ariz. May 4-5, 1973.

Smith, D.d., and W.H. Wischmeier. 1957. Factors affecting sheet and rill erosion. Trans. Amer. Geophys. Union 38: 889-896.

Smith, D.D., and W.H. Wischmeier. 1962. Rainfall erosion. Adv. Agron. 14: 109-148.

Soil Conservation Service, U.S. Dept. Agr. 1972. Procedure for computing sheet and rill erosion on project areas. Tech. Release No. 51, Geology.

Soiseth, R.J., J.R. Wight, and J.K. Aase. 1974. Improvement of panspot (solonetzic) range sites by contour furrowing. J. Range Manage. 27: 107-110.

Stall, J.B. 1972. Effects of sediment on water quality. J. Environ. Quality 1: 353-360.

Stall, J.B., and L.J. Bartelli. 1959. Correlation of reservoir sedimentation and watershed factors. Illinois State Water Survey Rep. No. 37.

Stephenson, G.R. 1976. Written Communication. USDA Agr. Res. Serv., Boise, Ida.

Stewart, B.A., D.A. Woolhiser, W.H. Wischmeier, J.H. Caro, and M.H. Frere. 1975. Control of water pollution from cropland. Vol. 1, A manual for guideline development USDA Agr. Res. Serv., Wash. D.C., ARS-H-5-1. 111 p.

Strand, R.I. 1975. Bureau of Reclamation procedures for predicting sediment yield. p. 10-15. *In:* Present and Prospective Technology for Predicting Sediment Yields and Sources. USDA Agr. Res. Serv., ARS-S-40.

Sturt, D.G., G.K. Bissonnette, T.D. Goodrich, and W.G. Walter. 1971. Effects of multiple us on water quality of high mountain watersheds. Bacteriological investigations of mountain streams. J. Appl. Microbiology 22: 1048-1054.

Suhr, W.E. 1967. Watershed action program—Tonto National Forest. Proc. Annu. Ariz. Watershed Symp. 11: 20-24.

Tatum, F.E. 1965. A new method of estimating debris-storage requirements for debris basins. p. 886-897. Fed. Inter-Agency Sedimentation Conf. Proc., 1963, USDA Agr. Res. Serv., Misc. Publ. No. 970.

Thomas, W.A. 1970. A digital model for simulating sediment movement in a shallow reservoir. *In:* Proc. of Seminar on Sediment Transport in Rivers and Reservoirs. U.S. Corps. of Engr., Davis Calif.

Thomas, D.B. 1975. The effects of gully plugs and contour furrows on erosion and sedimenttion in Cisco Basin, Utah. M.S. Thesis, Utah State Univ., Logan. 54 p.

Thompson, J.R. 1964. Quantitative effect of watershed variables on rate of gully head advancement. Trans. Amer. Soc. Agr. Engr. 7:54-55.

Toffaleti, F.B. 1968. A procedure for the computation of the total sand discharge and detailed distribution bed to surface. Comm. on channel stabilization. U.S. Army Corps of Engr. Tech. Rep. No. 5, Vicksburg, Miss.

Twenhofel, W.H. 1939. Principles of Sedimentation. McGraw-Hill Book Co., New York. 610 p.

Ursic, S.J. 1970. Hydrologic effects of prescribed burning and deadening upland hardwoods in northern Mississippi. USDA Forest Serv., Southern Forest Exp. Station Res. Paper SO-54. 15 p.

U.S. Dept. Agriculture. 1966. Procedures for determining rates of land drainage, land depreciation, and volume of sediment produced by gully erosion. USDA Soil Conserv. Serv. Tech. Release No. 32.

U.S. Dept. Agr., Agr. Res. Serv. 1975. Present and prospective technology for predicting sediment yields and sources. Proc. Sediment-Yield Workshop. Oxford, Miss., Nov. 28-30, 1972. 285 p.

U.S. Dept. Int., Geological Survey. 1969. Mean annual precipitation in California, Basi Data Compilation.

Valentine, K.A. 1947. Effect of water-retaining and water spreading structures in revegetating semi-desert rangeland. New Mexico Agr. Exp. Sta. Bull. 341: 22 p.

Vallentine, J.F. 1971. Range Development and Improvements. Brigham Young Univ. Press, Provo, Utah. 516 p.

Vanoni, V.A. editor. 1975. Sedimentation Engineering. Amer. Soc. Civil Engr., New York, N.Y. 745 p.

Visher, S.S. 1954. Climatic Atlas of the United States. Harvard Univ. Press, Cambridge. 403 p.

Waddell, W.W., et al. 1973. A dynamic hydraulic and water quality model for river basins. Battelle Pacific Northwest Lab: 19 p.

Wadleigh, C.H. 1968. Wastes in relation to agriculture and forestry. USDA Misc. Publ. 1065, Washington, D.C. 112 p.

Wallis, J.R., and H.W. Anderson. 1965. An application of multivariate analysis to sediment network design. Int. Ass., Sci. Hydrology 67: 357-378.

Wallis, J.R., and L.J. Stevan. 1961. Erodibility of some California wildlands soils related to their metallic cation exchange capacity. J. Geophys. Res. 66: 1225-1230.

Wallis, J.R., and D.W. Willen. 1963. Variation in dispersion ratio, surface aggregation ratio and texture of some California surface soils as related to soil forming factors. Int. Ass. Sci. Hydrology 8: 48-58.

Walter, W.G., and R.P. Bottman. 1967. Microbiological and chemical studies of an open and a closed watershed. J. Env. Health 30: 157-163.

Ward, G.H., Jr., and Witt Espey, Jr. 1971. Estuarine modeling: an assessment. Report to Environ. Protection Agency, Austin, Texas.

Weber, J.E., M.M. Fogel, and L. Duckstein. 1976. The use of multiple regression models in predicting sediment yield. Water Resour. Bull. 12: 1-17.

Wein, R.W., and N.E. West. 1972. Physical microclimates of erosion control structures in a salt desert area. J. Applied Ecology 9: 703-719.

Wein, R.W., and N.E. West. 1973. Soil changes caused by erosion control treatments on a salt desert area. Soil Sci. Amer. Proc. 37: 98-103.

Willrich, T.L. 1967. Animal wastes and water quality. Dep. Agr. Engl, Iowa State Univ., Zmes, Iowa. 16 p.

Wischmeier, W.H., and D.D. Smith. 1958. Rainfall energy and its relationship to soil loss. Trans. Amer. Geophys. Union 39: 285-291.

Wischmeier, W.H., and D.D. Smith. 1960. A universal soil-loss equation to guide farm planning. 7th Int. Cong., Soil Sci. Proc. 418-425.

Wischmeier, W.H., and D.D. Smith. 1965. Predicting rainfall-erosion losses from cropland east of the Rocky Mountains. Agr. Handbook 282, USDA Agr. Res. Serv., Wash., D.C.

Wischmeier, W.H., C.B. Johnson, and B.V. Cross. 1971. A soil erodibility nomograph for farmland and construction sites. J. Soil & Water Conserv. 26(5): 189-192.

Woodward, D.E. 1975. Discussion of "Estimation of Rainfall Erosion Index" by J.K.H. Ateshian (Am. Soc. Civil Engr., 100(IR3): 298-307, 1974). J. Irrig. and Drain. Division, Amer. Soc. Civil. Engr. 101(IR3): 245-247.

Woolhiser, D.A., and P. Todorovic. 1971. A stochastic model of sediment yield for ephemeral streams. p. 295-308. Proc. Symp. on Statistical Hydrology. USDA Agr. Res. Serv., Misc. Publ. 1275, issued June, 1974.

Workman, J.P., and J.E. Keith. 1971. Economic considerations of the Cisco soil treatment study. Unpublished manuscript, Utah St. Univ., Logan. 11 p.

Wright, H.A., F.M. Churchill, and W.C. Stevens. 1974. Effect of prescribed burning on sediment, water yield, and water quality from dozed juniper lands in central Texas. Texas Tech. Univ. Office of Water Res. & Technol., Tech. Coml. Rpt. (unnumbered):d 23 p.

Chapter 7

Evaporation and Transpiration

For most rangelands, evapotranspiration (ET), evaporation plus transpiration, represents by far the largest water loss from land surfaces with runoff and ground-water recharge being relatively minor losses (Figure 7-1). Shortages of water supplies for domestic, agricultural, and industrial uses in arid and semiarid climates have stimulated considerable interest in the reduction of evapotranspiration as a means of decreasing these water shortages.

Evaporation is the process by which a liquid is changed into vapor, and rate of evaporation is governed by vapor pressure differences. Evaporation occurs when vapor pressure of the atmosphere is lower than vapor pressure at the water surface. The principle, Dalton's law, is expressed by the formula

$$E = C(P_w - P_a) \qquad \textbf{(7-1)}$$

in which E is evaporation rate in inches per day, P_w is maximum vapor pressure corresponding to surface water temperature, P_a is vapor pressure in the air above the water surface, and C is a coefficient dependent on baromet-

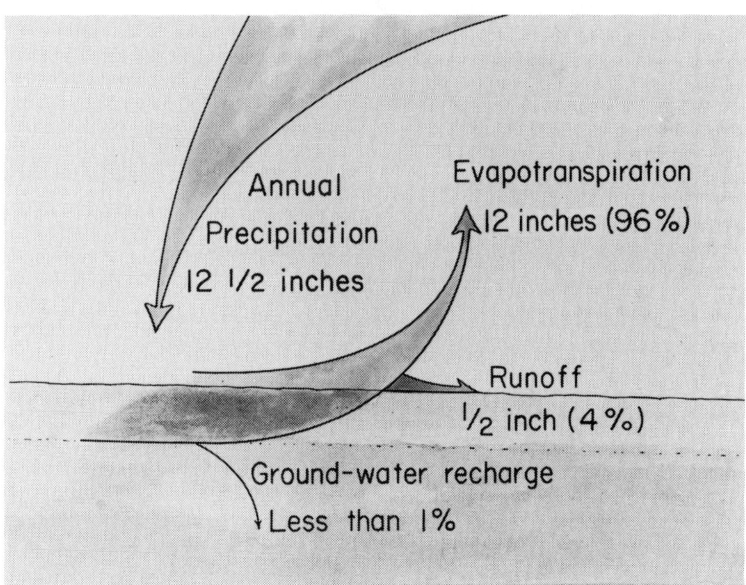

Fig. 7-1. Diagrammatic disposition of 318 mm (12.5 inches) precipitation as evapotranspiration, runoff, and ground-water recharge in a semiarid climate (from Branson 1976).

 ric pressure, wind velocity, water quality, and possibly other factors.

Transpiration may be considered a biological evaporation process and is affected by similar atmospheric factors but is more complex because of plant and soil influences (Figure 7-2). Transpiration rates are plant responses to air and soil temperatures, relative humidity, long- and shortwave-radiation balances, available soil moisture, stage of growth, kind of plant, leaf surface area, diffusion resistance of leaves, barometric pressure, wind movement, and depth and extent of rooting. Methods used to measure transpiration by plants and plant parts include psychrometric techniques (Slatyer and Bierhuizen 1964) and infrared gas analyzers (Decker and Wein 1960).

Evapotranspiration (ET) includes evaporation from soil and water surfaces and transpiration from plants (Figure 7-3). The relationship can be expressed as

$$ET = T + I + E \quad \textbf{(7-2)}$$

where T is transpiration loss, I is total interception loss, and E is evaporation from other soil and water surfaces not included under the other terms.

A major determinant of quantities of water evaporated or transpired is incoming solar energy (Figure 7-4). Evaporation involves a change from

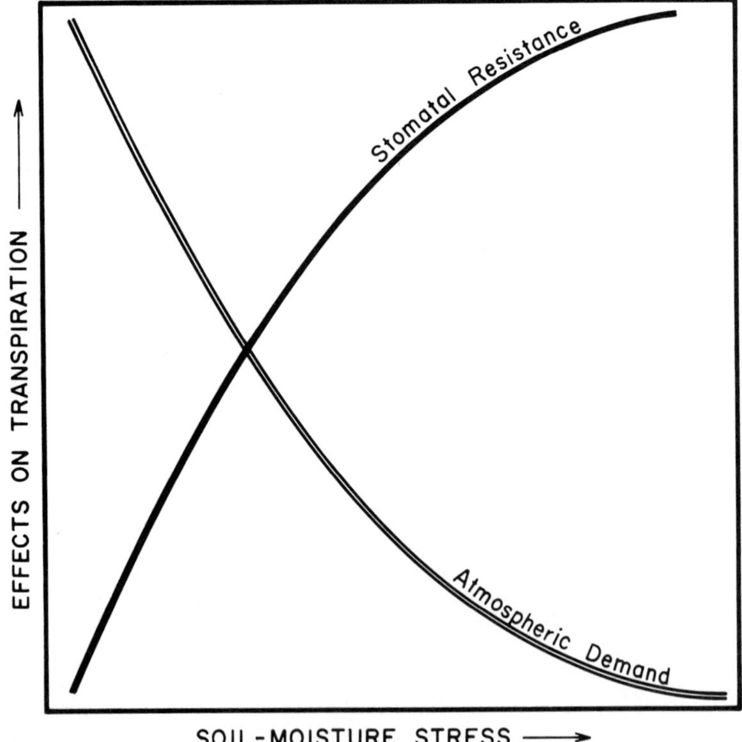

Fig. 7-2. Diagram showing probable interrelationships of soil-moisture stress, atmospheric demand, and plant effects (stomatal resistance, morphologic features, etc.) as they affect transpiration (from Branson 1976).

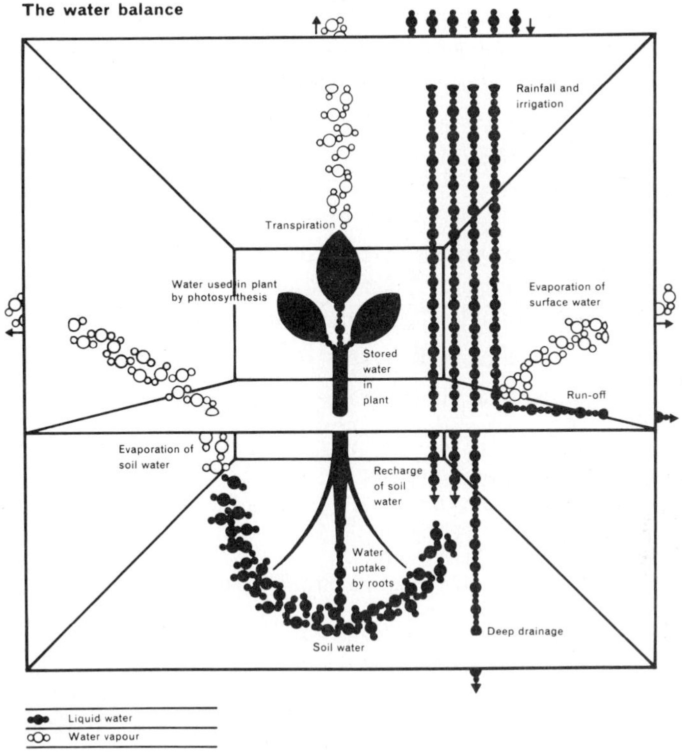

Fig. 7-3. Diagram of the water balance in soil, plant, and atmosphere (from CSIRO 1974).

liquid to vapor, and energy is required to supply the heat of vaporization. The estimated disposition of incoming radiation on southeast England (Penman 1948) was evapotranspiration 39%, back radiation 34%, reflection 20% warming the air 4%, warming the soil 2%, and photosynthesis 1%. In the Great Plains of the United States, 80 to 85% of the available energy may be used in evapotranspiration when ample water is available (Thornthwaite 1948). In deserts, most of the incoming radiation warms the air during most of the year. Keller (1971) states the following:

> Water is generally regarded as the limiting factor in forage production and arid rangeland. If 363.2 kilograms (800 lbs) is taken as the water requirement for a pound of range forage, 300 mm (12 inches) as the average precipitation, and 181.6 kilograms (400 lbs) as the average forage production/acre, only 12.5% of the precipitation, or 1.5 in, is used in producing the forage crop. If we estimate that, in addition, 12 mm (½ inch) is lost to deep percolation, 25 mm (1 in) to over-the-surface runoff, and 25 mm (1 inch) to undesirable vegetation, we account for 100 mm (4 inches). Thus, the remainder, two-thirds of the total precipitation, is lost evaporation, without benefit to man.

Measuring and Estimating Actual and Potential Evapotranspiration

Most early and many modern measurements of actual evapotranspiration

 (AET) have used variations of the water-budget or water-balance method where differences between water applied and water lost are a measure of ET. The water-balance equation is normally written

$$AET = P - RO - D \pm \Delta S \quad \textbf{(7-3)}$$

where P, RO, and D are precipitation, runoff, and deep drainage, ΔS is change in soil-moisture storage, and AET is actual evapotranspiration. This equation can be applied to land areas of any size. The water-balance equation can be solved for various components and can be expanded to include the number of significant variables that can be measured. In a study of the effects of removal of saltcedar on evapotranspiration on the Gila River in Arizona (Hanson, Kipple and Culler 1972), the following equation was used.

$$ET = Q_i - Q_o + Q_t \pm \Delta C + P \pm \Delta Z + G_i - G_o + G_b \pm G_s \text{ where:} \quad \textbf{(7-4)}$$

ET = evapotranspiration within the area defined by the boundaries of the cleared area,
Q_i = surface inflow from the Gila River,
Q_o = surface outflow of the Gila River,
Q_t = surface inflow from tributaries bordering the area,
ΔC = change in channel storage in the Gila River,
P = precipitation on the ET area,
ΔZ = change in moisture in the saturated and unsaturated soil zones within the ET area,
G_i = groundwater inflow through the saturated alluvium,
G_o = groundwater outflow through the saturated alluvium,
G_b = groundwater inflow into the alluvium from the basin fill, and
G_s = groundwater outflow across the side boundaries of the alluvium defined by the ET area.

Lysimeters, devices that usually consist of plants growing in soil and these in containers fitted with instruments that permit precise measurement of water loss or gain, also utilize the water-balance concept. Three kinds of lysimeters in common use are: (1) non-weighing, (2) weighing, and (3) floating. Non-weighing lysimeters were first used as measurements of percolation and ground-water recharge with tests by Dalton in Manchester, England, and by Maurice in Geneva, Switzerland in 1796 (Meinzer 1942). The earliest lysimetric records date to 1688 in France (Harrold 1966). Some of the oldest in the United States are those constructed by the U.S. Forest Service at Sierra Ancha, Arizona in 1935 (Rich 1959) and one of the world's largest lysimeter installations at San Dimas, California (Patric 1974). At both installations, surface runoff, percolation (groundwater discharge), and precipitation were measured and at San Dimas soil moisture was determined. Evapotranspiration could be determined for both sets of lysimeters by use of the water-balance equation. At Sierra Ancha, effects of different intensities of grazing (ungrazed, moderately grazed, and heavily grazed), and at San Dimas, bare soil plus seven kinds of vegetation were studied.

Plastic-lined lysimeters have been used at several locations in the southwestern United States to measure evapotranspiration by phreatophytes (riparian vegetation) (McDonald and Hughes 1968, Robinson 1970, and van Hylckama 1974).

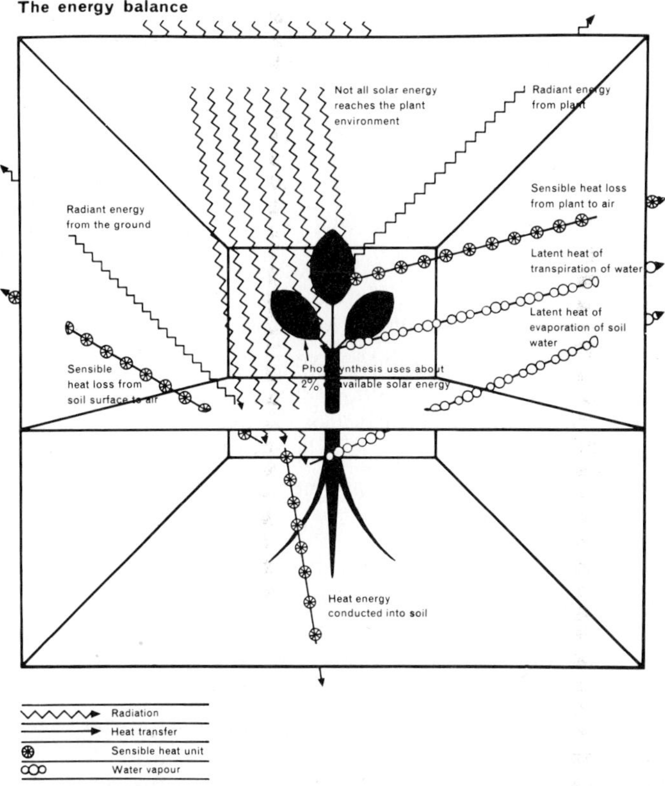

Fig. 7-4. Diagram of the energy balance of atmosphere, plant, and soil (from CSIRO 1974).

Numerous weighing lysimeters have been constructed (Harrold and Dreibelbis 1967, van Bavel and Reginato 1965, and Pruitt and Angus 1960). Advantages of weighing over non-weighing include: (1) short-term changes in water balance can be made, and (2) more precise measurements can be made. Soil monolith lysimeters at Coshocton, Ohio (Harrold and Dreibelbis 1967) record weight changes at 10-minute intervals and, by means of a strain gage and resistance device in Arizona (van Bavel and Reginato 1965), measurements at time intervals as short as 5 minutes were possible.

Floating lysimeters are similar in design to weighing lysimeters with the exception that fluid displacement instead of weight is used to measure ET or precipitation. Lysimeter tanks are placed on bags containing fluid. A tube (manometer) from the bags extends above ground and can be calibrated to permit direct readings of ET or precipitation (Hanks and Shawcroft 1965).

Lysimeters give a direct measure of evapotranspiration and have been used to determine the accuracy of estimate methods. Although lysimeters provide much useful information, they are also subject to numerous sources of error that should be minimized to duplicate natural conditions. Only a few of the many lysimeters in use in the United States and elsewhere are listed. More complete listings are found in papers by Harrold (1966), Jensen (1973), and Gangopadhyata (1968).

More recently used is the energy-budget method which, by measuring the components of incoming and outgoing energy, the energy available for evapotranspiration is determined (Penman 1963). The difference between incoming and outgoing energy, usually known as net radiation, represents energy absorbed by the surface under study, and is very important hydrologically. An equation can be expressed

$$Rn = LE + A + S + P + R \tag{7-5}$$

where Rn is net radiation, LE is latent heat of vaporization of water (about 590 cal/g water or roughly 600 cal/cm^3 water), A is transfer of measurable heat to air, S is heat storage in soil and vegetation, P is energy of photosynthesis, and R is energy of respiration. Terms that represent minor amounts of energy (e.g., S, which shows little net change from day to day, P, which usually represents a 2 to 4% energy sink, and R) are often omitted from the equation. Thus, the simplified equation may be written

$$RN = LE + A \tag{7-6}$$

Net radiation (Rn) is the difference between total incoming and outgoing radiation, and is measured with available instruments. When moisture is readily available, Rn gives a reasonable estimate of ET. Soil temperatures are easily measured and this heat-storage component is usually included in the simplified equation (Decker 1964). Evapotranspiration for a cornfield in Missouri varied seasonally from 56 mm/day (2.2 inches/day) in early July to 2.5 mm/day (0.1 inch/day) in early September (Decker 1964). The proportion of net radiation used in evapotranspiration was 70 to 80% during most of the summer. Under more arid conditions where moisture is limiting and plant cover is much less than 100%, the proportion of net radiation used in ET is considerably less (Aase and Wight 1970).

An indirect measure of vapor flux is obtained by the mass transfer method which employs the Dalton equation. A variation of the Dalton equation used by Eisenlohr (1966) in the pothole area of North Dakota is

$$ET = N[M(e_o - e_a)] \tag{7-7}$$

where,

N = a coefficient determined experimentally for a given area and arrangement of measuring instruments,

M = wind speed at a given height above the surface,

e_o = saturation vapor pressure corresponding to the temperature of the evaporating and transpiring surface, and

e_a = vapor pressure of the air.

For water bodies without emergent vegetation, the coefficient N remains a constant; with vegetation present, N varies in response to the changing development and volume of vegetation and has been used by Eisenlohr (1966) to compute separately the evaporation and transpiration components of ET. Use of the mass-transfer method for ET measurements on vegetated land surfaces has been questioned because the assumption that e_o is the saturation vapor pressure at the surface is not always valid (Slatyer 1967). The primary evaporative surface in plants is in the mesophyll cell walls, which are exposed to intercellular air spaces, and the temperature of these cell walls cannot be measured accurately.

An additional method that has considerable promise but has not seen wide use because of instrumentation problems is known as the eddy-correlation, eddy-fluctation, or eddy-transfer method (Sweinbank 1965). The method requires measurement of vertical changes in wind and vapor over short-time intervals of, at most a few seconds. Instrumentation must be extremely sensitive and recording systems elaborate to print and process the enormous amounts of data that are accumulated over short-time periods.

Numerous empirical formulas have been developed to compute *potential evapotranspiration* (PET) from readily available climatological data. Potential evapotranspiration is the water loss which would occur if there is no deficiency of water in the soil for use by vegetation. The concept has application to irrigated crops and shallow groundwater situations but has limited application to upland range. In an evaluation by Cruff and Thompson (1967) of methods developed by Thornthwaite (1948), U.S. Weather Bureau (a modification of the Penman [1948] method developed by Kohler, Nordeneon, and Fox 1955), Lowry and Johnson (1942), Blaney and Criddle (1950), Lane (1964), and Hamon (1961), the Blaney and Criddle method was considered the most practical of the six methods for estimating potential evapotranspiration.

As one example of these methods, the equation used by Blaney and Criddle (1950) is the following:

$$U = K\Sigma \frac{T \times p}{100} \quad \textbf{(7-8)}$$

where,

U = consumptive use, in inches, during the growth of the crop,

K = empirical consumptive use coefficient that is dependent on the type and location of the crop,

p = monthly percentage of total daytime hours in the year, and

T = mean monthly temperature, in degrees Fahrenheit.

The literature on evapotranspiration is voluminous. For more extensive coverage of the subject, the reader is referred to a recent review by Sosebee (1976), Jensen (1973), Burman, Rechard, and Munari (1975), and the extensive abstract bibliography by Horton (1973).

Evapotranspiration Suppression

A large variety of chemicals have been used to reduce transpiration and evaporation in attempts to improve water yields or water-use efficiency. Although research on the subject spans more than a decade, many problems have been encountered and such chemicals are not widely used. Antitranspirants can be classed in three broad groups (Davenport, Hagen, and Martin 1969): (1) reflecting materials that reduce the heat load on leaves, (2) film-forming materials that hinder the escape of water vapor from leaves, and (3) stomatal-closing chemicals that increase stomatal resistance. Most studies have shown that antitranspirants reduce water losses but these are accompanied by reductions in plant yields and the chemicals are expensive.

Evapotranspiration Research in Rangeland Evironments

It is unfortunate that we know so little about the hydrologic function of rangeland vegetation and the effects of possible vegetation conversion practices. To achieve the most efficient use of the water resource on rangelands, much more hydrologic information is needed. Selected studies concerning infiltration rates or runoff in different rangeland environments have already been covered; in this section, we will discuss evapotranspiration in relation to various plant communities.

Aspen Type

Studies dealing with water use on various aspen rangeland sites usually show a trend similar to that in Figure 7-5. Data for the illustration are from three adjacent plots at approximately 7,700 ft (2,347 m) elevation near the headwaters of Parrish Canyon, near Bountiful, Utah. Water loss is greatest from the deeper rooted aspen *(Populus tremuloides)* plots, intermediate from the herbaceous plot (aspen removed), and least from the bare plot. Studies by Croft and Monninger (1953) and more recently by Tew (1967), Johnston et al. (1969), and Johnston (1970) show that removal of aspen can reduce average transpiration by about 15 cm/year (6 inches/year) from a 3 m (10 ft) soil

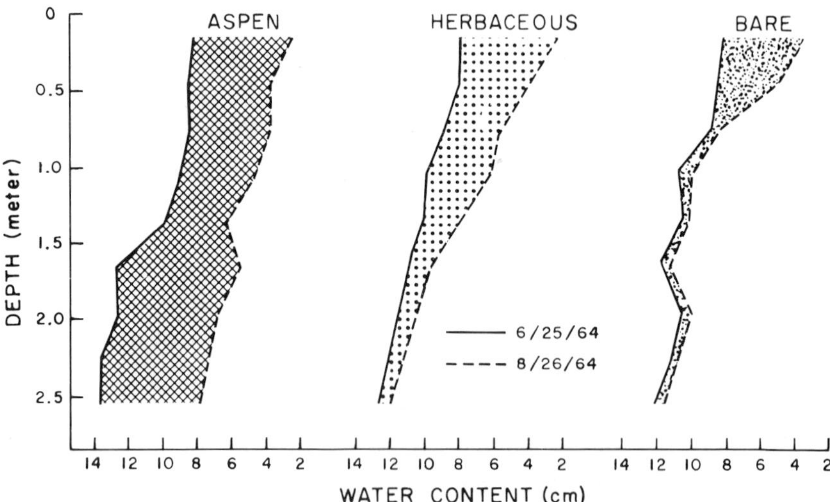

Fig. 7-5. Soil-moisture profiles under three cover conditions on one site at beginning and end of a growing season. The plots are located near Bountiful, Utah (from DeByle et al. 1969). (Multiply by 3.28 to obtain ft. Divide cm by 2.54 to obtain inches.)

profile. Where sprout stands become reestablished, a decrease in moisture-savings results.

Johnston et al. (1969) have documented, through a series of plot studies, water use by several important plant communities in the Intermountain Region. These data, illustrated in Figure 7-6, suggest the magnitude of evapotranspiration losses which may be reduced by converting from one vegetation type (aspen) to another (grass-forb).

Brown and Thompson (1965), in western Colorado, found that grassland sites used about 23 cm (9 inches) of water compared with about 46 cm (18 inches) for adjacent aspen sites (spring and fall measures for 3 years). Sampling was to a depth of 2.4 m (7.9 ft). Differences in water use were due largely to differences in moisture withdrawal by aspen below 1.2 m (4 ft).

Chaparral and Oak-Brush Type

Depending on site conditions, soil depth, energy available for evapotranspiration, etc., increases in soil moisture, and often increased streamflow, are noted following removal of woody brush species and conversion to grass. Some very striking examples involving tree removal are to be found in literature covering the varied aspects of forest hydrology (Hibbert 1967, Lull

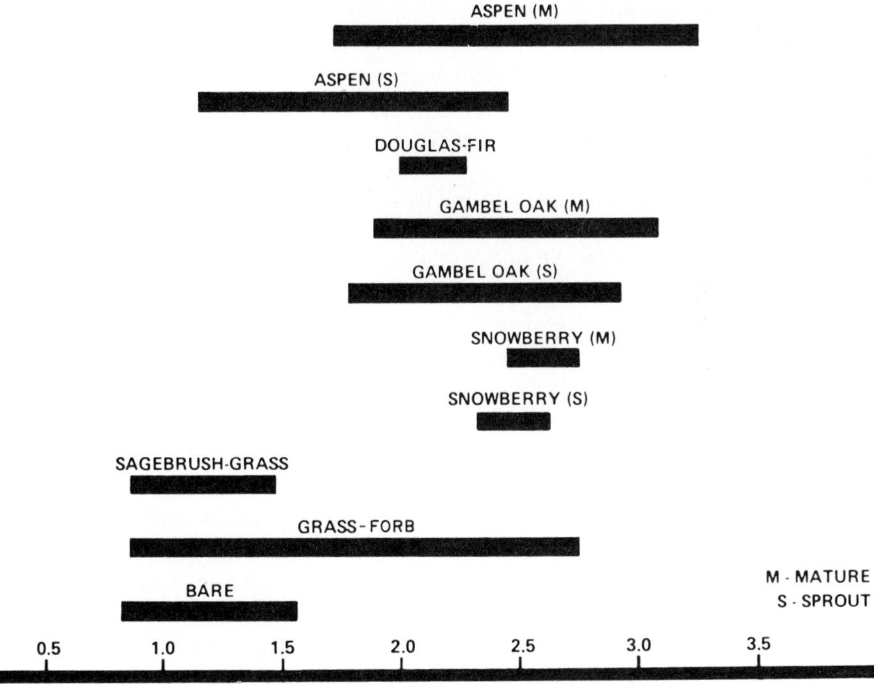

Fig. 7-6. Water-use values for several important plant communities in the Intermountain Region (data from Johnston et al. 1969). (Multiply inches/ft by 8.33 to obtain cm/m.)

and Reinhart 1967). Veihmeyer (1951, 1953) has found that extraction of water from burned chaparral plots in northern California depends upon the ability of reestablished plants to extend their roots to the full depth of soil and upon the persistency of the plants throughout the growing season. Water savings, determined from plot measurements, ranged up to about 5 cm (2 inches) as a result of burning. Where the chaparral was not burned, the soil moisture was reduced to the permanent wilting point to the full depth sampled (107 cm, 42 inches).

Conversely, Pillsbury, Osborn, and Pelishek (1963), in studying conversion of chaparral to grass in San Diego and Riverside Counties in southern California, found no residual moisture in soils under chaparral plots or in soils under plots converted to grasses and forbs. Measurements were made to a depth of 4.4 and/or 6.3 m (14 and/or 21 ft). In this instance the lower rainfall received (381 mm, 15 inches or less) wet the soil to depths of only 1.25 m (4.1 ft) or less.

Rowe and Reimann (1961) at San Dimas in southern California found that water yield cannot be appreciably increased in years of low rainfall or from shallow soils (less than 1 m or 3.28 ft deep) by converting oakbrush cover to grass. On deeper soils and during years of moderate rainfall, sufficient to wet through the soil and satisfy current evapotranspiration losses, a water savings of approximately 20 cm (8 inches) was realized by conversion to grass provided regrowth of brush and invasion of forbs and other deep-rooted plants were prohibited.

Water yield improvement studies in the chaparral type have been conducted in Arizona by the U.S. Forest Service for several years. Reference to some of these studies may be found in Chapter 5, *Runoff and Water Harvesting,* and an excellent review of these studies is available in Ffolliott and Thorud (1975). Most of the studies do not concern themselves with ET measurements per se but rather with water yield although reduced ET is frequently a source of saving.

Tew (1966, 1967) has studied soil-moisture depletion by gambel oak *(Quercus gambellii)* in central and northern Utah. In years of normal rainfall the oak depleted moisture from the entire soil profile (1.9 to 2.5 m[6.2 to 8.2 ft] sampling depth). Removing oak and allowing regrowth of a vigorous sprout stand reduced soil-moisture depletion nearly an inch during the year following cutting, but by the end of the third year sprout stands were using up to an inch more soil moisture than mature stands. In studying replacement of gambel oak with grass, a reduction of soil-moisture depletion by 7.8 cm (3 inches) the first year and 6.0 cm (2.4 inches) the second year was noted (Tew 1969a). As before, 2-year-old sprouts used as much water as mature oak stands.

Pinyon-Juniper Type

Water use studies focussing on the pinyon-juniper (*Pinus-Juniperus)* type have not shown great difference between natural stands and sites on which the trees have been removed and grass established. Studies at the Beaver Creek

Watershed near Flagstaff, Arizona, involving Utah Juniper *(J. osteosperma)* removal by cabling and burning the slash, and alligator juniper falling with power saws and leaving the debris in place, have after several years shown little change in water yield (Brown 1970). It should be recognized that water-yield increases (or decreases) are a function of both subsurface changes in soil-moisture storage patterns as well as possible treatment influences on surface hydrologic phenomenon (infiltration, overland flow, etc.).

Recent studies (Gifford and Shaw 1973; Gifford, Buckhouse, and Busby 1976) in both southeastern and southwestern Utah indicate that chained pinyon-juniper sites with debris-in-place have significantly more soil moisture than either the natural woodland or chained sites with windrowed debris. On a site in southeastern Utah, first-year soil moisture patterns were not significantly altered as a resulted of either grazing or burning a previously chained pinyon-juniper site. Gifford (1975) constructed some approximate annual water budgets of two pinyon-juniper sites in Utah and found that from 63 to 97% of the incoming precipitation (202 to 290 mm., 7 to 8 inches) was utilized as evapotranspiration. Evapotranspiration from natural woodland was actually slightly less than from chained (or cleared) areas because of greater interception rates. Interception actually represents an evaporative loss, but was included as a separate item in the above study.

Skau (1964b) measured soil-moisture changes under natural and cleared stands of alligator and Utah juniper at Beaver Creek to a depth of 60 cm (24 inches) for 1-½ years. His investigations indicated that clearing of juniper will have little effect on water yields insofar as they are influenced by soil water storage in the uper 60 cm. Any increased water availability apparently went into increased forage production but this does not constitute a waste in range as does evapotranspiration.

Sagebrush Type

Artificial vegetation manipulation practices are common in the big sagebrush type, yet few studies have evaluated soil-moisture changes or evapotranspiration as a result of such practices. Since big sagebrush *(Artemisia tridentata)* may transpire year around (Gifford 1968a), it is of interest to determine how replacement of sagebrush with grass species may influence evapotranspiration from the given site. Cook and Lewis (1963) compared soil moisture samples taken to a 1-meter depth from sprayed and unsprayed areas on 12 sample locations for three summers in Utah. Soil-moisture content was greater on sprayed plots at the 60 and 90 cm (24 to 36 inches) depths the first two growing seasons after spraying. Similar results were obtained by Shown, Lusby, and Branson (1972) in western Colorado on four small watersheds, two of which had been converted from big sagebrush to beardless bluebunch wheatgrass *(Agropyron inerme).* Evapotranspiration was about 5 cm (2 inches) greater the first year and 2.5 cm (1 inch) greater the second year from the sagebrush watersheds than from the grass watersheds. Differences thereafter were small. Sturges (1973) has suggested much the same.

Gifford (1968b) has compared soil-moisture patterns to a depth of 1.4 m

 (4.6 ft) under six cultural treatments in west-central Nevada. Treatments were 1)rip and drill, 2)spray and drill, 3)spray and contour deep furrow drill, 4)plow and drill, 5)plow and contour deep furrow drill, and 6) control. One year of data from each treatment at two locations indicated the two spraying treatments were most favorable toward soil moisture and accumulation. At one site the soil profile under spray treatments had a mean of from 2.8 to 5.0 cm (1.1 to 2 inches) more water than control plots on each sampling date, and at the other site from 1.8 to 2.0 cm (0.7 to 0.8 inch) more water than control plots. Soil-moisture change below 1 m (6.6 ft) soil depth were minimal.

Tabler (1968) measured soil-moisture changes to depth of 2 m following spraying of big sagebrush northeast of Dubois, Wyoming. Three years of records indicated that soil-moisture depletion rate was greater on control plots than on sprayed plots. Evapotranspiration differed by 4.4 ± 2.3 cm (1.7 to 0.8 inches) between sprayed and unsprayed plots the second year following treatment. Most of the difference accrued between June and late August, and about 75% of the difference in total moisture depletion occurred within the 1 to 2 m (3.3 to 6.6 ft) soil depth (Figure 7-7).

Studies covering several years at the Squaw Butte Experiment Station in Eastern Oregon have provided similar findings (Hedrick et al. 1966). Soil moisture depletion following sagebrush control by spraying or rotobeating

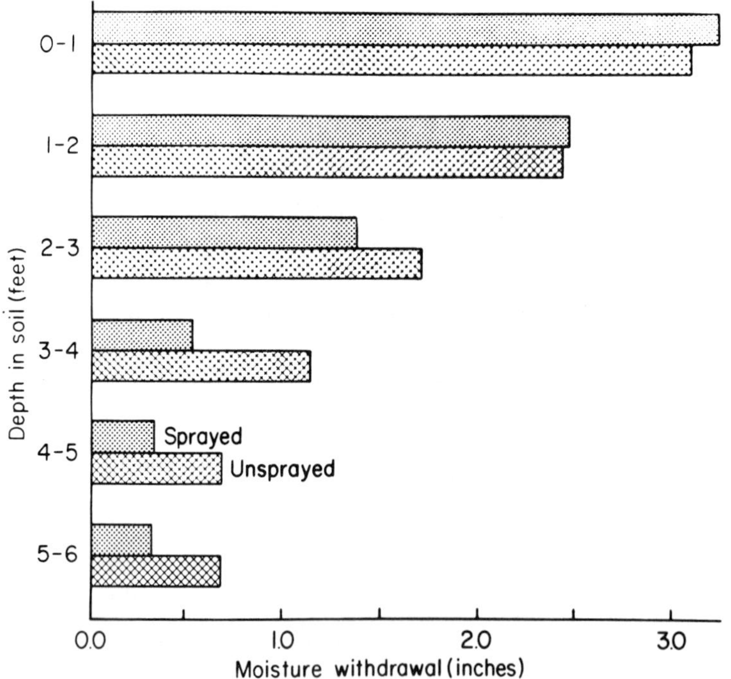

Fig. 7-7. The reduction in moisture use after sagebrush removal is concentrated in soil below the major rooting zone of replacement vegetation. About 75 to 80% of the savings accrue in soil between 1 and 2 m deep (from Sturges 1975). (Divide feet by 3.28 to obtain meters. Multiply inches by 2.54 to obtain centimeters.)

was less rapid on treated than on untreated plots at shallower depths in the first few years after treatment and after 8 years at the 76 cm (30 inch) level. Additional moisture is utilized by desirable perennial forage grasses on fair to good condition ranges, while on poor condition ranges most of the available moisture is used by annual plants or new sagebrush.

Figure 7-8 shows estimated relative quantities of water evapotranspired by the 12 plant communities near Grand Junction, Colorado, studied by Branson, Miller, and McQueen (1976). Values in the figure were obtained by subtracting minimum soil-water storage (August, 1973) from maximum storage (March, 1973). Values shown for the *Chrysothamnus nauseosus, Artemisia tridentata,* and *Sarcobatus vermiculatus* communities may be under-estimates. All three sites receive run-in moisture and all but the *Artemisia* site had groundwater available to the plant roots at least part of the growing season. The method for estimating ET in this instance does not measure groundwater use.

Average annual soil moisture curves for the top 1 m (3.28 ft) of soil of three networks of soil moisture access tubes which represent a range of soils, elevations, and precipitation zones within the big sagebrush type on Idaho's Reynolds Mountain Watershed have been presented by Rawls, Zuzel, and Schumaker (1973). Belt (1970), on the same area, measured springtime evaporative flux rates for low sagebrush *(Artemisia arbuscula)* range. He found daily ET rates ranged from 0.13 to 0.30 cm (0.05 to 0.12 inch) under differing conditions of soil moisture and radiant energy supply. On all but one of the 6 days of observation, more energy was partitioned into sensible heat than latent heat indicating moisture availability, not energy availability, was the

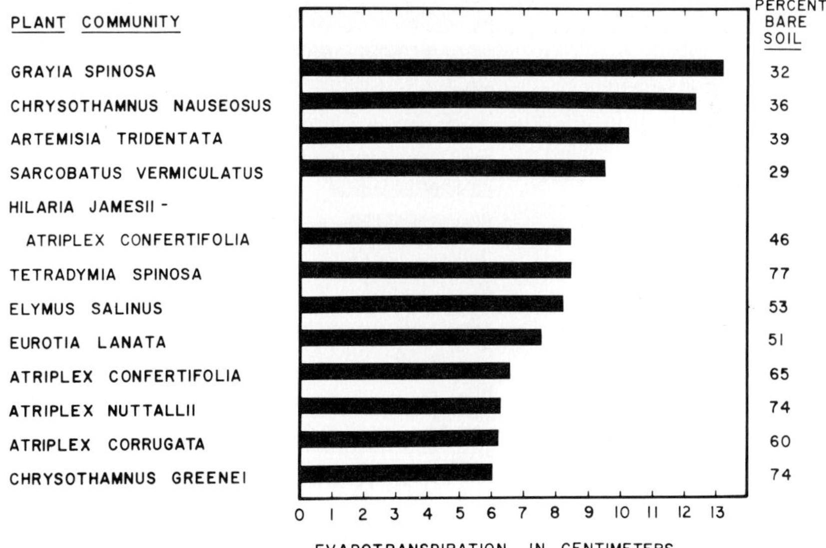

Fig. 7-8. Relative quantities of water evapotranspired and percent bare soil for 12 plant communities. Note the general inverse relationship (r = –0.78) between bare soil and water use (from Branson, Miller, and McQueen 1976).**

 limiting factor during the spring period.

Grassland Type

Seasonal accumulation and depletion of soil moisture on three soils of two native range pastures near Miles City, Montana, were studied by Houston (1968). One pasture had been heavily grazed and the other lightly grazed since 1932. The study was established in 1958. Significant differences between grazing intensities were strongly influenced by soil texture and vegetation composition, and to some extent by precipitation levels. Interactions existed and tended to mask grazing effects. On sandy soils there were no grazing effects but some differences were found on finer soils. During a favorable moisture year, on loam soil, a substantial accumulation of moisture occurred under heavy grazing at 45 cm (18 inches) of depth and below. During a dry year, this difference between stocking levels largely disappeared. On silty clay soil, moisture accumulation at 76 cm (30 inches) and below was greatly reduced under heavy grazing. This was attributed to reduced rates of soil moisture infiltration and depth of infiltration from some combinations of soil compaction and sealing from livestock trampling, reduced root channels through the soil, and less litter cover on the soil from the lower-producing, more shallow-rooted species present.

Buckhouse and Coltharp (1976) have reported that extreme clipping treatments (complete denudation) at 2,500 m (8,202 ft) elevation on Utah's Wasatch Plateau resulted in significantly less soil moisture withdrawl than on

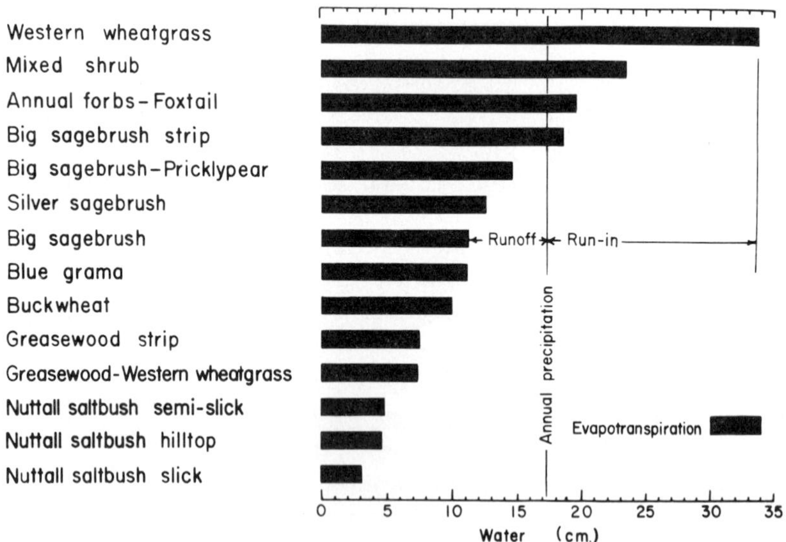

Fig. 7-9. Soil-moisture storage for 14 sites in northeastern Montana. Values were obtained by subtracting soil-moisture storage at the end of the growing season from maximum storage at the beginning of the growing season and adding water quantities contributed to the soils by summer season precipitation (from Branson, Miller, and McQueen 1970).

unclipped grassland. Other clipping intensities showed reductions in soil moisture withdrawal though differences were not always statistically significant. No significant differences were found among clipping treatments at a subalpine site at 3,300 m (10,800 ft) elevation.

In northeastern Montana, Branson, Miller, and McQueen (1970) found that soil moisture storage exceeded precipitation applied on four of 14 sites studied (Figure 7-9). Stated differently, four sites were characterized by "run-in" moisture and 10 produced differing amounts of runoff. The western wheatgrass site having the greatest water stored (recharge and discharge) as soil moisture also had one of the lowest infiltration rates.

Tew (1969b) has found that grasses best suited to high elevation sites in northern Utah also depleted the greatest amount of water from the soil profile. During the second growth season, available soil moisture was completely depleted to at least 1 meter. Brown (1973) studied transpiration rates of Kings fescue *(Hesperochloa kingii)*, smooth brome *(Bromus inermis)* and intermediate wheatgrass *(Agropyron intermedium)* at an elevation of 2,896 m (9,500 ft) in the Wasatch Mountains of northern Utah. Maximum transpiration rates ranged from 9.8 to 7.3 $g/dm^2/hr$ for smooth brome and intermediate wheatgrass respectively to 6.8 $g/dm^2/hr$ for Kings fescue, the native species.

Soil-moisture conditions under pastures of cool-season and warm-season grasses were compared by Conrad and Youngman (1965) near Lincoln, Nebraska. Measurements were made several times each year, from 1956 to 1961, to a depth of 1.6 m (5.25 ft). The amounts of water under warm-season grasses ranged from 25 to 240% more than under cool-season grasses in midspring each year. As a result, cool-season pastures suffered from midsummer drought more often than did the warm-season pastures.

Both dry-matter production and water-use efficiency were greater on grassland plots with 50% cover than on 25%, 75%, or 100% cover near Sidney, Montana (Aase and Wight 1970). The treatment applied was rotary tillage of 0%, 25%, 50%, 75%, and 100% of the native mixed-grass cover. Pan evaporation was found to be a poor estimator of ET from rangeland in this climatic region.

Aase, Wight, and Siddoway (1973) recently tested a model for estimated soil-water content on a mixed prairie site near Sidney, Montana. The model is based on the Penman combination method for estimating PET, and includes factors to account for crop development, limiting soil-water content, and increased evaporation after rain. The model gave reasonable estimates of actual soil-water conditions within a 15% limit suggested as being practical for rangeland management purposes.

Riparian, Phreatophytic, and Hydrophytic Vegetation

One means of reducing consumptive use of water by phreatophytes (plants that use ground water) is by converting phreatophyte-infested floodplains to forage plant production. It has been estimated by Robinson (1958) that phreatophytes cover about 6.4 million hectares (15.6 million acres) in the 17 western states and discharge as much as 30.8 billion cubic meters (25 million acre-ft) of water into the atmosphere annually by transpiration; this is considered by many as intolerable waste, particularly during drought years. The annual use of water by phreatophytes ranges from a few hundred cubic meters to more than 8,631 cubic hectare-meters (hm^3) (7 acre-ft/acre). Some of the more common phreatophytes listed by Robinson (1958) include saltcedar *(Tamarix pentandra)*, greasewood *(Sarcobatus vermiculatus)*, willows (*Salix* spp.), cottonwoods (*Populus* spp.), saltgrass *(Distichlis stricta)*, rabbitbrushes (*Chrysothamnus* spp.), and pickleweed *(Allenrolfea occidentalis)*.

Ffolliott and Thorud (1975) have tabulated the results of riparian vegetation water-use studies conducted in the Southwest. They state that, when all the source data from the various studies are considered, a trend in the water use by species is apparent. Dense stands of saltcedar, arroweed, cottonwood, and hydrophytes (tules [*Scirpus* spp.] and cattails [*Typha* spp.]) are the heaviest users of water (annual use of between 1.3 and 2.6 m [4 and 8.5 ft] of water). Intermediate users of water are seepwillow *(Baccharis glutinosa)*,

Table 1. Physical gain or loss of resource products resulting from mechanical or chemical conversion treatments in riparian zones (from Ffolliott and Thoroud 1975). (Divide acre/ft by 3.28 × 10_{-4} to obtain hm^3.

Resource outputs	Physical gain or loss
Water	Under the most favorable conditions, water yields of up to 2 acre-feet per acre annually may be obtained. However, some of this water will be used by reinvading vegetation.
Wood	At this time, riparian species have little or no commercial value except for cottonwood. However, fence posts, firewood, or chips for particle board and mulch are potential products.
Range	Areas which have shallow water tables (5 feet or less) and higher precipitation (10 inches–15 inches) may produce herbage following artificial seeding, but the amount of production has not been adequately determined. If grasses become established naturally, it will take about 5 years, and they will consist of weeds and annuals.
Wildlife	The quality of habitat for fish and wildlife will probably decline, particularly for doves, small game species, songbirds, javelina, and migratory water-fowl. Bristow (1968) placed a value of $150 per acre on phreatophyte lands as a result of hunting days lost and the price which the Arizona Game and Fish Dept. is paying per acre for phreatophyte land. Mitigation of these conflicts in use becomes more difficult as phreatophyte lands for fish and wildlife habitat decline in area. A positive factor may be use of cleared areas for wintering water fowl.
Flood prevention	Removal of riparian vegetation prevents blockage of streamflow and the subsequent speading of flood flows on adjacent floodplains which are sometimes developed.
Recreation and aesthetics	The aesthetic value of the treatment area is lowered in the initial stages. However, cattle grazing on riverine grass land and a view of flowing water may be aesthetically superior to riparian vegetation for some observers.
Other	Erosion rates increase between the initial removal of riparian vegetation and the establishment of grasses.

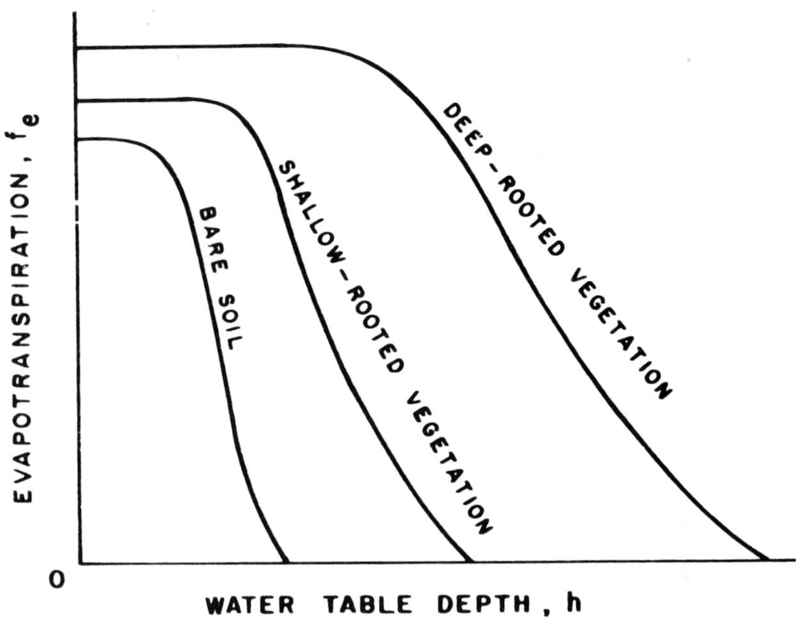

Fig. 7-10. General nature of relations between evapotranspiration (f_e) and water table depth for deep- and shallow-rooted vegetation and bare soil (from Bouwer 1975).

mesquite *(Prosopis juliflora),* quailbrush *(Atriplex lentiformis),* fourwing saltbush *(Atriplex canescens)* and greasewood (annual use of between 0.6 and 1.6 m [2 and 5 ft] of water). The least amount of water is transpired by grasses such as saltgrass *(Distichlis stricta),* Bermuda grass *(Cynodon dactylon),* sedges *(Carex* spp.) and from bare soil (annual use of less than 1 m of water). Some physical gain or loss of resource products resulting from mechanical or chemical conversion treatments in riparian zones in the Southwest are given in Table 7-1.

The general nature of evapotranspiration rates and depth of the water table is shown in Figure 7-10. Bouwer (1975) indicates that the deeper the roots, the greater the depth of the water table at which the vegetation can maintain essentially *potential evapotranspiration* rates. In developing a procedure for calculating seepage from a stream due to uptake of groundwater by vegetation or evaporation from soil in the floodplain, Bouwer states the following:

> Application of the procedure to a hypothetical stream and floodplain shows that replacing a deep-rooted vegetation with bare soil or shallow-rooted vegetation causes a rise in the water table of the floodplain. The percentage reduction in seepage losses from the stream by phreatophyte control is significant only if the depth from which groundwater can be removed by evaporation is much smaller after removal than before. The percentage reduction in channel seepage due to phreatophyte control tends to increase, and the actual amount of water saved to decrease with increasing vertical distance between the floodplain and the water level in the stream.

Information presented in this section indicates that some potential exists for reducing transpiration and evaporation water losses. In many instances, replacement of woody species with grasses has resulted in decreased evapotranspiration and some increases in runoff. However, additional information would be desirable before operational application of most of the results presented is practical. Operational procedures should include considerations of economics, aesthetics, wildlife responses, erosion rates, flood hazards, and additional costs or benefits that might be associated with application of the treatments.

Table 7-2. Water requirements (g water/g dry weight) of native or seeded species within the pinyon-juniper type (from Gifford 1976).

Crested wheatgrass *Agropyron cristatum* or *Agropyron desertorum* or *A. cristatum* x *A. desertorum*	573–616 (Bleak & Keller 1973)	705 (Shantz & Piemeisal 1927)	880–1,024 (Dillman 1931)
Siberian wheatgrass *A. sibiricum*	549–612 (Bleak & Keller 1973		
Russian wild rye *Elymus junceus*	598 (Bleak & Keller 1973)		
Intermediate wheatgrass *Agropyron intermedium*	126–180 (Baker & Hunt 1961)		
Western wheatgrass *Agropyron smithii*	429–701 (Bailey 1940)	1,076 (Shantz & Piemeisel 1927)	1,247-1,420 (Dillman 1931
Blue grama *Bouteloua gracilis*	596 (McGinnies & Arnold 1939)		
Black grama *Bouteloua eriopoda*	572 (McGinnies & Arnold 1939)	836–1,141 (Dwyer & DeGarmo 1970)	
Tobosa *Hilaria mutica*	916–1,127 (Dwyer & DeGarmo 1970)		
Russian thistle *Salsola kali*	98–222 (Dwyer & Wolde-Yohannis 1972)	107–263 (Wallace 1970)	
Fourwing saltbush *Atriplex canescens*	1,544–1,946 *(Dwyer &* DeGarmo 1970)		
Mormon tea *Ephedra* spp.	1,946 (McGinnies & Arnold 1939)		
Snakeweed *Gutierrezia sarothrae*	2,578–5,954 (Dwyer & DeGarmo 1970)		

Water-use Efficiencies

The quantity of water a plant requires for the production of a gram of dry matter, exclusive of roots, has been termed by Briggs and Shantz (1913, 1914) as its water requirement. Although it is recognized that many factors influence water requirements of plants (available water, nutrient availability, genotypic variation, environmental demands, stage of growth, rooting depth, length of growing season, soil temperature, etc.), it would appear logical that water requirement per unit of plant growth would be at least one of the several considerations, and perhaps most important generally, in ultimately deciding among several species which might be seeded on a given site. Gifford (1976) has recently compiled a list of the water requirements of some native and seeded species frequently encountered within the pinyon-juniper type (Table 7-2). The brief listing indicates significant variability within a species, but a general trend of higher water-use efficiencies may be noted for the grasses as compared to the shrub species (last three on the list). Gifford has concluded, however, that this type of information, under present conditions, does not appear to lend itself as a worthwhile guideline for selecting species useful in seeding.

Literature Cited

Aase, J.K., and J.R. Wight. 1970. Energy balance relative to plant covery in a native community. J. Range Manage. 23: 252-255.

Aase, J.K., J.R. Wight, and F.H. Siddoway. 1973. Estimating soil water content on native rangeland. Agr. Meteorol. 12: 185-191.

Bailey, L.F. 1940. Some water relations of three western grasses. I. The transpiration ratio. Amer. J. Bot. 27: 122-128.

Baker, J.N., and O.J. Hunt. 1961. Effects of clipping treatments and clonal differences on water requirements of grasses. J. Range Manage. 14: 216-219.

Belt, G.H. 1970. Spring evapotranspiration from low sagebrush range in southern Idaho. Water Resources Res. Inst. Res. Project Tech. Completion. Rep. A-014-Ida. 44 p.

Blaney, H.F., and W.D. Criddle. 1950. Determining water requirements in irrigated areas from climatological and irrigation data. U.S. Dep. Agr., Soil Conserv. Serv. Tech. Paper 96. 48 p.

Bleak, A.T., and W. Keller. 1973. Water requirement, yield, and tolerance to clipping of some cool-season semiarid range grasses. Crop Sci. 13: 367-370.

Bouwer, H. 1975. Predicting reduction in water losses from open channels by phreatophyte control. Water Resources Res. 11: 96-101.

Branson, F.A. 1976. Water use on rangelands, p. 193-209. *In:* Watershed Management on Range and Forest Lands, Proc. 5th U.S./Australian Rangelands Panel, Boise, Idaho, 1975.

Branson, F.A., R.F. Miller, and I.S. McQueen. 1970. Plant communities and associated soil and water factors on shale-derived soils in northeastern Montana. Ecology 51: 391-407.

Branson, F.A., R.F. Miller, and I.S. McQueen. 1976. Moisture relationships in twelve northern desert shrub communities near Grand Junction, Colorado. Ecology 57: 1104-1124.

Briggs, L.J., and H.L. Shantz. 1913. Water requirements of plants. Investigation in the Great Plains in 1910 and 1911. U.S. Dep. Agr. Bur. Plant. Industry Bull. 284.

Briggs, L.J., and H.L. Shantz. 1914. Relative water requirement of plants. J. Agr. Res. 3: 1-63.

Bristow, B. 1968. Statement of Arizona Fish and Game Department on phreatophyte clearing projects, p. 41-43. *In:* Arizona Watershed Symp., Vol. 12. Phoenix, Ariz.

Brown, H.E. 1970. Status of pilot watershed studies in Arizona. Amer. Soc. Civil Eng. Proc., J. Irrigation and Drainage Div. (Paper 7129).

Brown, R.W. 1973. Transpiration of native and introduced grasses on a high-elevation harsh site, p. 467-481. *In:* R. Hutnik and C. Davis, ed., Ecology and Reclamation of Devastated Land, Gordon and Breache Sci. Pub., L.T.D., London.

Brown, H.E., and J.R. Thompson. 1965. Summer water use by aspen, spruce, and grassland. J. Forest. 63: 756-760.

Buckhouse, J.C., and G.B. Coltharp. 1976. Soil moisture response to several levels of foliage removal on two Utah ranges. J. Range Manage. 29: 313-315.

Burman, R.D., P.A. Rechard, and A.C. Munari. 1975. Evapotranspiration estimates for water right transfers, p. 173-195. *In:* Specialty Conf. Amer. Soc. Civil Eng., Irrigation and Drainage Div.

Conrad, E.C., and V.E. Youngman. 1965. Soil moisture conditions under pastures of cool-season and warm-season grasses. J. Range Manage. 18: 74-78.

Cook, C.W., and C.E. Lewis. 1963. Competition between big sagebrush and seeded grasses on foothill ranges in Utah. J. Range Manage. 16: 245-250.

Croft, A.R., and L.V. Monninger. 1953. Evapotranspiration and other water losses on some aspen forest types in relation to water available for stream flow. Trans. Amer. Geophys. Union 34: 563-574.

Cruff, R.W., and T.H. Thompson. 1967. A comparison of methods of estimating potential evapotranspiration from climatological data in arid and subhumid environments. U.S. Geol. Surv. Water-Supply Pap. 1839-M. 28 p.

C.S.I.R.O. 1974. The Pye laboratory. Division of environmental mechanics, C.S.I.R.O. Commonwealth Sci. and Ind. Res. Organ., Canberra, Aust. 13 p.

Davenport, D.C., R.M. Hagen, and P.E. Martin. 1969. Antitranspirants research and its possible application to hydrology. Water Resources Res. 5: 735-743.

DeByle, N.V., R.S. Johnston, R.K. Tew, and R.D. Doty. 1969. Soil moisture depletion and estimated evapotranspiration on Utah watersheds. Int. Conf. on Arid Lands in a Changing World. Tucson, Ariz. June 3-13. 14 p.

Decker, W.L. 1964. Total energy budget of the plant canopy and its relationship to evapotranspiration from corn. Univ. Missouri Agr. Exp. Sta. Res. Bull. 854. 22 p.

Decker, J.P., and J.D. Wein. 1960. Transpirational surges in tamarix and eucalyptus as measured with an infrared gas analyzer.. Plant Physiol. 35: 340-343.

Dillman, A.C. 1931. The water requirement of certain crop plants and weeds in the northern Great Plains. J. Agr. Res. 42: 187-238.

Dwyer, D.D., and H.C. DeGarmo. 1970. Greenhouse productivity and water-use efficiency of selected desert shrubs and grasses under four soil-moisture levels. New Mexico State Univ. Agr., Exp. Sta. Bull. 570. 15 p.

Dwyer, D.D., and K. Wolde-Yohannis. 1972. Germination, emergence, water use, and production of Russian thistle. Agron. J. 64: 52-55.

Eisenlohr, W.S., Jr. 1966. Water loss from a natural pond through transpiration by hydrophytes. Water Resources Res. 2: 443-453.

Ffolliott, P.F., and D.B. Thorud. 1975. Water yield improvement by vegetation management focus on Arizona. Nat. Tech. Inform. Serv. Rep. PB-246 055.

Gangopadhyaya, M. (ed.). 1968. Measurement and estimation of evaporation and evapotranspiration. World Meteorol. Organization. Tech. Note 83. 121 p.

Gifford, G.F. 1968a. Apparent sap velocities in big sagebrush as related to nearby environment. J. Range Manage. 21: 266-268.

Gifford, G.F. 1968b. Influence of various rangeland cultural treatments on runoff, sediment production, and soil moisture patterns in the big sagebrush type, Eastgate Basin, Nevada. PhD. Diss. Utah State Univ. 138 p.

Gifford, G.F. 1975. Approximate annual water budgets of two chained pinyon-juniper sites. J. Range Manage. 27: 73-74.

Gifford, G.F. 1976. Vegetation manipulation—a case study of the pinyon-juniper type, p. 141-148. *In:* Watershed Management on Range and Forest Lands, Proc. 5th U.S./Australian Rangelands Panel, Boise, Idaho, June 15-22, 1975.

Gifford, G.F., J.C. Buckhouse, and F.E. Busby. 1976. Hydrologic impact of burning and grazing on a chained pinyon-juniper site in southeastern Utah. Center for Water Resources Res., Utah State Univ., Logan. Completion Rep. PRJNR012-1. 22 p.

Gifford, G.F., and C.B. Shaw. 1973. Soil moisture patterns on two chained pinyon-juniper sites in Utah. J. Range Manage. 26: 436-440.

Hamon, W.R. 1961. Estimating potential evapotranspiration. Amer. Soc. Civil Eng., Hydraulics Div. J. 87. HY3: 107-120.

Hanks, R.J., and R.W. Shawcroft. 1965. An economical lysimeter for evapotranspiration studies. Agron. J. 57: 634-366.

Hanson, R.L., F.P. Kipple, and R.C. Culler. 1972. Changing the consumptive use on the Gila River floodplain, southeastern Arizona, *In:* Age of changing priorities for land and water. Amer. Soc. Civil Eng. New York.

Harrold, L.L. 1966. Measuring evapotranspiration. *In:* Evapotranspiration and its role in water resources management, p. 28-33. Amer. Soc. Agr. Eng. Conf. Proc. Amer. Soc. Agr. Eng. St. Joseph, Michigan.

Harrold, L.L., and F.D. Dreibelbis. 1967. Evaluation of agricultual hydrology by monolith lysimeters. U.S. Dep. Agr. Tech. Bull. 1367. 123 p.

Hedrick, D.W., D.N. Hyder, F.A. Sneva, and C.E. Poulton. 1966. Ecological response of sagebrush-grass range in central Oregon to mechanical and chemical removal of Artemisia. Ecology 47: 432-439.

Hibbert, A.R. 1967. Forest treatment effects on water yield, p. 527-543. *In:* Int. Symp. Forest Hydrol. Penn. State Univ. Proc. Aug. 29-Sept. 10, 1965.

Horton, J.S. 1973. Evapotranspiration and watershed research as related to riparian and phreatophyte management. An abstract bibliography. U.S. Dept. Agr., Misc. Pub. 1234. 192 p.

Houston, W.R. 1968. Soil moisture on native grazing lands in the semiarid northern plains of USA. Ann. Arid Zone 7: 230-234.

Jensen, M.E. (ed.). 1973. Consumptive use of water and irrigtion water requirements. Amer. Soc. Civil Eng. New York 227 p.

Johnston, R.S. 1970. Evapotranspiration from bare, herbaceous, and aspen plots: A check on a former study. Water Resources Res. 6: 324-327.

Johnston, R.S., R.K. Tew, and R.D. Doty. 1969. Utah mountain watersheds. U.S. Dept. Agr., Res. Serv. INT 67. 13 p.

Keller, W. 1971. Limits on western range forage production—water or man. J. Range Manage. 24:243-247.

Kohler, M.A., T.J. Nordenson, and W.E. Fox. 1955. Evaporation from ponds and lakes. U.S. Weather Bur. Res. Pap. 38. 21 p.

Lane, R.K. 1964. Estimating evaporation from isolation. Amer. Soc. Civil Eng., Hydraulics Div. J. 90, No. HY5: 33:41.

Lowry, R.L., and A.F. Johnson. 1942. Consumptive use of water for agriculture. Amer. Soc. Civil Eng. Trans. 107: 1243-1266.

Lull, H.W., and K.G. Reinhart. 1967. Increasing water yield in the Northeast by management of forested watersheds. U.S. Dept. Agr., Res. Serv. Pap. NE-66. 45 p.

McDonald, C.C., and G.H. Hughes. 1968. Studies of consumptive use of water by phreatophytes and hydrophytes near Yuma, Arizona. U.S. Geol. Surv. Prof. Pap. 486-F. 23 p.

McGinnies, W.G., and J.F. Arnold. 1939. Relative water requirement of Arizona range plants. Univ. Ariz. Agr. Exp. Sta. Tech. Bull. 80. 79 p.

Meinzer, O.E. (ed.). 1942. Hydrology. McGraw-Hill Book Co., Inc. New York. 712 p.

Patric, J.H. 1974. Water relations of some lysimeter-grown wildland plants in southern California. U.S. Dept. Agr., Forest Serv. Northeastern Forest Exp. Sta., Upper Darby, Pa. (Mimeo.) 117 p.

Penman, H.L. 1948. Natural evaporation from open water, bare soil, and grass. Royal Soc. Proc. A193: 120-145.

Penman, H.L. 1948. Vegetation and hydrology. Commonwealth Agr. Bur. Franham Royal, England. 124 p.

Pillsbury, A.F., J.F. Osborn, and P.E. Pelishek. 1963. Residual soil moisture below the root zone in southern California watersheds. J. Geophys. Res. 68: 1089-1091.

Pruitt, W.O., and D.E. Angus. 1960. Large weighing lysimeter for measuring evapotranspiration. Trans. Amer. Soc. Agr. Eng. 3: 13-35.

Rawls, W.J., J.F. Zuzel, and G.A. Schumaker. 1973. Soil moisture trends on sagebrush rangelands. J. Soil & Water Conserv. 28: 270-272.

Rich, L.R. 1959. Hydrologic research using lysimeters of undisturbed soil blocks. *In:* Symposium of Hanoversch-Munden. Int Sci. Hydrol. Ass. Pub. 49. 2: 139-145.

Robinson, T.W. 1958. Phreatophytes. U.S. Geol. Surv. Water-Supply Pap. 1423. 84 p

Robinson, T.W. 1970. Evapotranspiration by woody phreatophytes in the Humbolt River valley near Winnemucca, Nevada. U.S. Geol. Surv. Prof. Pap. 491-D. 41 p.

Rowe, P.B., and L.F. Reimann. 1961. Water use by brush, grass, and grass-forb vegetation. J. Forest. 59: 175-181.

Shantz, H.L., and L.N. Piemeisel. 1927. The water requirement of plants at Akron, Colorado. J. Agr. Res. 34: 1093-1190.

Shown, L.M., G.C. Lusby, and F.A. Branson. 1972. Soil-moisture effects of conversion of sagebrush cover to bunchgrass cover. Water Resources Bull. 8: 1265-1272.

Skau, C.M. 1964b. Soil water storage under natural and cleared stands of alligator and Utah juniper in northern Arizona. U.S. Forest Serv. Res. Note RM-24. 3 p.

Slatyer, R.O. 1967. Plant-water relationships. Academic Press, New York. 336 p.

Slatyer, R.O., and J.F. Bierhuizen. 1964. A differential psychrometer for continuous measurements of transpiration. Plant Physiol. 39: 1051-1056.

Sosebee, R.E. 1976. Hydrology: The state of the science evapotranspiration, p. 95-104. *In:* Watershed Management on Range and Forest Lands, Proc. 5th U.S./Aust. Rangelands Panel, Boise, Idaho. 1975.

Sturges, D.L. 1973. Soil moisture response to spraying big sagebrush the year of treatment. J. Range Manage. 26: 444-447.

Sturges, D.L. 1975. Hydrologic relations on undisturbed and converted big sagebrush lands: The status of our knowledge. U.S. Forest Serv. Res. Paper RM-140. 23 p.

Sweinbank, W.C. 1965. The measurement of exchanges between the atmosphere and the underlying surface, p. 82-86. *In:* F.E. Eckardt (ed.). Methodology of plant eco-physiology. Montpellier Symp. Proc. UNESCO. Arid Zone Res. XXV.

Tabler, R.D. 1968. Soil moisture response to spraying big sagebrush with 2,4-D. J. Range Manage. 21: 12-15.

Tew, R.K. 1966. Soil moisture depletion by gambel oak in northern Utah. U.S. Dept. Agr., Res. Serv. Note INT-54. 7 p.

Tew, R.K. 1967. Soil moisture depletion by aspen in central Utah. U.S. Dept. Agr., Res. Serv. Note INT-73. 8 p.

Tew, R.K. 1969a. Converting gambel oak sites to grass reduces soil moisture depletion. U.S. Dept. Agr., Res. Serv. INT-104. 4 p.

Tew, R.K. 1969b. Water use, adaptability, and chemical composition of grasses seeded at high elevations. J. Range Manage. 22: 280-283.

Thornwaite, C.W. 1948. An approach toward a rational classification of climate. Geog. Rev. 38: 55-94.

van Bavel, C.H.M., and R.J. Reginato. 1965. Precision lysimetry for direct measurement of evaporation flux, p. 129-135. *In:* F.E. Eckhardt (ed.), Metholdogy of Plant Eco-physiology. Montpellier Sym. Proc. UNESCO, Arid Zone Res. XXV.

van Hylckama, T.E.A. 1974. Water use by saltcedar as measured by the water budget method. U.S. Geol. Surv. Prof. Pap. 491-E. 30 p.

Veihmeyer, F.J. 1951. Hydrology of rangelands as affected by the presence or absence of brush vegetation, p. 226-234.*In:* Int. Union Geod. Geophys. Int. Ass. Hydrol. Sci. Assemblee Generale de Bruxelles. 1951.

Veihmeyer, F.J. 1953. Use of water by native vegetation versus grasses and forbs on watersheds. Trans. Amer. Geophys. Union 34: 201-212.

Wallace, A. 1970. Water use in a greenhouse by *Salsola kali* grown at different soil temperatures and at limiting soil moisture. Soil Sci. 110: 146-149.

Chapter 8
Snow Hydrology and Snowpack Management on Rangelands

Rangelands (as defined previously) encompass geographic areas of diverse topography, climate, and vegetation cover. For this reason it is difficult to characterize the hydrologic processes of a rangeland snowcover in general terms. Since rangeland can be found at nearly all elevations and all latitudes, characteristics of the rangeland snowpack vary from highly infrequent and ephemeral snowfalls to deep, persistent, seasonal snowpacks having pockets of perennial snow and ice. For the most part, rangelands are characterized by a highly diverse vegetal cover and a winter climate that includes moderate to strong winds. It is this physical regime that gives much of our rangelands unique patterns of snow accumulation and redistribution.

Growing season precipitation, particularly on semiarid range, may be highly variable and undependable. Consequently, forage production has been shown to be strongly correlated with winter precipitation. This phenomenon is ultimately expressed in the vegetation species gradation which exists in snowdrift areas. On our higher ranges, "snowbank aspen", snowberry, and bitterbrush occupy heavy snow accumulation sites. Even on the shortgrass prairie, lee slopes often support colonies of grasses more mesic than the surrounding vegetation.

On most range sites snow probably assumes more ecological importance than hydrologic importance. Some opportunities may exist for improving water yields through rangeland snowpack management, particularly in terms of stock water supplies, but in most cases our objectives will likely be the optimization of plant-available soil water.

The occurrence of snow on rangelands is especially important at higher elevations, an example of which is given in Figure 8-1 for Wyoming. In general, snow accumulation at the higher elevations begins during approximately mid-October and extends through perhaps mid-April, after which the melt season begins (Figure 8-2). At lower elevations the season of accumulation may be much shorter in duration and accumulated quantities of snow much less (Figure 8-3). Most of the perennial streams that either originate on or traverse rangeland landscapes are the end product of the accumulation and melt processes associated with snowpacks. Forecasts of the runoff expected

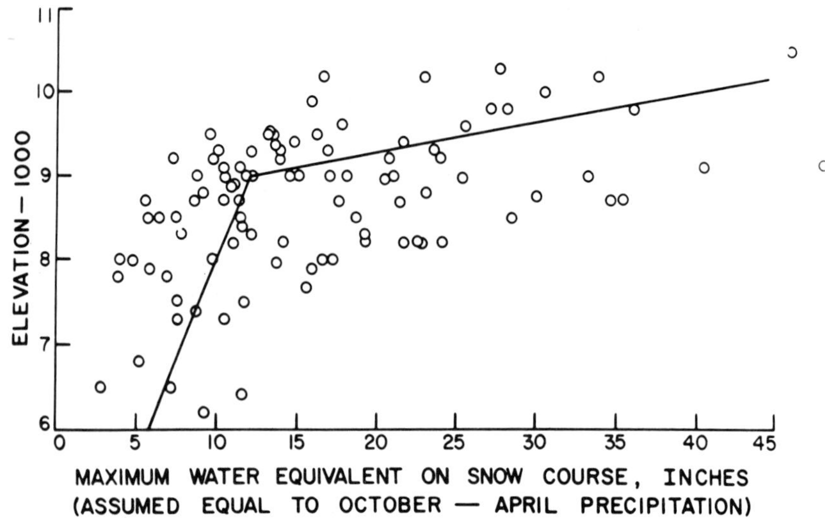

Fig. 8-1. Approximate relationship of winter precipitation to elevation in feet in Wyoming (from Rechard 1973). (Divide ft by 3.28 to obtain m, multiply inches by 2.54 to obtain cm)

from these snow accumulations are made by the Water Supply Forecasting Units of the Soil Conservation Service based on snow course surveys in the Western United States. Published reports of the predictions for individual major river basins are available from individual state Soil Conservation Service offices.

Rangelands are somewhat unique in their ability to collect and hold snow. For instance, possibilities for manipulating vegetative cover to improve water yields are often difficult, if not impossible. Actual measurements of snow is

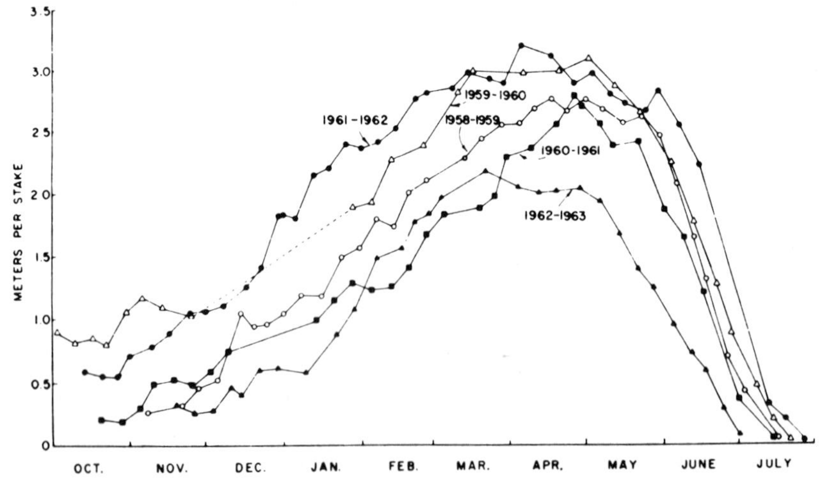

Fig. 8-2. Snow depths in Loveland Basin, Colorado, for 5 winters. Each point represents the average snow depth at 29 stakes (from Martinelli 1975). (Multiply m by 3.28 to obtain ft.)

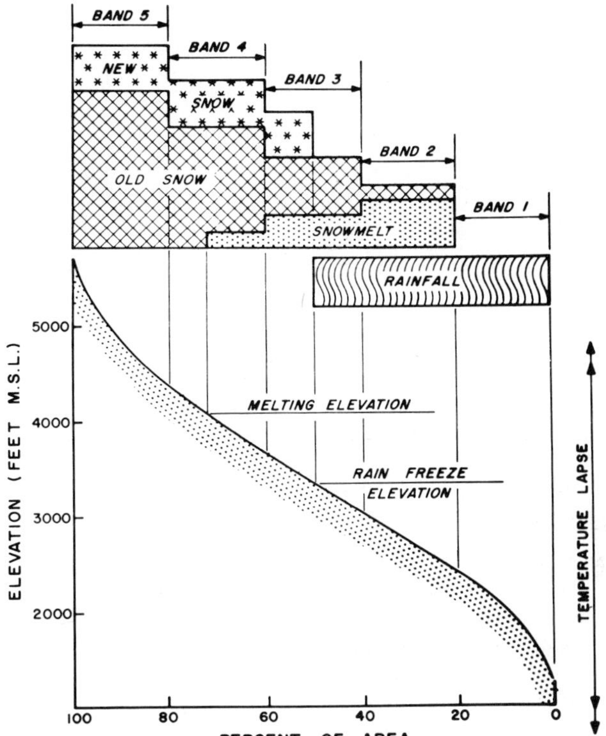

Fig. 8-3. Various conditions of precipitation (rain or snow) and snowmelt that may occur simultaneously as a function of elevational bands within a basin (Army Corps of Engineers 1971). (Divide ft by 3.28 to obtain m.)

difficult, the difficulties arising because of the physical characteristics of snow, the influence of wind, and the interaction of numerous other physical and topographic factors such as slope, distance to moisture, distance to air flow barriers, geographic location, etc.).

Physical Properties of Snow

Snowcover is a general term which includes the ground accumulation of snow, ice pellets, various forms of frost, glaze, liquid water and various pollutants. Therefore, its structure is complex and is undergoing constant change depending on prevailing weather, storm characteristics, etc.

Snow Density

The specific gravity (or density) of a snowpack is a vertically averaged value which integrates the varied effects of winds, periodic thaws, percolation, etc. on the component layers of snow. McKay (1968) has indicated that density changes in snow may result from any of the following processes:

1. Heat exchange, due to convection, condensation, radiation and conduction.

 Table 8-1. Snowpack densities.[1]

Snow Type	Density
Wild snow	0.01 to 0.03
Ordinary new snow, immediately after falling in still air	0.05 to 0.065
Settling snow	0.07 to 0.19
Settled snow	0.2 to 0.3
Very slightly-wind-toughened, immediately after falling	0.063 to 0.08
Average wind toughened snow	0.28
Hard wind slab	0.35
New firn snow[2]	0.4 to 0.55
Advanced firn snow	0.55 to 0.65
Thawing firn snow	0.6 to 0.7

[1]From McKay (1968)
[2]Snow partly consolidated into ice.

2. The pressure of the overlying snow.
3. The wind.
4. The temperature and water variation within the pack.
5. Percolation of melt water.

Typical densities of snow in its various states are given in Table 8-1. The impact of time at select locations on snow densities is illustrated in Figure 8-4. Higher densities are found in windy rangeland environments than are found in forested regions.

Snow Albedo

The albedo, reflectivity, of the snowpack varies appreciably. New fallen snow may reflect 80% or more of the incident insolation whereas a ripe granular snowpack may reflect as little as 40%. Albedo is primarily a function

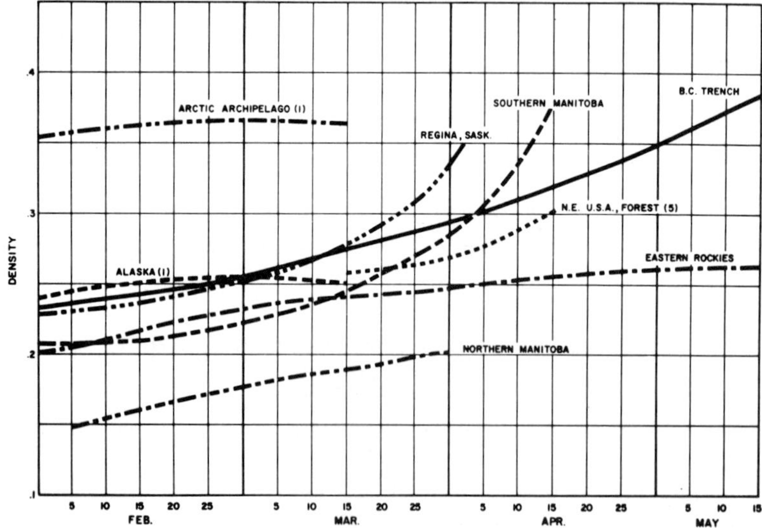

Fig. 8-4. Seasonal variation in typical snow densities for various geographic areas (from McKay and Thompson 1967).

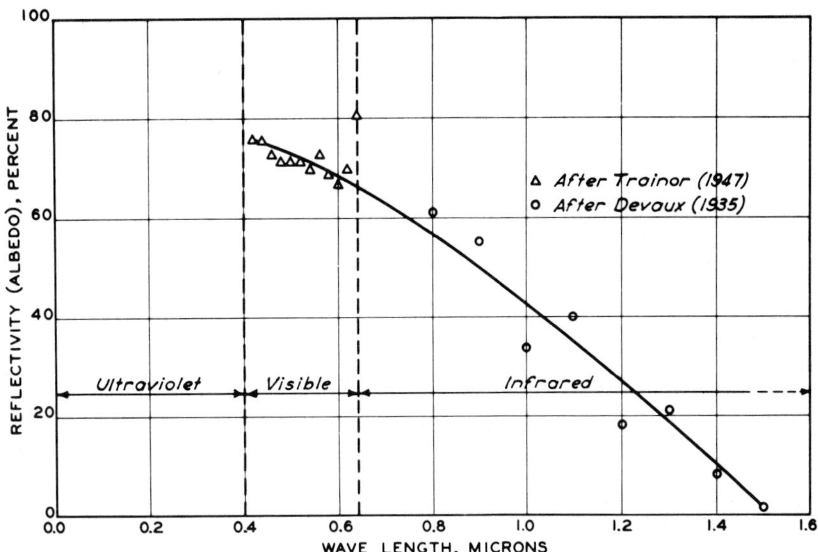

Fig. 8-5. Approximate spectral reflectivity of melting snow (from U.S. Army Corps of Engineers 1956).

of the condition of the surface layers of the snowpack. Figure 8-5 and 8-6 indicate the approximate spectral reflectivity of melting snow and variation in albedo with time during both the accmulation and melt season. Of particular interest is the fact that spectral measurements of snow albedo have shown it to be reasonably constant through the visible range, as is obvious from the extreme whiteness of the snow surface. Progressing into the near infrared the

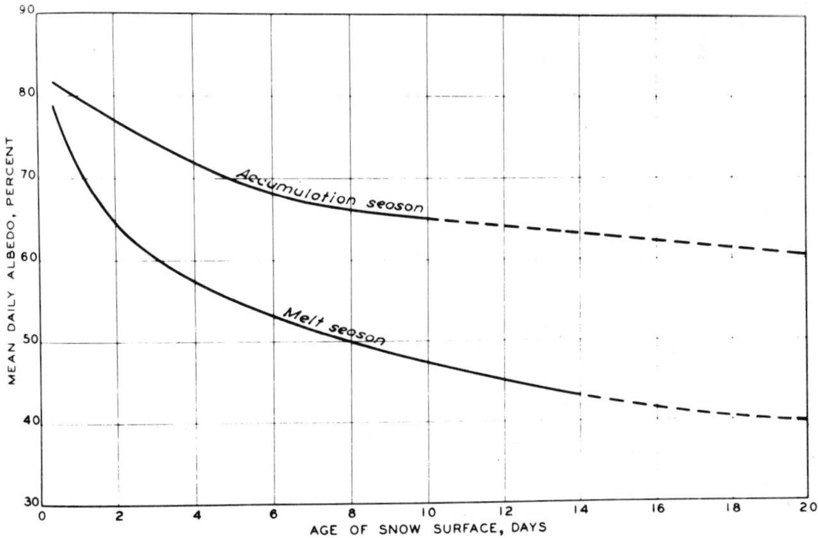

Fig. 8-6. Variation in albedo with time (from U.S. Army Corps of Engineers 1956).

 reflecting power decreases rapidly. Beyond the near infrared region snow acts very much as a perfect black body. Because of solar angles, albedo is highly dependent on the slope aspect and is higher in the morning and evening that it is at noon. Under diffuse sky radiation conditions, this effect is absent.

Snow Porosity Relationships

The relationship of airflow and gas exchange characteristics of snow to snow metamorphism and snowpack evaposublimation is important. As pointed out in a review by Van Haveren (1975), porosity directly influences the air permeability characteristics of snow, as shown in Figure 8-7. Table 8-2 gives permeabilities for various materials with a wide range of porosities and particle sizes. Van Haveren states the following: ". . . it is the width, continuity, shape, and tortuosity of the conducting snow pores which control flow rates of air through snow. Pore geometry changes considerably as a snow stratum undergoes the aging process. The result is that new snow has an initial high air permeability because of the large void space to solids ratio. Conducting pores are small and their pathways tortuous but there are many of them.

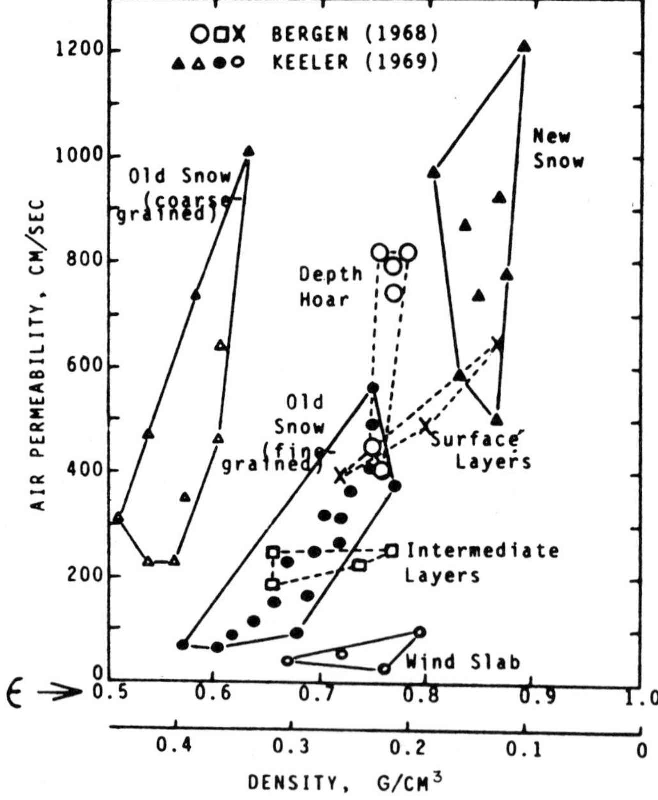

Fig. 8-7. Relationship between air permeability and porosity (E) for various types of snow (after Bergen 1968, and Keeler 1969). (Divide cm/sec. by 2.54 to obtain inches/sec. Multiply g/cm³ by 62.4 to obtain lb/ft³).

Table 8-2.[1] Air permeabilities and porosities of some common porous media.

Material	Porosity	Air Permeability cm/sec
Coarse sand (0.75 mm)	.40	6
Very coarse sand (1.1 mm)	.39	28
2-mm gravel	.40	75
Wind-packed snow	.67–.80	25–100
Old snow (fine-grained)	.60–.77	100–500
Old snow (coarse-grained)	.50–.78	200–1000
New snow	.80–.87	500–1000
10-mm gravel	.40	1885
Straw	.97	5400

[1]From Van Haveren (1975).

The sintering process reduces these pores in size, closing some of them, and air permeability decreases. This reduction in permeability is evident in the clusters of fine-grained old snow and intermediate layers show. . . As temperature-gradient metamorphism advances to the depth hoar stage, pore sizes increase, tortuosity decreases, and the pore geometries are such that air permeability increases. Permeabilities in depth hoar often equal or exceed the values for new snow. . ." Van Haveren concluded that airflow and gas exchange processes in snow exert a strong control over the water vapor balance of a snowpack as well as over the concentrations of other gases found in the void spaces of a seasonal snowpack.

Snow Movement and Accumulation

Snow Transport

Snow, because of its relatively low density, is easily moved about by the wind. This is particularly true in rangeland circumstances since significant vegetation barriers to wind movement are often lacking. Land use practices are important factors in the accumulation process, but only until the snow has filled available storage and the ground becomes effectively smooth.

The amount of snow in transport over a friable surface is a function of the wind speed. The relation between speed and load is shown in Figure 8-8. Where loose snow conditions prevail and the winds entering and leaving an area are measured, it is possible to estimate the net snow transport using relationships shown in Figure 8-8. Sturges (1975) indicates that newly fallen snow begins to drift when winds reach 5.3 m/sec (17.6 ft/sec); greater speeds are required for transport of metamorphosed snow.

As the wind blows, a considerable portion of the snow is relocated in a horizontal direction around natural barriers or vegetative windbreaks. A considerable portion of the transported snow may also sublimate, i.e., the mass of snow that returns to vapor as the air-snow mixture moves downward over a horizontal snow surface. Schmidt (1972) has recently derived a model for sublimation of wind-transported snow, critical parameters being particle speed, relative humidity, air temperature, and total insolation. Tabler and

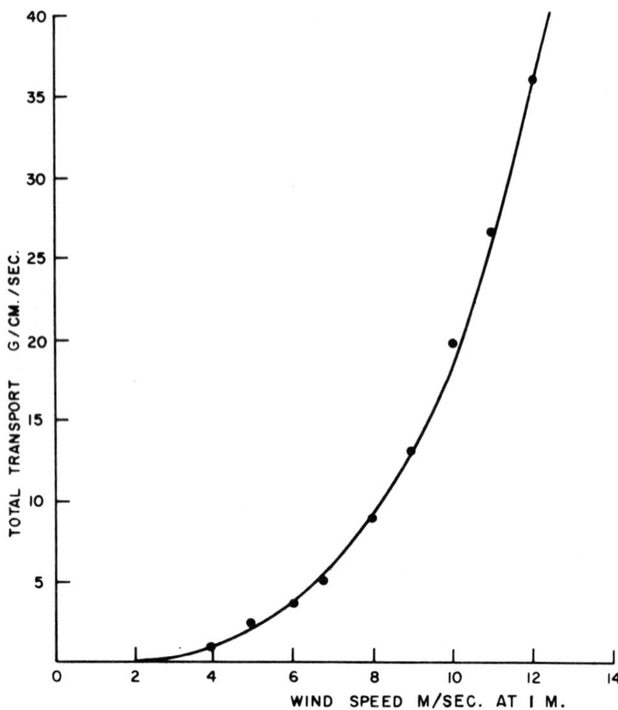

Fig. 8-8. Total snow transport (after Komarov 1954). (Multiply m/s by 3.28 to obtain ft/s. Divide g/cm/sec by 14.88 to obtain lb/ft/sec. Multiply m by 3.28 to obtain ft.)

Schmidt (1973) studied weather conditions that determine snow transport distances, based on the above model at a site in Wyoming. Figure 8-9 shows snow transport distance (before the particle completely sublimates) as a function of particle diameter and Figure 8-10 shows percentage distribution of total transport per select factor during all drifting events during their study. Such determinations provide the basic data necessary for design of snow fences for increasing water yield from wind-swept areas.

Elevational Effects on Snow Accumulation

Meiman (1968) has summarized 12 studies which include the effects of elevation on snow accumulation. This summary is given in Table 8-3 (see also Figure 8-1). Meiman calls attention to the range of values in column (4) and (5), and comments that these values reflect the great variation encountered between major physiographic areas as well as the spatial-temporal variations within a given area. His review indicates that other land surface factors as well as atmospheric conditions (year to year as well as within year) are factors that interact strongly with elevation.

Aspect Effects on Snow Accumulation

Meiman (1968) also reviewed the impact of aspect on snow accumulation,

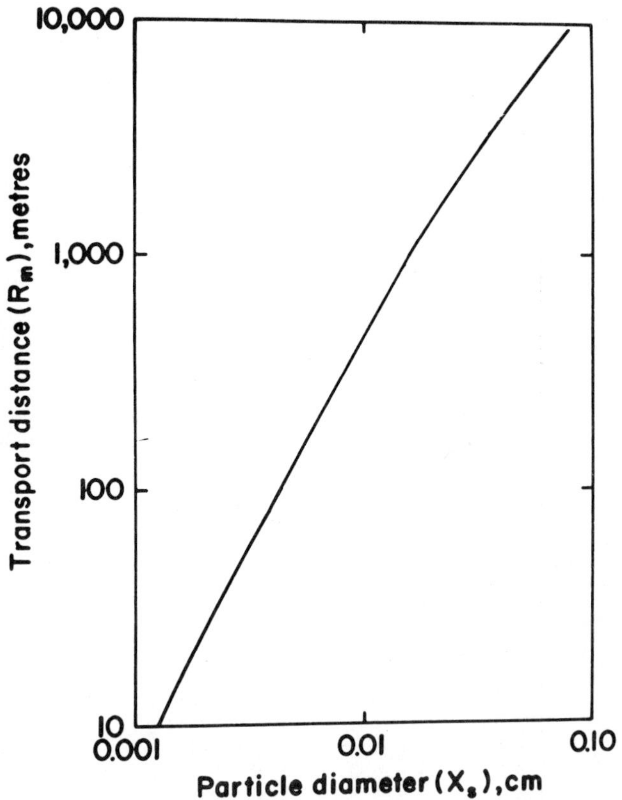

Fig. 8-9. Snow transport distance as a function of particle diameter, using the mean values for winter conditions in Wyoming during drifting over the 1970-71 winter (from Tabler and Schmidt 1973).

and selected results are presented in Table 8-4. Apparently the effect of aspect on accumulation in most areas tends to be considerably less than that of elevation and the effect appears to be predominantly a melt influence.

Snowmelt

In considering snowmelt on rangelands, it is necessary to examine the heat exchange between atmosphere, vegetation, soil, snowpack. The total heat exchange may be considered an interaction of the following components: (1) shortwave (solar) radiation, (2) longwave radiation, (3) convected and advected energy (sensible heat), (4) condensation, (5) conduction from underlying soil, and (6) warm rain.

Gray and O'Neill (1974) indicate that the energy budget approach is an appropriate framework for the development of a snowmelt model in that the

Table 8-3. Selected studies on the effect of elevation on snow accumulation.[1,2]

		Results			
(1) Source location	(2) Procedure	(3) General and descriptive	(4) Average increase in inches per inches per 1000 ft	(5) Percentage increase from site per 1000 ft	(6) Reviewer's remarks
Anderson & Pagenhart (1957) Calif.	Multiple regression analysis of 32 snow courses as measured on April 10, 1951.	Elevation most important of the six terrain and three forest variables used.	20.5 (6800–8600)	114	Strong influence of melt. Values for columns (4) and (5) estimated from text Fig. 1 (See Meiman 1968.)
Anderson & West (1965) Calif.	March 1 readings on 163 snow courses, 1958 thru 1960 anayzed by principal component regression and factor analysis	No decline with elevation; 1- and 2-inch greater increase in going from 7000 to 8000 ft as compared to 6000 to 7000 ft.	7 to 10	—	Strong influence of melt. Column (4) estimated from text text Fig. 1 (See Meiman 1968.)
Church (1912) Calif.	Readings taken between April 27 and May 5, 1910.	Mount Tallac sparse timber at lower elevation; unforested at higher.	10.9 (8000–9000)	44	Melt influence. Church gave figures to indicate potential for study; changes in vegetation as well as elevation.
Court (1963)	Eighteen snowcourses analyzed for period 1951-60; readings at end of March.	Increase with elevation "presumably a response to the decrease of temperature with height."	1.8 (8200–11200)	7	Used the 8 "north" courses only.
Mixsell et al (1951) Calif.	Long-term records for 13 snow courses having generally similar terrain; (April 1)	Linear correlation with elevation was highly significant, $r = 0.84$	22 (5800–7200)	—	—
Curry & Mann (1965) Alberta, Canada	Divided region into 6 areas judged to differ climatically on basis of gage records and topography; records since early 1950's.	Values of r^2 for four areas (8 to 17 gages per area) for mid-Oct to mid-May precipitation vs. elevation were from 0.71 to 0.83.	5.4 (4600–7400)	54	Sacramento storage gages. Values for columns (4) and (5) estimated from text Fig. 4 for area 3 (17 gages) (See Meiman 1968).

Stanton (1966) Alberta, Canada	Compared two north-facing courses of 20 points each within spruce-fir forest on April 22, 1965.	Course #2	24.5	15.5 (5000–5500)	97	
		Course #3	16.0			
Grant & Schleusener (1961) Colorado	Daily observations from Feb. 19 to May 5, 1960 at sites spaced at 1-mile intervals over Fre- ıt mont pass.	Snowfall increase with elevation to top of 11,300′ pass.		2.7 (9300–11300) (N slope)	73	Values for columns (4) and (5) estimated from text Fig. 2 (See Meiman 1968). These are additive values from daily snowboard measurement.
				5.7 (10200–11300) (S slope)	204	
Leaf (1962) Colorado	Daily observations on 15 sites during Feb. 20 to May 5, 1960 and Jan. 1 to April 27, 1961; also used 14,000′ winds over Denver and Grand Junction, Colorado	Wind Direction (clockwise)		—	—	
		WNW-ENE		2.177		
		WSW-NNW		0.313		
		SSE-WNW		-0.004		
U.S. Soil Conservation Service (1965, 66, 67) Colorado	Snow Survey results from 14 course 1965-1967; maximum snowpack water equivalents regardless of date.	**1965**		**24.8**	**62**	
		1966		15.1	67	
		1967	26.5	86 (8180-9600)		
		1965		15.3	83	
		1966		8.2	75	
		1967		12.8 (8180–9600)	102	
Gary & Coltharp (1967) New Mexico	Eight plots, one north and one south in each cover type; 10 points at 10 ft intervals on each plot	Douglas fir (S) to spruce fir (s)		4.8 (9900–11150)	63	
		Aspen (S) to grass(S)		5.4 (9900–11150)	70	
		Douglas fir (N) to spruce fir (N)		3.6 (9900–11150)	38	
		Aspen (N) to grass (N)		0.7 (9900–11150)	5	

Table 8-3. (continued) Selected studies on the effect of elevation on snow accumulation.[1,2]

		Results			
(1) Source location	(2) Procedure	(3) General and descriptive	(4) Average increase in inches per inches per 1000 ft	(5) Percentage increase from site per 1000 ft	(6) Reviewer's remarks
Packer (1962)	Sampled 950 acres WS at 10-day intervals 1949-1952 for a total of 1242 station years. Used maximum water equivalent regardless of date.	Elevation had greatest effect of factors measured. Rates of accumulation be- 3000 and 4000 ft less than at lower or higher elevation. Elevational differences not affected by forest canopy differences.	10.75 (2700–5500)	154	Some melt involved; author indicated 3000 to 4000 ft effect might be result be resut of inverstion.

[1]Values are snowpack water equivalents determined by standard snow tube measurements unless indicated otherwise in column (6).
[2]Data from Meiman (1968). (Divide ft by 3.28 to obtain m. Multiply in by 2.54 to obtain cm. Multiply m by 1.6 to obtain km.)

Table 8-4. Selected studies on the effect of aspen on snow accumulation.[1,2]

		Results				
(1) Source and location	(2) Procedure	(3) General and descriptive (elevations in feet and snow snow depths in inches)			(4) Maximum difference (inches)	(5) Difference as % of lesser
Gary & Coltharp (1967) New Mexico	See Table 1 (See Meiman 1968).	North vs. South comparisons:	N	S		
		Doug-fir	9.4	7.6	1.8	24
		Aspen	13.1	7.7	5.4	70
		Spruce-fir	13.9	13.6	0.3	2
		Grass	14.0	14.5	0.5	4

Grant & Schleusener (1961) Colorado	Accumulated daily snowboard water equivalents for Feb. 19, to May 5, 1960 at 1-mile intervals over 11,300 ft pass.	From text Fig. 2:						
		Elev.			N	S		
		≃10250			6.55	2.75	3.8	138
		≃10900			7.30	5.14	2.2	43
Haupt (1951) Idaho	Measurements 1935 thru 1937 at 258 stations in ponderosa pine forest at 4000 to 5000 ft	March 1 readings:						
		N(55 stations)			9.58			
		E(52 stations)			9.23	2.5	35	
		W(76 stations)			8.77			
		S(75 stations)			7.11			
Packer (1962) Idaho	Snow courses sampled at 10-day intervals, 1949-1952; aspects ranged from 160° clockwise to 50° from north.	From text Figs. 1 and 3 (See Meiman 1968).						
		Elev.			165°	11°		
		2700			5.6	10.0	4.4	78
		3657			9.9	19.1	9.2	93
		5500			29.3	48.1	18.8	64
Mixsell et al. (1951) Calif.	Results presented here include only 1947 data on the 21 snow courses for March 12-15 and April 14-17.		N	E	W	S		
		Mar	36.8	34.9	35.9	32.1	4.7	15
		Apr	39.6	34.5	33.2	30.6	30.6	9.0
Stanton (1966) Alberta, Canada	Paired snow courses, 1 in cut area and 1 in adjacent forest on N, S, and E aspects; cut over areas ranged from 8H to 16H.[3]			N	S	E		
		Forest		16.0	16.0	15.5	0.5	3
		Cut		17.5	21.0	25.5	8.0	46
Leaf (1962) Colorado	See Table 1 (Meiman 1968). Individual storm comparisons using regression equation	Wind direction			Aspect effect by 10° classes from mag. N.			
		WNW-NNE			+0.006			
		WSW-NNW			+0.012		—	—
		SSE-WNW			+0.012			

[1]Values are snowpack water equivalents determined by standard snow tube measurements except for snowbird determinations of Grant and Schleusener, and Leaf.
[2]Data from Meiman (1968).
[3]H indicates tree height.

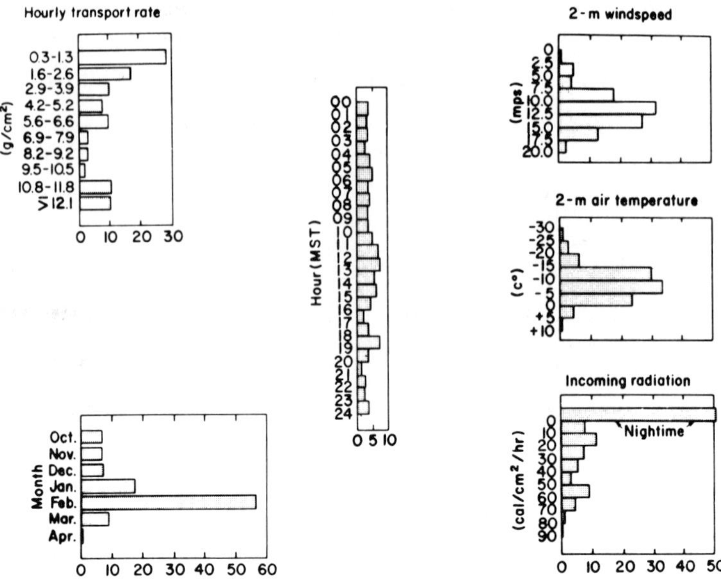

Fig. 8-10. Percentage distribution of total snow transport per select factor during all drifting events over the 1970-71 winter (from Tabler and Schmidt 1973). (Multiply g/cm² by 2.05 to obtain lb/ft². Multiply cal/cm²/hr by 6.45 to obtain cal/inch²/hr.)

process may be explained on a physical basis and the melt quantity, M, expressed as:

$$M = \frac{Q_M}{L} \tag{8-1}$$

where

Q_M = the amount of heat available for the melt process and
L = the latent heat of fusion.

Partitioning of Q_M into its components requires examination of the relative importance of radiative, turbulent, conductive and advective heat transfer processes in the melt sequence. The budget for a nonisothermal snowpack, wholly or partly below O° C may be written as:

$$Q_s = Q_{IS} - Q_{RS} - Q_{GS} + Q_{LD} - Q_{LU} + Q_H + Q_E + Q_V + Q_G - Q_M \tag{8-2}$$

where:

Q_S = the increase in internal energy storage within the snowpack,
Q_{IS} = the energy supplied by incoming solar radiation,
Q_{RS} = the energy loss due to the reflection of incoming solar radiation,
Q_{GS} = the energy transferred to the underlying soil surface by penetration of solar radiation through the pack,
Q_{LD} = the energy supplied by longwave radiation to the snow surface,
Q_{LU} = the energy loss by emission of longwave radiation from the snow surface,

Q_H = the gain of energy by turbulent transfer of sensible heat from the air to the snow surface,

Q_E = the gain of energy by turbulent transfer of latent heat (evaporation/condensation or sublimation) from the air to the snow surface,

Q_V = the gain of energy by all vertical advective processes (e.g., rain, the supply of heat in condensate and the mass removal by evaporation/sublimation), and

Q_G = the supply of energy to the snowpack from conduction from the underlying ground and percolation of melt-water and vapour transfer between snow and ground.

When continuous melt is in progress, a snowpack will rapidly reach an isothermal condition at a temperature of 0° C and therefore Q_S may be assumed negligible. Similarly, it is often assumed that Q_{GS} is negligible or can be included in Q_G. Thus, Equation 2 reduces to:

$$Q_M = Q_N + Q_H + Q_E + Q_V + Q_G \quad \textbf{(8-3)}$$

in which Q_N = the net all-wave radiation supplied by the snowpack. Equation

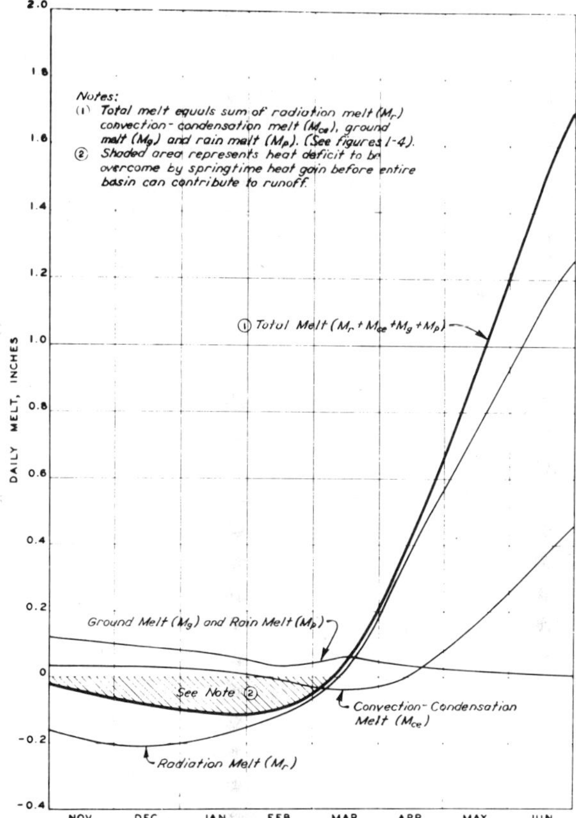

Fig. 8-11. Snowmelt summary (from U.S. Army Corps of Engineers (1956). (Multiply inches × 2.54 to obtain cm).

 8-3 has been found to be generally satisfactory for synthesizing melt at a point.

Figure 8-11 represents the various heat fluxes averaged over a 5-year period (1946-51) at the Central Sierra Snow Laboratory in California. As may be may be seen, radiation (shortwave plus longwave) is extremely important. It will be noted that for the months November through February, the total heat flux to the snowpack is negative (snowpack is actually cooling). During March the total flux becomes positive, yet the amount of heat is small. During April there is an acceleration in the heat supply to the snowpack which continues through June. By considering these heat fluxes, it can thus be seen why the snowpack accumulates through March and melts thereafter.

In general then, potential rate of snow melt increases rapidly as spring advances. The observed melt rate is, however, controlled not only by atmospheric conditions, but also by the condition of the snowpack and by its environment. As shown in Figure 8-12 the effect of season and environment on melt rates for the Edson area of Alberta is of particular importance. For example, snowcover may persist within a forest long after it disappears from the adjacent rangeland. Differences in the energy balance, and particularly the increased solar radiation received in the open, can account for these differences, even though total accumulation in the forest may be less than in the adjacent rangeland.

In some instances it may be possible to use snow and ice melt as indicators of certain hydrologic conditions. For example, Heinemann, Myers, and Moore (1972) have studied snow and ice melt patterns as surface expressions of subsurface characteristics in eastern South Dakota. They were interested in

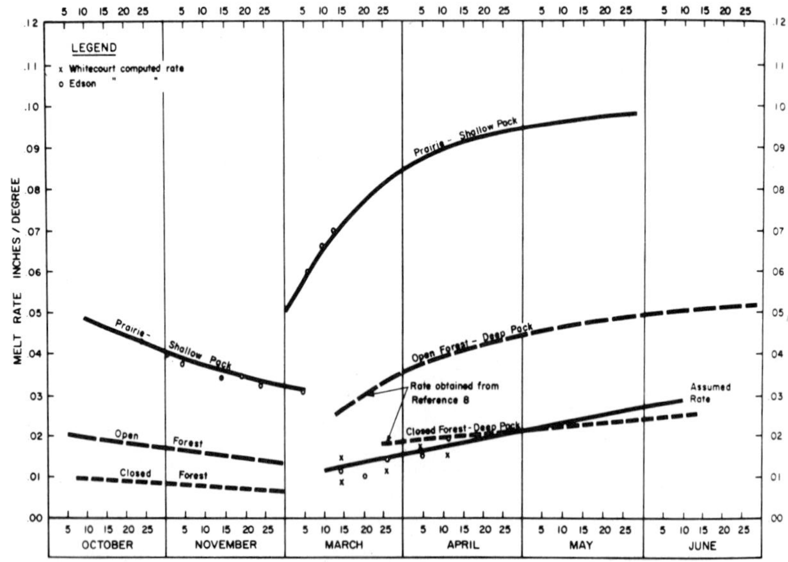

Fig. 8-12. Melt rates, in inches of water per degree day above 32° F. One degree-day (or one "heat unit") is an average temperature of 33° F for 24 hours (from McKay 1968; multiply in × 2.54 to obtain cm).

transient thermal conditions created by intermittent aquifers which might possibly involve a differential heat transfer to the surface of the snow. Using a heat sink concept (greater heat storage in saturated materials), remote sensing of snow and ice melt patterns aided in delineation of shallow aquifers in glacial drift and the location of ground water movement into rivers.

Ignoring for the moment losses from the snowpack due to evaporative processes, snowmelt per se generally either goes to recharge soil moisture deficits or becomes a part of the groundwater system. In some instances a portion of the snowmelt may appear either as surface runoff or interflow. Several indices of snowmelt recharge efficiency (recharge to the soil water system) or snowmelt runoff efficiency have been generated recently. For instance, Greb (1975) introduced the concept of the "snowmelt storage efficiency" and defined this index as the net gain in soil water within a specified soil profile from a given water equivalent of snowfall as measured by core sampling on the ground. Another similar index, the "snowmelt recharge efficiency", is defined as the ratio (in %) of soil water recharge due to snowmelt to the storm precipitation as measured with shielded precipitation gages. Van Haveren (1974) defined another index, the "snowmelt recharge ratio", as the ratio of the net soil water recharge received between growing seasons. Soloman, Ffolliott, and Thorud (1975) have defined a "snowmelt runoff efficiency" as follows:

$$PE(\%) = \frac{R}{(P-S_w)} 100 \qquad \textbf{(8-4)}$$

where

R = surface runoff,

PE = precipitation efficiency,

P = precipitation in inches, and

S_w = Δ snowpack water-equivalent for the period of measurement.

For additional information regarding snowmelt processes, the reader is referred to Wilson (1941), U.S. Army Corps of Engineers (1956), Anderson and Crawford (1964), and Eggleston, Israelson, and Riley (1971).

Snowfall Measurement

Standard recording and non-recording (storage) rain gages are generally used for measuring snow. However, because of the influence of wind it is necessary to protect the gages either with natural barriers (buildings, trees) or with some sort of artificial barrier (Nipher or Alter shields, snow fences). One example of a solution to this problem is the Reynolds Creek precipitation network which consists of dual-gage sites with a density of one site per 5.2 km^2 (2 sq. miles). A site consists of two recording gages, mounted on posts, with collectors at 3.04 m (10 ft). One gage is shielded with a modified Alter shield, with the baffels constrained at 30 degrees from vertical to maintain a constant airflow across the collector. The second gage is unshielded. The dual-gage network (Figure 8-13) is intended to be used for calculating "true" precipitation in areas of snowfall, since the wind and snow combination is a major

Fig. 8-13. Picture of a shielded and unshielded raingage as used on the Reynolds Creek Experimental Watershed near Boise, Idaho.

error source in gage catch. The method used at Reynolds Creek for computing "true" precipitation was developed by Hamon (1972) and requires the unshielded gage catch, shielded gage catch, and an empirically determined calibration coefficient as input parameters.

Other examples of attempts to overcome adverse wind effects in measuring snowfall or blowing snow include those of Rechard and Larson (1971) and Jairell (1975a). In the first instance (Figure 8-14) snow fences are suggested as a means for shielding precipitation gages and in the second instance (Figure 8-15) an improved recording gage is described for estimating total snow transport during drifting events.

Measurement of Snow on the Ground

Snowcover measurements are required generally to determine the water equivalent (snow density × depth) or an index of the water equivalent over an area. McKay (1968) has reviewed some of the problems associated with measuring and evaluating snow cover, and Table 8-5 is taken from his paper. The techniques shown in Table 8-5 are currently the most common, but recent efforts have shown that airborne measurement of passive terrestrial gamma radiation (from natural radioisotopes in the soil) (Bissell and Peck 1973), snow density profiling through backscattering of X rays (Blincow and Domi-

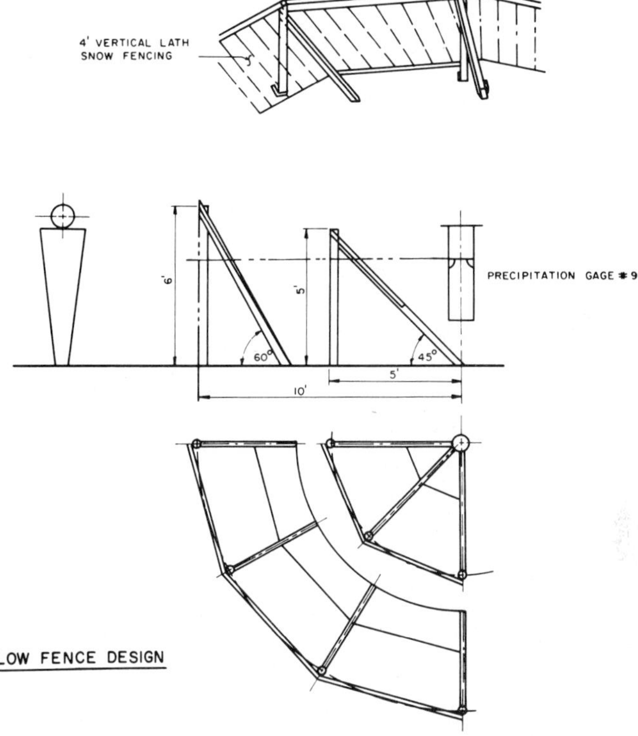

Fig. 8-14. Raingage installed inside two concentric circles of snow fences inclined at 45° and 60° angles from the horizontal. Snow fence material is 1.3 m (4 ft) vertical-lath highway-type fencing, with about 50% density (from Rechard and Larson 1971).

ney 1974), and measurements of attenuation of highly penetrating cosmic radiation (Bissell and Burson 1974) may eventually become operational as techniques for measuring the snowpack. It should be emphasized that instruments are not available which give areal measurements of the snowcover water equivalent. Statistical techniques using point measurements appear at present to be the best method of obtaining areal water equivalents. It is important to know whether particular measurements will be used as an index or as an estimate of areal distribution of snow water equivalent. For example, Soil Conservation Service snowcourses are designed to be indices used in forecasting water yields. However, most watershed simulation models in use today require some estimate in time and space of snow water equivalent over the entire watershed surface.

Measuring snow depths and snow distribution by aerial photogrammetry has been shown to be a practical method for obtaining detailed information on the snow distribution in areas of complex relief (Cooper 1965). This photogrammetric method for measuring snow consists of initially making a topographic map of the desired area including establishing the horizontal and vertical at specified points. Then, at the desired intervals during the snow

Table 8-5. Methods of snow measurement[1].

	Type of measurement						Limitations and potential					
	Point values				Areal values							
	Depth	W.E.[2]	Structure	Quality	Depth	Cover	Telemetry	Access	Measurement error	Weather sensit.	Syst. error	Other
Snow tubes	X	X						X	±12%		X	Site selection
Stakes	X							X	±1″		X	Site selection
Aerial markers	X							X	±6″	X	X	Site selection
Photography								X		X		Affected by unseasonal snowcover
Photogrammetry	X				X	X			Function of area, equipment, and technique	X		Shallow cover, steep slopes, forests
Radioisotope		X					X	X	±5% within 50″			Cost, hazard
Twin Υ probe	X	X	X				X	X	±2% (density)			Cost, profiles soil as as well . 1/2″ increments, 5 secs/in
Microwave	X	?		X			X					In development— cost
Pressure pillow		X					X	X	Function of pack character		X	Calibration, ice layers, leakages, response time related related to ice layers
Υ Backscatter	X	X	X						Density 20–40% only			Geometry of source and detector a factor, ice layer shielding, 16″ sphere a function of snow composition

[1]From McKay (1968).
[2]Water equivalent.

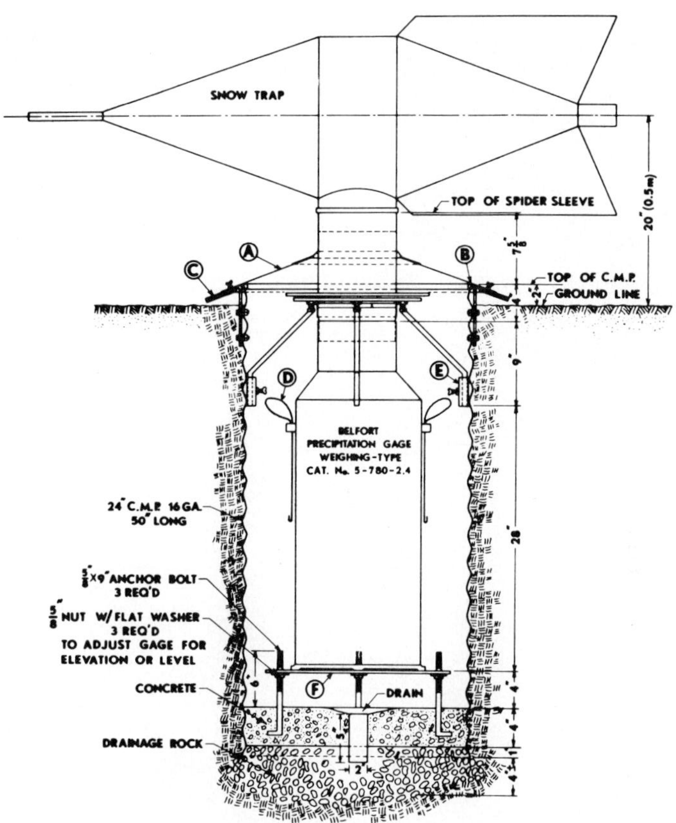

Fig. 8-15. Diagram of pit gage installation for a 0.5 m orifice height (from Jairell 1975a; used by permission, Amer. Geophys. Union).

season, the control stations are remarked, the snow depth measured at each of the control stations, and the area reflown. The snow elevations are determined from the aerial photographs and the volume is snow is computed. Although the snow depth over an area may be accurately determined by this method (Figure 8-16 and 8-17), considerable error could be introduced when calulating the total water content of the snow cover because of the variation in snow density over the study basin.

Continued study of photogrammetric snow measurements on the Reynold Mountain Study Basin (Northwest Watershed Research Center of the Agricultural Research Service, Boise, Idaho) have indicated that grid spacing could be increased from 25 feet to 100 feet or 325 feet (7.6 m to 30.5 m or 99.1 m) with a loss of accuracy in total volume of snow of only 2.5 percent and 10 percent, respectively. The above change in grid spacing enabled the number of points processed to be decreased by 94 and 99%, respectively. Continued evaluation of snow density on the watershed had indicated that density varies according to aspect and drift location. Some drift locations may have snow

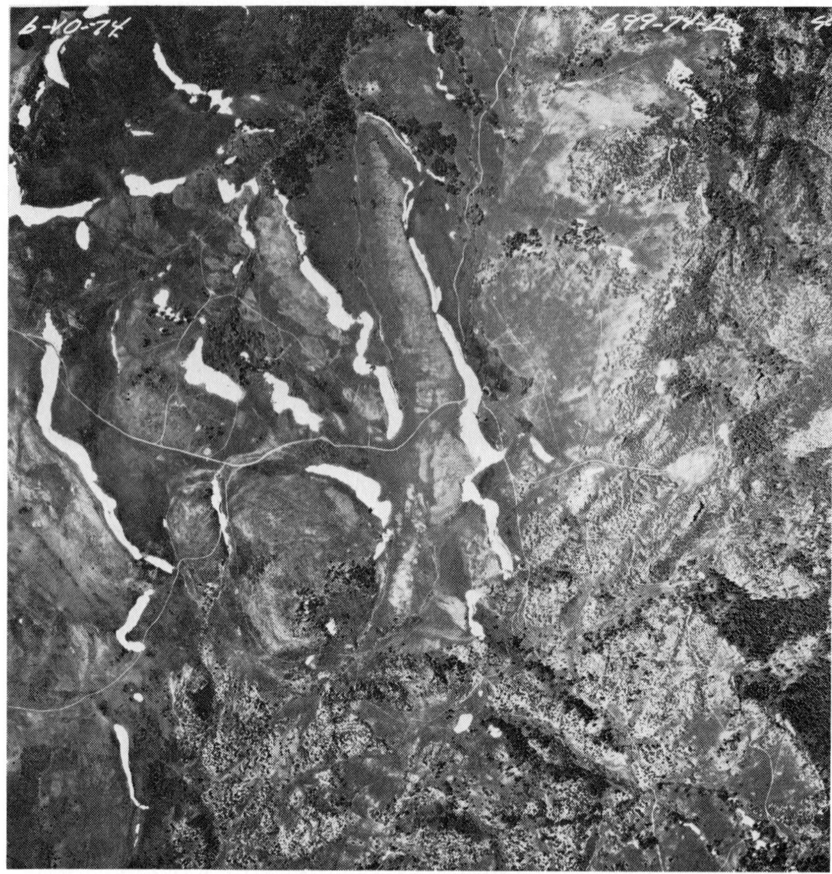

Fig. 8-16. Snowdrift sites, 6/10/74. Discontinuous snow storage as massive drifts is a unique feature on each watershed (Renard and Brakensick 1975).

density 10% greater than other sites.

Information regarding additional snow-related problems may be found in Leaf and Konner (1972), Haeffner and Barnes (1972), Hasholt (1972), Logan (1972), Bartos and Rechard (1973), Steppuhn and Dyck (1974), Young (1973), Young (1974), Larson, Ffolliott, and Moessner (1974), Tabler (1975a), Jairell (1975b), Steppuhn (1976) and Kerr (1976).

Evaluating Snowmelt Estimates

The vitality of the economy of many sections of the country largely depends on water derived from snow, which finds its way into streams and reservoirs in spring and early summer. Zuzel and Ondrechen (1975) state:

> Reliably forecasting the quantity and timing of snowmelt runoff are essential to the optimum control and management of the water resource, since these forecasts are used to make watershed management decisions concerning agriculture, conservation, municipal

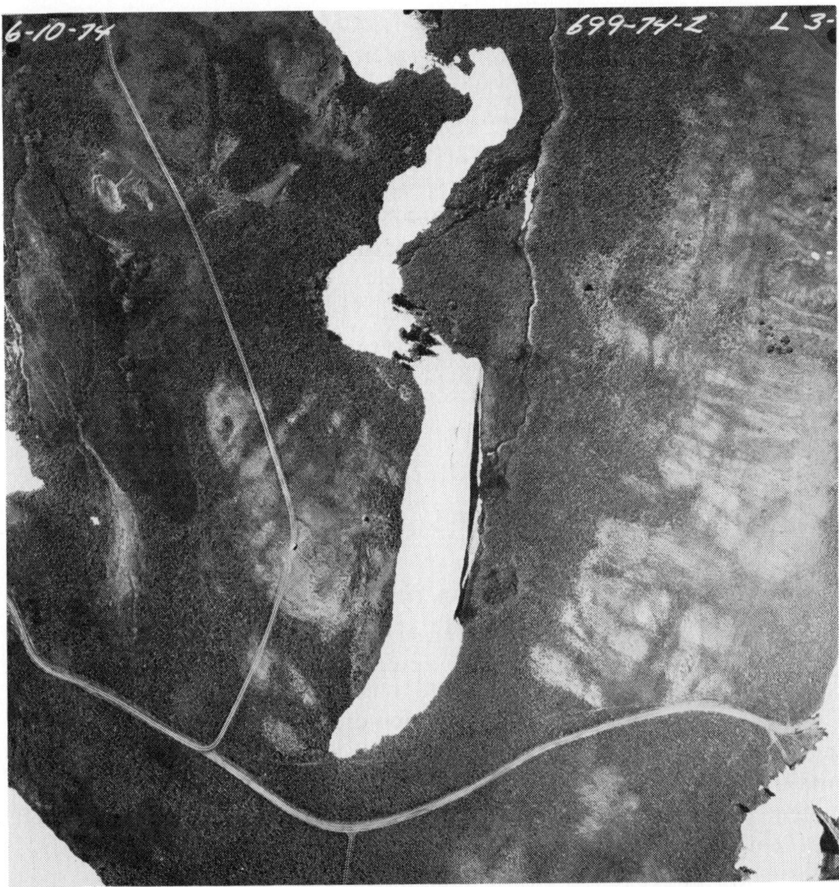

Fig. 8-17. Segment of Fig. 8-16 on 6/10/74. Lower fringes up to trees contain 23 acre-ft of water (28,383 m³) (Renard and Brakensiek 1975).

water, power generation, recreation, navigation, flood and pollution control.

For example, the annual January 1 flow forecast for the Columbia River of the Dalles, Oregon, is in error by about 12%. An estimated improvement of only 1% in the Columbia Basin forecast could return $6.2 million annually from improved management of power production and irrigation alone.

Presently forecasts are made about the first of each month, from January to June, by the Soil Conservation Service, National Weather Service, Bureau of Reclamation, Corps of Engineers, and the Bonneville Power Administration, as well as privately owned power companies. The forecast models now in use generally utilize snow water content at selected snow courses, and October through March precipitation totals from stations at low elevations in multiple regression equations. Since these forecasts are prepared by the various agencies they often differ greatly, and great uncertainties in reservoir operations result. These uncertainties naturally produce a less than acceptable management of the water resource.

The foregoing conclusions by Zuzel and Ondrechen (1975) were derived from a comparison of a forecast model developed by the Agricultural Research Service in Idaho (ARS-IDWR model) on the Boise River (drainage area of approximately 11,000 km² (4247 sq. miles) with forecasts made by three other models. The Corps of Engineers model is a complex procedure utilizing snow-water content from five snow course sites and six precipitation gages in a multiple linear regression scheme. The Bureau of Reclamation model also uses multiple linear regression with snow course water content, precipitation total and the antecedent natural flow of the Boise River. The Soil Conservation Service model is a simple linear regression, utilizing snow-water content from a selected snow course or the sum of water contents from several courses as the independent variable.

The ARS-IDWR model is a linear model of the type

$$Y = \alpha + \beta_1 X_1 = \beta_2 X_2 + \ldots. \beta_n X_n = \alpha + \sum_{i=1}^{n} (\beta_i)(X_i) \qquad \textbf{(8-5)}$$

where

$Y =$ runoff volume for a specific forecast period

$\alpha =$ a fitting coefficient

$X_1 \ldots X_n =$ snow water content at each of n snow courses

$\beta_1 \ldots \beta_n =$ coefficients

A general optimization program developed by the Tennessee Valley Authority (Green 1970) was used to generate beta coefficients and alpha values for various forecast periods. Tables 8-6 and 8-7 show typical errors in the forecsts of the four models for different periods within the water year. They also presented analyses of the models used in Table 8-6 for non-typical runoff years and found that the ARS-IDWR model had greater accuracy in both low and high runoff years. Allowing the coefficients applied to water content and precipitation values to vary with the forecast seems logical and appeared to improve the model predictions. They concluded that "further refinement will probably require additional inputs for which adequate data networks also do not exist."

Table 8-6. Comparison of Soil Conservation Service (SCS) published forecast and ARS-IDWR forecast errors for 1969 through 1972 for the Boise River at Lucky Peak Dam (Zuzel and Ondrechen 1975).

Forecast period	SCS published average error (%)	ARS-IDWR average error (%)
April–September	7 (3–16)	6 (1–12)
May–September	11 (3–21)	7 (2–11)

[1] Numbers in parenthesis indicate range of forecast errors.

Solomon, Ffolliott, and Thorud (1975) state:

> Many attempts have been made to model the snowmelt process with energy budget approaches (Federer 1968; Leaf & Brink 1973; Rantz 1964; Willen, Shumway and Reid 1971). A problem with many of these models is maintaining usability without requiring data inputs that are difficult to obtain. Ideally, a model should require as few input variables as possible, yet be capable of simulating the snowmelt process. A computer model developed to simulate Colorado snowmelt characteristics appeared to meet the requirements of easily measurable input, a desirable degree of simulation, and sensitivity to "key" variables (Leaf

Table 8-7. Comparison of average forecast error for forecast periods (Zuzel and Ondrechen 1975).

Forecast period	Years	Average forecast error in percent		
		Bureau of Reclamation	Corps of Engineers	ARS-IDWR model
January–July	17	20	22	17
February–July	25	14	15	12
March–July	25	12	11	10
April–July	25	9	9	6

and Brink 1973). However, as the model was originally designed to simulate Colorado snowmelt, some modifications were necessary to simulate the intermittent snowpack conditions that are common in Arizona. Thus, many of the subroutines in the original model have been altered or eliminated, and new subroutines have been added.

The modified computer program is dependent on four daily input variables: maximum temperature, minimum temperature, precipitation, and solar radiation. In addition, the following initializing values are required: initial snowpack temperature, a solar radiation

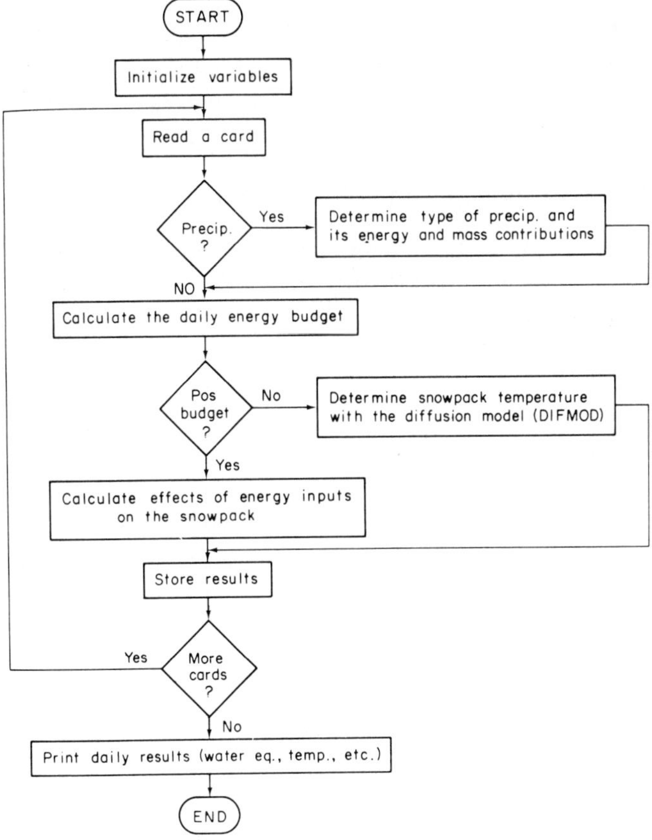

Fig. 8-18. General flow pattern for modified computer model of snowmelt process simulation (Solomon, Ffolliott, and Thorud 1975).

transmissivity coefficient, forest cover density, initial snowpack water-equivalent, a threshold value for calculating reflectivity, the average slope and aspect of the watershed, latitude, an atmospheric absorption coefficient, and a time interval for hour-angle solar changes.

A general flow pattern for the modified computer model is shown in Figure 8-18. With this computer model, Solomon, Ffolliott and Thorud (1975) estimated daily snowmelt runoff efficiencies and concluded:

> No one distinct pattern of snowmelt runoff efficiency throughout a season was apparent for the experimental watershed studied. Every year was unique, with some patterns demonstrating gradual transitions from the beginning to the end of the runoff season, while others possessed sharp breaks between the beginning, middle, and the end of the runoff season. But, it appeared that all patterns started with a greater rate of snowpack ablation than runoff, and the rate of runoff increased more rapidly than the rate of snowpack ablation as the season progressed, until about 70 to 80% of the snowpack had ablated, then the rate of runoff compared to the rate of ablation again decreased.

Schematically their finding is shown in Figure 8-19.

Solomon, Ffolliott and Thorud (1975) concluded for watersheds in Arizona that—although snowmelt runoff efficiency patterns were unique for each of the experimental watersheds, there did exist a general pattern of relatively low efficiencies at the beginning and end of a snowmelt-runoff season, with relatively high efficiencies occurring between 40 and 60% of seasonal snowpack ablation. Patterns of efficiencies throughout a season did not graduate in a smooth pattern, but rather fluctuated from day-to-day. These daily fluctuations were found to be positively correlated with daily snowmelt and

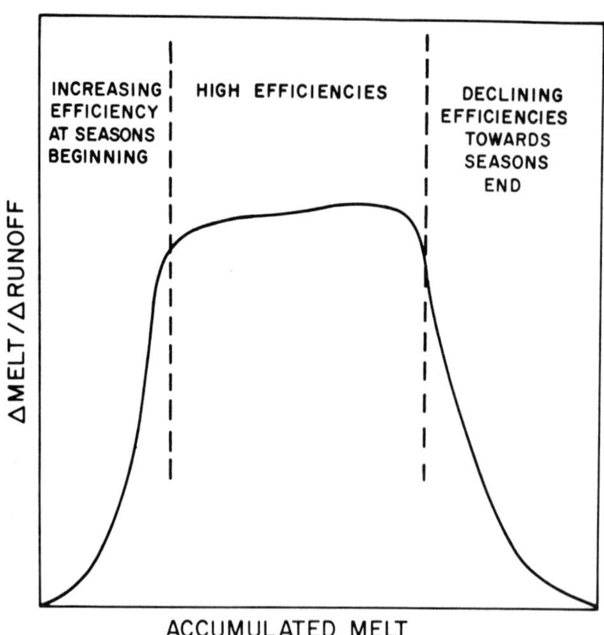

Fig. 8-19. A schematic diagram of the changing pattern of a normalized snowpack ablation-runoff curve (Solomon, Ffolliott and Thorud 1975).

runoff, with greater efficiencies being associated with high snowmelt rates and high volumes of runoff. Further, significant correlations between snowmelt runoff efficiencies and antecedent soil moisture, peak snowpack accumulation, and duration of runoff were found. These independent variables were significantly correlated with snowmelt runoff efficiencies within a watershed from year-to-year and among watersheds within a snowmelt-runoff season. Regression equations were developed with the use of these and other variables to provide a basis for predicting efficiencies from year-to-year.

Opportunities for Snowpack Management

Vegetation Manipulation

As previously indicted, the potential for manipulating vegetation to enhance snow accumulation on rangeland landscapes does not usually exist, except perhaps in the form of windbreaks. In those plant communities where snow is important, the vegetation can be effective in accumulating snow, but only until the surface becomes smooth again. Hyder and Sneva (1956) found that live sagebrush *(Artemisia tridentata)* plants and skeletal remains of sprayed plants were more effective than herbaceous vegetation in retaining snow. Near Dubois, Wyoming, Hutchinson (1965) found that a stand of sagebrush about 50 cm (20 inches) high contained an additional inch of water by the time the brush was covered compared with adjacent grassland. Sturges (1975) indicates that sagebrush conversion reduces snow accumulations most strongly on windward slopes, since the loss of sagebrush foliage opens the site to the scouring action of wind. Once vegetation is covered, further deposition becomes primarily a function of prevailing wind direction in combination with topographic locations.

Within the snow zone of the Sierra Nevada west side, Wilken (1967) found that brush fields fields of greenleaf manzanita *(Arctostaphylos patula)* have the potential for filling from the ground up to form fairly solid packs. Voids or open spaces make up a large part of newly deposited brush field snowpacks, but tend to disappear in interstorm periods of consolidation.

Swanson and Stevenson (1971) have studied the management of snow accumulation and melt under leafless aspen *(Populus tremuloides)* on rangelands of southern Alberta which are generally mixed aspen-willow and grass. They found that small openings in the leafless canopy were effective snow traps, with an average of one-third more snow accumulated in small openings than under the surrounding leafless canopy. Contrary to most results in coniferous forests, snow within the openings ablated 30% slower than under the leafless canopy. They speculate that it should be possible to enhance groundwater recharge through manipulation of aspen stands growing on recharge areas to increase snow accumulation and retard ablation.

Grazing has an impact on vegetation and as a result may have an impact on snow accumulation and ultimate soil moisture recharge. Wight, Neff, and Siddoway (1975), citing unpublished data, reported on a study of snow accumulation on plots simulating grazing intensities created by 1) clipping a native grass to a uniform 7.6 cm (3 inches) height (smooth), 2) an intermediate roughness with patches of natural grass interspersed with patches of grass clipped down to 5 cm (2 inches), and 3) a natural grass cover 30 to 40 cm (12 to 16 inches) tall. Snow depth measurements following two storms indicated greater snow depths and a greater total snow cover on the untreated (rough) plots as compared to the intermediate and smooth plots. However, no measurable relationship was noted between snow accumulation and changes in soil water content from fall to spring. Galbraith (1971) and Van Haveren and Striffler (1976) have studied various aspects of the water balance of ungrazed and grazed watersheds on the shortgrass prairie in northeastern Colorado. In both instances, increased snow retention on ungrazed rangeland, with its greater vegetation height, was responsible for increased soil water storage under this treatment. Magnitudes of the treatment differences, however, were not great.

Modifying the Energy Budget

Snow albedo modifications

Examination of the snowpack energy budget reveals that atmospheric radiation, sensible heat exchange, and emitted radiation (longwave) are not readily susceptible to manipulation by man. However, control of latent heat exchange and shortwave radiation (through modification of the albedo) are aspects that may be modified. Slaughter (1969) has reviewed the literature in this respect, and the reader is referred to that publication for details. In essence, modification of snow albedo is usually done for the purpose of increasing snowmelt rates. The modification results in some artificial darkening of the snow surface (reduced albedo) thereby enhancing melt rates. Most of these techniques have been applied with respect to urban and agricultural problems with little practical application thus far to wildland situations.

Modifying evaporation from snow.

Evaporation from the snow surface involves a change-of-state. The snow, upon reaching 0° C, changes from a solid to a liquid and then to a gas (water vapor). Snow at below freezing temperatures may change directly from the solid to the vapor state, this process being referred to as sublimation. Snowpack evaposublimation has been noted by Van Haveren (1975) to include five different mechanisms summarized as:

1. Sublimation of ice grains at the snowpack surface,
2. Evaporation of hygroscopic meltwater at the snowpack surface,
3. Diffusion of water vapor from within the snowpack,
4. Free convection of water vapor, and
5. Forced convection of water vapor.

All five processes are probably common for the dry, porous snow of continental areas. Diffusion is probably the slowest process and results in the lowest water vapor loss from the snowpack. Free convection occurs when masses of void air are displaced upwards from the lower layers of the snow cover, transporting both heat and water vapor. Displacement is caused by very steep temperature gradients in the snowpack which results in instability of the snowpack air mass. Forced convection is the mass flux of snowpack air which occurs at and just below the snow surface, and is caused by pressure fluctuations which originate above the surface.

As indicated in a review by Slaughter (1970), evaporation seemingly varies from "insignificant" upwards to 40% of the total melt, depending on study location, methods used, etc. Tabler (1973) has shown in Wyoming that upwards of 83% of drifting snow returns to the atmosphere when natural or artificial barriers are spaced at 3.0 times the average transport distance (the average distance required for a drifting snow particle to sublimate). During the ablation period evaposublimation from snowdrifts under oasis conditions in Wyoming approximates 20 to 30% of the water equivalent volume of the drift at the start of ablation (Rechard and Raffelson 1974). Meiman and Grant (1974) found in Colorado that evaporation losses from mountain snowpacks amounted to approximately 60% of the snow season precipitation in an alpine setting and to approximately 45% in both a forest and a forest opening. Despite research extending back into the 1870's, Slaughter states the following: "Although progress has been made and is being made in this field, there is still essentially as much latitude for research on snow evaporation as was noted three decades ago."

Compounds most commonly used in evaporation retardation by monolayers are hexadecanol and octadecanol. Neither compound is soluble in water, and both are straight-chain primary alkanols which, on a water surface, become oriented in a monomolecular layer with the polar (hydroxl) ends toward the water (Koberg, Cruse, and Shrewsbury 1963). However, as pointed out by Slaughter (1970), control of evaporation from snow has received little attention in the literature. An example of the potential for evaporation reduction from snow is that of Anderson et al. (1963), some of their results being given in Table 8-8. In their case, hexadecanol was applied as a 2-½% emulsion at 11 kg/ha (10 lb/acre). As can be seen from the table, reduction of evaporation was greatest where the evaporation potential was considered to be the highest, i.e., the exposed ridge site.

Avalanching

Avalanching is a technique whereby snow is concentrated in valley bottoms or cirques for the purpose of making a fresh storage area for new snow, for possible treatment of the accumulated snow to reduce evaporative losses, for concentrating eventual melt waters, and for prolonging the melt season. Though some of this takes place indirectly through avalanche control measures on winter recreation sites, the technique has not been used to any extent as a management tool. Meiman and Grant (1974) studied the technique as a management tool in the Colorado Rockies and concluded that periodic removal of ridgeline deposits with the snow being relocated in cirques (in this instance) offered a good potential for increasing water yields (primary through reduced vapor losses from the snowpack).

 Table 8-8. Evaporation from snow with and without hexadecanol (HD)[1] central Sierra Nevada, California[2].

Site and treatment	Evaporation of water[3] 8 am–5 pm (mm)	5 pm–8 am (mm)	Total	Reduction by hexadecanol (%)
12 days totals: 12–19 April, 2, 31 May and 2–3 June 1962				
Center of large open meadow				
Without HD	5.3	0.1	5.4	
With HD	1.9	-0.2	1.7	
Reduction	3.4	0.3	3.7	68
Center of opening in forest, 2 tree heights across:				
Without HD	5.2	-0.9	4.3	
With HD	2.3	-1.1	1.2	
Reduction	2.9	0.2	3.1	73
4-day totals: 29–31 May, 3 June 1962				
Exposed ridge site				
Without HD	2.0	0.0	2.0	
With HD	0.5	-0.3	0.2	
Reduction	1.5	0.3	1.8	90
Center of large open meadow:				
Without HD	2.3	-0.5	1.8	
With HD	0.5	-0.3	0.2	
Reduction	1.8	-0.2	1.6	87

[1]Hexadecanol applied at 2-½% emulsion, 11 kg/ha (12 lb/acre); snow changed daily at 5 p.m. in plastic pans set into the snowpack.
[2]From Anderson, et al. (1963).
[3]Minus sign indicates condensation.

Snow Fencing

The possibilities of using snow fencing as a management tool for enhancing snow accumulation and reducing evaporative losses from snow on rangelands are increasing, though current usage is primarily for protecting highways, railroads, etc., from excessive drifting. Since snow transport has already been discussed, this section will deal primarily with fence design. Martinelli (1973) indicates there are three basic types of snow fenches: 1) collecting fences—solid or porous barriers that slow the wind and allow the snow to settle out, 2) blower fences—solid barriers that accelerate the wind in local deposition areas so the blowing snow will settle elsewhere, and 3) guide walls—vertical solids barriers aligned to steer snow-laden winds past objects to be protected from snowdrifts. Only collecting fences are of interest at this point.

In general, when snow-bearing wind meets an obstacle, such as a snow fence, eddies are formed on the leeward side of the obstacle. The snow particles passing either over or through the fence follow these eddies. As mean forward velocity is reduced, the particles are deposited to the leeward side of the fence. The resulting drift increases in size until it fills the eddy area, after which the fence is saturated, and further particles will be deposited beyond the eddy area. The length of the eddy area depends essentially on the height of the fence and on its density ratio, but is nearly independent of wind speed at naturally occurring speeds. Density ratio is defined as the ratio of frontal area of the solid part of the fence to the total frontal area, a close-boarded fence having a density ratio of 1.0. Figure 8-20 shows typical eddy patterns around a solid fence and an open fence and Figure 8-21 shows the effect of density ratio of fence on profiles of snow drifts at four stages in their growth. The optimum density ratio appears to be about 0.5 (Hogbin 1970).

In designing a snow fence system, Martinelli (1973) states that the first step is to formulate the objective of the project. He goes on to describe fence design and layout, and the following is extracted from his paper:

> Once the basic objectives have been set, it is helpful to estimate the amount of snow blowing past the site. This may be done from measurements of snowdrifts near the site plus an estimate of the trapping efficiency of the site or from equations of snow transport as a function of precipitation, contributing distance, and terrain features. . .
>
> Useful estimates of the amount of blowing snow arriving at a given site can be made from equations such as the following:
>
> $$Q = 1/2\ \theta P R_m - q_s$$
>
> Where Q is snow transport in cubic meters of water per lineal meter of cross-wind distance, θ is snow transport coefficienct (amount of precipitation redeposited by wind), P is water equivalent of winter precipitation (meters), R_m is maximum distance (meters) the average snow particle travels before sublimating completely (maximum contributing distance) (Figure 8-22), and q_s is the amount of snow (m^3 water/m distance) trapped by upwind terrain features.
>
> A theoretical analysis by Schmidt (1972) has shown R_m to be related to particle size, particle speed, relative humidity gradient near the snow surface, air temperatures, and total insolation. . .
>
> After objectives have been set and blowing snow estimates made, fence design and layout can be determined from a sizable fund of knowledge relating certain fence variables to snow accumulation. Some of the most helpful of this information is summarized below. . .
>
> 1. *Fence height (H)*—Combines with fence density, bottom gap, and terrain features to determine drift size and shape. Price (1961) states that drift profiles at saturation are approximated by part of one petal from a rose equation using fence height and density.

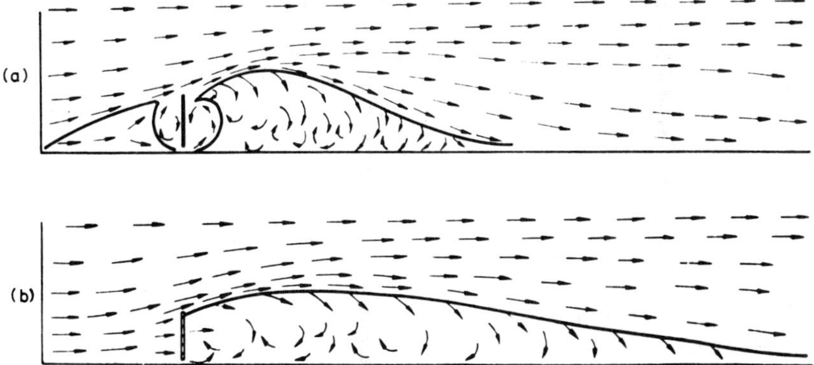

Fig. 8-20. Typical eddy patterns and drift profiles about (a) a solid fence (b) an open space (from Hogbin 1970).

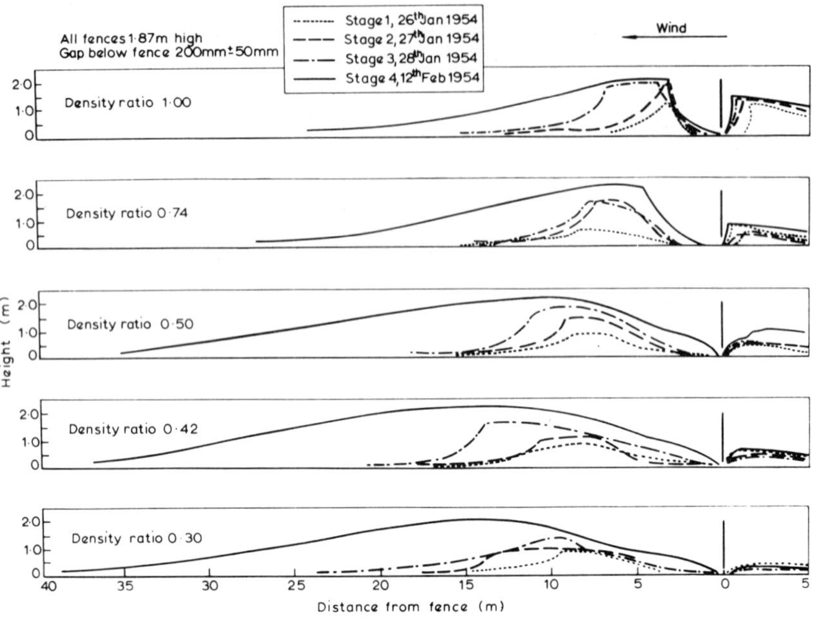

Fig. 8-21. Effect of density ratio of fence on profiles of snow drifts at four stages in their growth (from Hogbin 1970). (Multiply m by 3.28 to obtain ft. Divide mm by 25.4 to obtain inches).

Tabler[1] confirms this and suggests drift cross-section (A) prior to saturation is related to H by a sigmoidal growth function of the form

$$A = a\,(1 - e^{-bH})^{c}$$

where e is the base of natural logarithms and a,b, and c are parameters that control the upper asymptote and rate of approach to this asymptote. They can be related to weather, fence, and site conditions. There is some evidence that trapping efficiency improves with H up to about 4-½ meters which is near the practical upper limit of H.

2. *Fence density*—This is expressed as the percent of total frontal area of the fence made up of solid material. Solid barriers (100 percent density) give short, deep drifts. More permeability fences give longer, shallower drifts. Optimum density in terms of the area of lee drift is between 40 and 60 percent. Limited studies suggest the size and shape of the openings in permeable fences are unimportant so long as the openings and solid parts are about equal in size and do not exceed 25 cm in width.

3. *Bottom gap*—This gap between the fence and ground reduces windward drift, keeps the lee drift away from the fence, and makes the lee drift longer. Optimum gap size for smooth level terrain is 0.1 H to 0.15 H, where H is fence height. Larger gaps may be needed in irregular terrain to avoid windward drifts that plug the gap.

4. *Length of the lee drift*—Function of fence height, fence density and gap size. For a fence with a 0.1 H gap, the saturated lee drift will be about 10 H long for a fence with 100 percent density and 15 H to 17 H for 50 percent density.

5. *Maximum depth of lee drift*—Varies from 0.9 H to 1.1 H. Moves downward as fence density decreases from 100 percent and as gap width goes from zero to 0.15 H. Located 2 H to 3 H behind fence for 100 percent density and about 8 H for 40 percent density fence with gap of 0.1 H.

6. *Fence length and continuity*—Fence should be at least 20 H long and should have no breaks because they act as nozzles that erode the lee drift.

[1]Tabler, R.D., personal communications.

7. *Inclination of the fence from vertical*—Although fences can be tilted as much as 30° from vertical without marked disadvantage. . .most are built vertical because of the ease of construction and economy of materials. In some types of terrain, a 10° to 15° downward tilt improves stability enough to be worthwhile.

8. *Cumulative effect of a set of tandem fences*—There appears to be a slight gain over single fence efficiency for two fences set 2 H apart. . . There is a loss from single fence efficiency for two or more fences set 10 H apart. Single fence efficiency is maintained for fences set 20 H or more apart.

9. *Terrain effect*—In gently rolling terrain, fence storage capacity can be increased by placing the fence upward of a topographic depression. In steep terrain, fences on windward slopes have less storage capacity, while those on lee slopes have more storage capacity than comparable fences in level terrain. Fences on lee slopes, however, are quite subject to burial and damage from snow settlement.

10. *Contributing distance*—Theoretically, this can vary from a few meters to infinity. Recent studies in rolling Wyoming grasslands, however, show average values from 1,000 to 2,000 meters. . .

Martinelli (1973) gives an example of how the above design criteria were actually used on a snow fence project on Mount Bethel, 96 km (60 miles) west of Denver, Colorado:

The objective of this project was to trap blowing snow on avalanche-free terrain and thus reduce the amount of snow blown into the avalance starting zone. About 300 meters of collecting

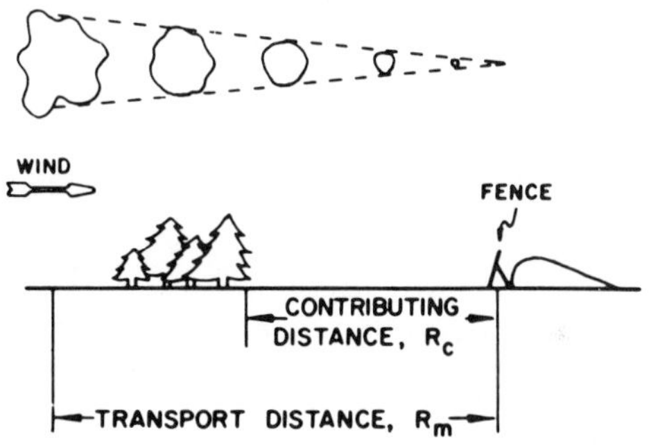

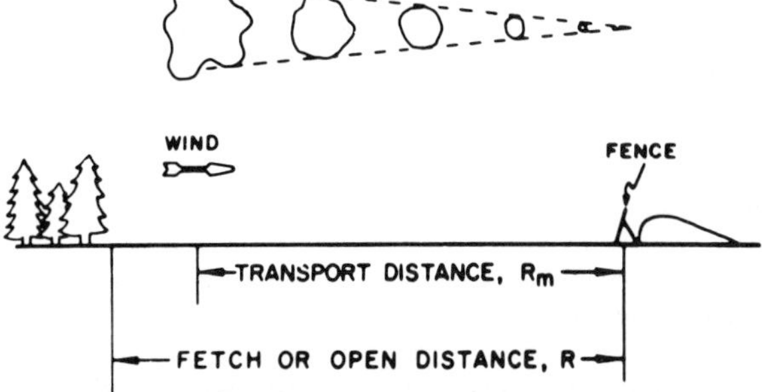

Fig. 8-22. Diagram of transport distance concept used to estimate sublimation loss from wind transported snow (from Rechard 1973).

fences was to be used in the initial effort.

Snow transport (Q) at the site was estimated to be 300 m³ of water per lineal meter, based on an average November-April precipitation (P) of 0.6 m, contributing distance (M_m) of 1,000 meters and assuming snow transport coefficient (θ) to be 1 and upwind trapping (q_s) to be zero.

The fences were designed considering the following facts:

(1) the only practical access was by helicopter,

(2) the steep, rocky terrain and fragile plant cover argued against dug post holes and anchors,

(3) proximity to a heavily used highway posed an aesthetic problem, and

(4) strong winds demanded sturdy construction.

The fences used were 3.65 m tll, tilted 15° downwind, had a density of 50 percent, and a gap of 0.9 m, and the horizontal fence elements and openings were 33 cm wide. The large bottom gap was used to keep the fence from becoming engulfed in its own drift... an important consideration at this steep site. Panels 4.1 m long were assembled at a roadside site and heli-lifted to the mountain. Each panel had front and back cable guys and a stiff back brace that gave good support without any dug holes. Quick-rusting steel was used to help camouflage the fence from the highway.

Layout of the fences was another problem. Our estimates of drift snow potential (300 m³ of water/m of fence) and storge capacity of this type of fence 47 m³ of water/m of fence)... showed a tandem array of six fences each 250 to 300 m long would be needed to trap all the blowing snow in an average winter. Although blowing snow crosses the entire 350-meter windward edge of the starting zone, most avalanches start below the middle section of this windward edge. The best use

NATURAL MOUNTAIN CLOUDS

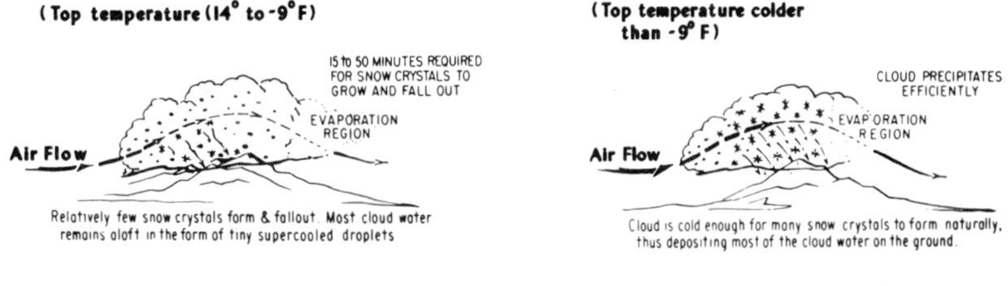

SEEDED MOUNTAIN CLOUDS

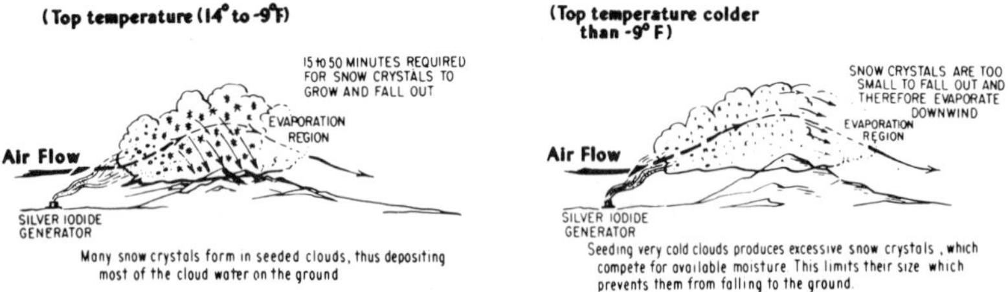

Fig. 8-23. The objective of orographic cloud seeding is to increase the number of snow crystals in naturally deficient clouds in order that all available cloud water will be utilized to grow the crystals. These will fall to the ground thus depositing most of the cloud water. The silver iodide particles transform many of the supercooled cloud droplets into ice crystals which then attract moisture from surrounding droplets (due to the vapor pressure gradient) and grow large enough to fall to the ground as snow (from Dep. of the Interior 1973).

of the limited amount of fencing seemed to be two tandem fences upward of the most serious avalance starting points, rather than a single fence along the entire windward edge. To allow for the larger-than-normal bottom gap and the likelihood of completely saturated drifts, the two fences should have been spaced about 75 meters (20 H) apart with the downwind fence about the same distance from the edge of the avalance. This spacing was possible at the lower end of the fences, but rock required shortening it to 45 m (12 H) at the upper end. The fence closest to the avalanche path was made a little longer than the other to allow for variations in local wind direction.

At maximum efficiency, these fences should trap about one-third of the snow normally blown into the avalanche starting zone.

Besides in Wyoming and Colorado, snow fencing has also been shown useful in central Oregon (Swank and Booth 1970) and in eastern Montana (Saulmon 1973) for increasing water yield through induced snowdrifts. Additional information regarding snow fencing may be obtained by consulting Mellor (1965), Schneider (1962), Pugh (1950), Tabler and Johnson (1971), Tabler (1968, 1971, 1975b), Berndt (1964), Dyunin (1959), and Schmidt (1970).

Contour Trenching, Furrowing, and Dikes.

The only study of the impact of contour trenches on snow accumulation has been that of Doty (1970) on the Davis County Experimental Watershed near Farmington, Utah. The trenches studied are on a windswept southwest exposure where snow redistribution by wind is important. The trenches increased snow accumulation slightly, which appeared to be more significant to revegetation than to water yield.

Contour furrowing was evaluated by Wight and Neff (unpublished data cited by Wight, Neff, and Sidoway 1975) as a snow management practice. They measured snow accumulation and subsequent soil water recharge on furrowed panspot and saline upland range sites in Montana. The contour furrows were 2.5 to 4 cm deep and 8 to 10 cm wide on 1.6 m centers. Over a 5-year period contour-furrowed areas trapped an average of 6.4 cm/yr snow water as compared with 4.1 cm/yr in nonfurrowed areas. Nearly all of the snowmelt was lost as runoff on nonfurrowed areas, whereas only about half was lost on furrowed areas.

In Canada, Hubbard and Smoliak (1953) evaluated contour dikes and single contour furrows as a method spreading water, particularly snowmelt runoff, on rangelands. Contour dikes were effective in spreading water and increasing forage production. However, the furrows, which were only 4 to 5 inches (10 to 13 cm) deep, became filled with ice and snow during the winter and were of no value in spreading water.

Windbreaks

Field windbreaks (single or multiple-row tree plantings) were originally established in the Great Plains to prevent wind erosion. However, the potential for snow spreading and perhaps increased production was soon noted. Frank and George (1975) indicate that to best manage snow in field windbreaks, the objective should be to trap snow over as wide a cropping area as possible for soil water recharge and for subsequent crop production. They cite several investigators who have described snowpack profiles adjacent to and within single and multiple-row windbreaks (Stoeckler and Dortignac 1941, George 1943, Potter, Longwell, and Mode 1952, Willis et al. 1961, Stoeckler 1962, George 1971, McMartin, Frank and Heintz 1974). Frank and George further comment that there is no simple solution to managing snow catch in

SEEDING EFFECTS (Small Clouds)

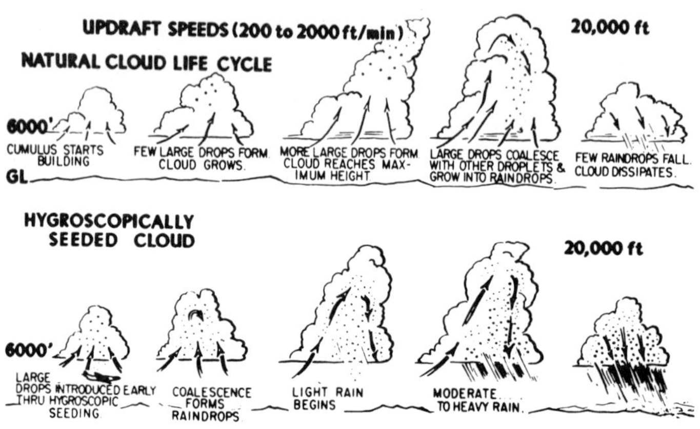

SEEDING EFFECTS (Large Storm)

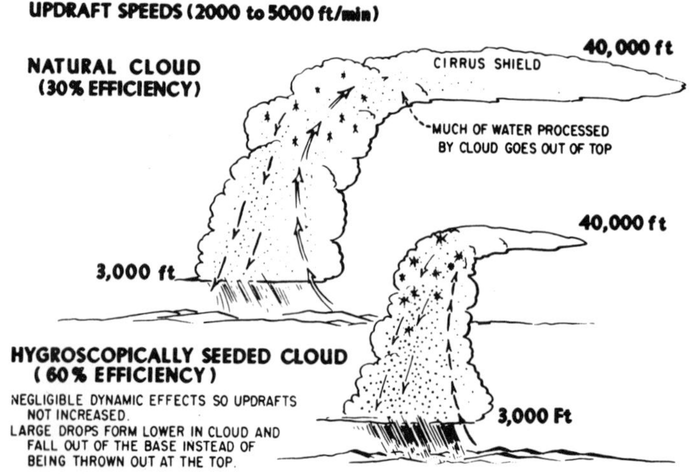

Fig. 8-24. Seeding effects on small clouds and large storms, both of a convective nature (from Dep. of the Interior 1973).

windbreaks. Large snowpack formation which creates the potential for water erosion under certain conditions and the inconvenience of late seeding of crops in the snowpack area can be lessened if: 1) windbreaks are not planted on land where snowmelt runoff may cause soil erosion and 2) species are selected that have a low winter density or self-pruning characteristic.

Weather Modification

The hydrologic impact of weather modification is potentially significant. Any changes in rainfall can affect not only soil moisture and reservoir storage, but peak rates of flow, sediment and chemical transport, dissolved solids, and possibly even the regime of a river. Seely and DeCoursey (1975) have provided a brief review of hydrologic impact studies to date, and they state the following:

> Several important points must be considered in evaluating the hydrologic impact of weather modification. The effect of weather modification on rainfall is not as precise as reports in newspapers and some scientific journals have indicated. Impact on "secondary" areas, such as hydrology, could be very important if the performance claims of many weather modification programs are true. All programs should have an impact statement to anticipate hydrologic effects, which could be severe, and such considerations should be included in planning the programs.
>
> The hydrologic impact should be estimated based on the region in which the program is targeted and on the likely effects of the program on rainfall. The need for such estimates at the planning stage and the time and cost involved will generally dictate use of mathematical hydrologic models either in planning or in providing extrapolations to other areas.
>
> Several important types of uncertainty should be considered—the uncertainty of the

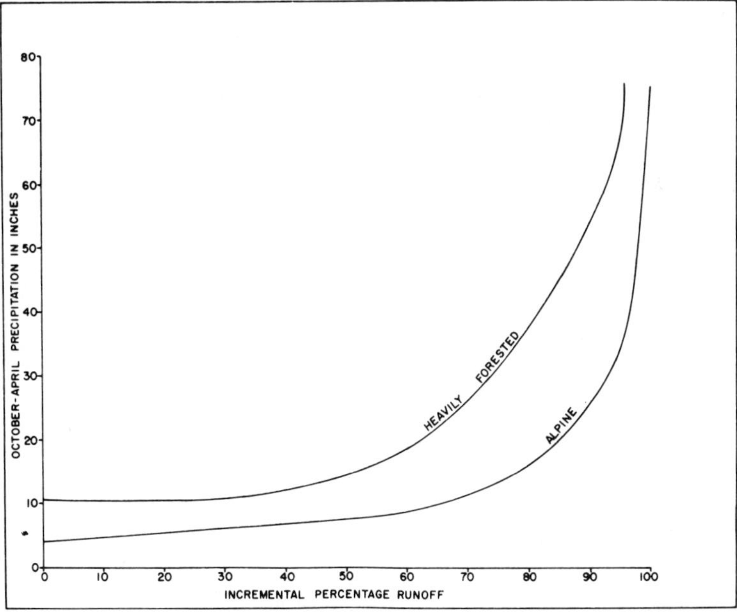

Fig. 8-25. Precipitation-runoff relationship for the Donner Mountain mass in the Sacramento major basin. The same general relationship holds for the Upper Colorado mountains (from Dep. of the Interior 1975).

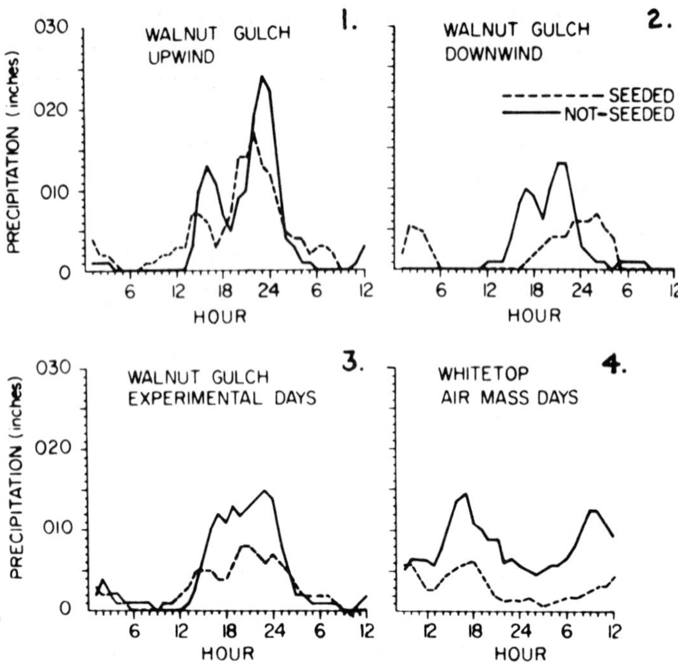

Fig. 8-26. Diurnal changes in hourly precipitation amounts: 3 hour moving averages (from Neyman and Osborn 1971). (Multiply inches × 2.54 to obtain cm).

effect that weather modification will have on rainfall; the uncertainty of the hydrologic effect of different strategies; and the uncertainty of conclusions based on hydrologic models, because of error in the model and their parameters.

Most weather modification efforts within the western United States concentrate on cold orographic cloud systems during the winter season in an effort to augment winter snowpacks (Figure 8-23 and 8-24). An example of this effort is the Winter Orographic Snowpack Augmentation program (WOSA) which offers a relatively inexpensive method of increasing the water supply in the Colorado River Basin (Stanford Research Institute 1974). For a cost of perhaps $5.4 million per year, the project is expected to generate about 2,837 million m^3 (2.3 million acre-ft) of runoff within the Basin and 1,480 million m^3 (1.2 million acre-ft) outside it. Runoff from incremental precipitation in the Upper Colorado mountains increases substantially with increases in elevation both as a result of increased precipitation subject to treatment and increased efficiency of runoff in alpine regions (U.S. Dep. of the Interior 1975) (Figure 8-25). Interestingly enough the environmental, economic, and social impacts of WOSA on the high mountain areas where it will be operated are uniformly minimal. The water produced in the mountains would be of no benefit to local residents but would be used by downstream desert states.

Chappel (1972) has indicated that snowfall augmentation ranging from 10–30% of the wintertime precipitation is possible over many of the mountain ranges of the western United States.

Weather modification research has also been conducted on summer cumulus type cloud patterns but the success in such endeavors has been minimal with reported instances of decreased downwind precipitation catch as far as 100 miles (161 km) from the target areas. Neyman and Osborn (1971) reported evidence of significant apparent effects of local seeding on precipitation at distant areas, found for Grossversuch III, for Whitetop Project, and for the Arizona experiments is a strong argument for more careful experimental design. They showed significant decreases on days where Walnut Gulch was downwind from the Santa Catalina Mountain seeding area (Figure 8-26). Another excellent treatment of the problems of experimental design in weather modification is Neyman and Scott (1972).

Literature Cited

Anderson, E.A., and N.H. Crawford. 1964. The synthesis of continuous snowmelt runoff hydrographs on a digital computer. Dep. of Civil Eng., Standford Univ., Tech. Rep. No. 36 103 p.

Anderson, H.W., and T.H. Pagenhart. 1957. Snow on forest slopes, p. 19-23. *In:* Proc. 25th Annu. Western Snow Conf., April 17-19., Santa Barbara, Calif.

Anderson, H.W., and A.J. West. 1965. Snow accumulation and melt in relation to terrain in wet and dry years, p. 73-82. *In:* Proc. 33rd Annu. Western Snow Conf. April 20-22. Colorado Springs, Colo.

Anderson, H.W., A.J. West, R. Ziemer, and F.R. Adams. 1963. Evaporative loss from soil, native vegetation and snow as affected by hexadecanol. Int. Ass. Sci. Hydrol. Pub. No. 62: 7-12.

Bartos, L.R., and P.A. Rechard. 1973. Snow sampling techniques on a small subalpine watershed, p. 52-61. *In:* Proc. Western Snow Conf., April 17-19, 1973, Grand Junction, Colo.

Bergen, J.D. 1968. Some measurements of air permeability in a mountain snow cover. Bull. Int. Ass. Sci. Hydrol. 13: 5-13.

Berndt, H.W. 1964. Inducing snow accumulation on mountian grassland watersheds. J. Soil and Water Conserv. 19: 196-198.

Bissell, V.C., and Z.G. Burson. 1974. Deep snow measurements suggested using cosmic radiation. Water Resources Res. 10: 1243-1244.

Bissell, V.C., and E.L. Peck. 1973. Monitoring snow water equivalent by using natural soil radioactivity. Water Resources Res. 9: 885-890.

Blincow, D.W., and S.C. Dominey. 1974. A portable profiling snow gage, p. 53-57. *In:* Proc. 42 Annu. Western Snow Conf., April 16-20, Anchorage, Alaska.

Chappell, C.F. 1972. Orographic cloud seeding as a water resource, p. 389-397. *In:* Age of Changing Priorities for Land and Water, Proc. Irrig. and Drainage Div. (Amer. Soc. Civil Eng.) Spec. Conf., Spokane, Wash., Sept. 26-28.

Church, J.E. 1912. The conservation of snow: its dependence on forests and mountains. Sci. Amer. Supp. 74: 152-155.

Cooper, C.F. 1965. Snowcover measurement. Photogrammetric Eng., July: 611-619.

Court, A. 1963. Snow cover relations in the Kings River Basin, California. J. Geophys. Res. 68: 4751-4761.

Curry, G.E., and A.S. Mann. 1965. Estimating precipitation on a remote headwater area of western Alberta, p. 58-66. *In:* Proc. 33rd Annu. Western Snow Conf. April 20-22, Colorado Springs, Colo.

Doty, R.D. 1970. Influence of contour trenching on snow accumulation. J. Soil & Water Conserv. 25: 102-104.

Dyunin, A.K. 1959. Fundamentals of the theory of snowdrifting. Nat. Res. Council of Canada, Ottawa, Tech. Transl. 952 (1961). 26 p.

Eggleston, K.O., E.K. Israelson, and J.P. Riley. 1971. Hybrid computer simulation of the accumulation and melt processes in a snowpack. Utah State Univ., Utah Water Res. Lab. Rep. PRWG65-1. 77 p.

Federer, C.A. 1968. Radiation and snowmelt on a cleared watershed, p. 28-42. *In:* Proc. Eastern Snow Conf. Boston, Mass. February.

Frank, A.B., and E.J. George. 1975. Windbreaks for snow management in North Dakota. p. 144-154. *In:* Snow Management on the Great Plains, Great Plains Agr. Council Publ. 73, Univ. Nebraska, Lincoln.

Galbraith, A.F. 1971. The soil water regime of a shortgrass prairie ecosystem. Ph.D. Diss., Colo. State Univ., Ft. Collins. 127 p.

Gary, H.L., and G.B. Coltharp. 1967. Snow accumulation and disappearance by aspect and vegetation type in the Santa Fe Basin, New Mexico. U.S. Forest Serv. Res. Note. RM-93.

George, E.J. 1943. Effects of cultivation and number of rows on survival and growth of trees in farm windbreaks on the northern Great Plains. J. of Forest. 41: 820-828.

George, E.J. 1971. Effect of tree windbreaks and slat barriers on wind velocity and crop yields. U.S. Dep. Agr. Prod. Res. Rep. 121. 23 p.

Grant, L.O., and R.A. Schleusener. 1961. Snow fall and snow fall accumulation near Climax, Colorado, p. 53-64. *In:* Proc. 29th Annu. Western Snow Conf. April 11-13, Spokane, Wash.

Gray, D.M., and A.D.J. O'Neill. 1974. Application of the energy budget for predicting snowmelt runoff, p. 108-118. *In:* Advanced Concepts and Techniques in the Study of Snow and Ice. Resources, Symp., Monterey, Calif., Dec. 2-6, 1973, Nat. Acad. of Sci. Washington, D.C.

Greb, B.W. 1975. Snowfall characteristics and snowmelt storage at Akron, Colorado, p. 45-64. *In:* Snow Management on the Great Plains, Great Plains Agr. Council Pub. 73, Univ. Nebraska, Lincoln.

Green, R.F. 1970. Optimization by the pattern search method. Tennessee Valley Authority Res. Pa. 7.

Haeffner, A.D., and A.H. Barnes. 1972. Photogrammetric determinations of snow cover extent from uncontrolled aerial photographs, p. 319-340. *In:* Amer. Soc. Photogramm. ASP Tech. Sess. Proc., Columbus, Ohio.

Hamon, R.W. 1972. Computing actual precipitation. Symp. World Meteorlogical Organization.. Int. Ass. Sci. Hydrol. Geilv, Norway. 1-15.

Hasholt, B. 1972. Random sampling techniques applied in measuring snow water equivalent in a drainage basin. Int. Symp. on the Role of Snow and Ice in Hydrol. Banff, Alberta, Canada. 7 p.

Haupt, H.F 1951. Snow accumulation and retention on ponderosa pine lands in Idaho. J. Forest. 49: 869-871.

Heinemann, L.R., V.I. Myers, and D.G. Moore. 1972. Snow and ice melt as indicators of hydrologic conditions—exploratory study. Remote Sensing Inst., S. Dak. St. Univ., Brookings. Interim. Tech. Rep. RSI-72-01. 19 p.

Hogbin, L.E. 1970. Snow fences. Road Res. Lab. (Ministry of Transport), Crowthorne, Berkshire, England, England. Rep. LR 362, 13 p.

Hubbard, W.A., and S. Smoliak. 1953. Effect of contour dykes and furrows on short-grass prairie. J. Range Manage. 6: 55-62.

Hutchinson, B.A. 1965. Snow accumulation and disappearance influenced by big sagebrush. U.S. Forest Serv. Res. Note RM-46. 7 p.

Hyder, D.N., and F.A. Sneva. 1956. Herbage response to sagebrush spraying. J. Range Manage. 9: 34-38.

Jairell, R.L. 1975a. An improved recording gage for blowing snow. Water Resources Res. 11: 674-680.

Jairell, R.L. 1975b. A study probe for measuring deep snowdrifts. U.S. Forest Serv. Res. Note RM-301.

Keeler, C.M. 1969. Some physical properties of alpine snow. U.S. Corps of Engineers, Cold Regions Res. and Eng. Lab., Hanover, New Hampshire. 67 p.

Kerr, W.E. 1976. Snow pillow experiences in a prairie environment, p. 39-47. *In:* Proc. West. Snow Conf., Calgary, Alberta.

Koberg, G.E., R.R. Cruse, and C.L. Shrewsbury. 1963. Evaporation control research, 1956-60d. U.S. Geol. Surv. Water-Supply Pap. 1962. 53 p.

Komarov, A.A. 1954. As quoted by Kuzmin in Snow Cover and Snow Reserves. Gidrometeorologicheskoe Isdatelsko, Leningrad, 1960. Translation Nat. Sci. Found., Washington, D.C., 1962 p. 34.

Larson, F.R., P.F. Ffolliott, and K.E. Moessner. 1974. Using aerial measurements of forest overstory and topography to estimate peak snowpack. U.S. Forest Serv. Res. Note RM-267.

Leaf, C.F. 1962. Snow measurement in mountain regions. MS Thesis, Colo. State Univ., Fort Collins, Colo.

Leaf, C.F., and G.E. Brink. 1973. Computer simulation of snowmelt within a Colorado subalpine watershed. U.S. Forest Serv. Res. Pap. RM-99.

Leaf, C.F., and J.R. Konner. 1972. Sampling requirements for areal water equivalent estimates in forested subalpine watersheds. Water Resources Res. 8: 713-716.

Logan, L.A. 1972. Basin-wide water equivalent estimation from snowpack depth measurements. *In:* Intern. Symp. on the Role of Snow and Ice in Hydrology. Banff. 22 p.

Martinelli, M., Jr. 1973. Snow fences for influencing snow accumulation, p. 1394-1398. *In:* The Role of Snow and Ice in Hydrology. Symp. on Measurement and Forecasting (WMO), Banff, Alberta, Canada, Sept. 1972.

Martinelli, M., Jr. 1975. Water yield improvement from alpine areas: the status of our knowledge. U.S. Forest Serv. Res. Pap. RM-138. 16 p.

McKay, G.A. 1968. Problems of measuring and evaluating snowcover, *In:* Snow Hydrology, Proc. Workshop Seminar, Univ. of New Brunswick, Feb. 28-29.

McKay, G.A., and H.A. Thompson. 1967. Snowcover in the Prairie Provinces. Paper presented at joint Can. Soc. Agr. Engr. and Amer. Soc. Agr. Eng. Meeting, Saskatoon, June 27-30.

McMartin, W., A.B. Frank, and R.E. Heintz. 1974. Economics of shelterbelt influence on wheat yields in North Dakota. J. Soil and Water Conserv. 29: 87-91.

Meiman, J.R. 1968. Snow accumulation related to elevation, aspect, and forest canopy, p. 35-47. *In:* Snow Hydrology, Proc. Workshop Seminar, Univ. of New Brunswick, Feb. 28-29.

Meiman, J.R., and L.O. Grant. 1974. Snow-air interactions and management of mountain watershed snowpack. Colo. State Univ. Environ. Resources Center Completion Rep. Series 57: 36 p.

Mellor, M. 1965. Blowing snow. U.S. Army Corps of Eng., Cold Regions Res. and Eng. Lab. Cold Regions Sci. and Eng. Part III, Sec. A3c. 79 p.

Mixsell, J.W., D.H. Miller, S.E. Rantz, and K.G. Brecheen. 1951. Influence of terrain characteristics on snowpack water equivalent. U.S. Army Crops of Engineers, Analytical Unit, Res. No. 2, San Francisco, Calif.

Neyman, J., and H.B. Osborn. 1971. Evidence of widespread effects of cloud seeding at two Arizona experiments. Proc. Nat. Acad. Sci. USA 68: 649-652.

Neyman, J., and E.L. Scott. 1972. Some current problems of rain stimulation research. p. 1167-1244. *In:* Proc. Intern. Symp. on Uncertainties in Hydrologic and Water Resource Systems. Univ. of Arizona, Tucson, Vol. III.

Packer, P.E. 1962. Elevation, aspect, nd cover effects on maximum snowpack water content in a western white pine forest. Forest Sci. 8: 225-235.

Potter, L.D., J. Longwell, and C. Mode. 1952. Shelterbelt snowdrifts. N. Dak. State Univ. Biomo. Bull. 14: 176-179.

Price, W.I.J. 1961. The effects of the characteristics of snow fences on the quantity and shape of the deposited snow, p. 89-98. *In:* Int. Ass. Sci. Hydrol., General Assembly, Helskinki, 1960, Int. Ass. Sci., Hydrol. Pub. 54.

Pugh, H.L.D. 1950. Snow fences. Depth of Sci. & Indus. Res., Road Res. Lab., London, Road Res. Tech. Paper 19.

Rantz, S.E. 1964. Snowmelt hydrology of a Sierra Nevada stream. U.S. Geol. Surv. Water-Supply Pap. 1779-r. 36 p.

Rechard, P.A. 1973. Opportunities for watershed management in Wyoming, p. 423-448. *In:* Proc. Irrig. & Drainage Div. (Amer. Soc. Civil Eng.) Spec. Conf. "Agricultural and Urban Considerations in Irrigation and Drainage", Ft. Collins, Colo., Augs. 22-24.

Rechard, P.A., and L.W. Larson. 1971. Snow fence shielding of precipitation gages. Amer. Soc. Civil Eng. Proc. (HY 9)97: 1427-1439.

Rechard, P.A., and C.N. Raffelson. 1974. Evaporation from snowdrift under oasis conditions, *In:* Advanced Concepts and Techniques in the Study of Snow and Ice Resources, Symp., Monterey, Calif., Dec. 2-6, 1973, National Academy of Sciences, Washington, D.C.

Renard, K.G., and D.L. Brakensiek. 1975. Precipitation on intermountain rangeland in the western United States. p. 39-59. *In:* Proc. 5th U.S./Aust. Range Sci. Workshop, Boise, Id.

Saulmon, R.W. 1973. Snowdrift management can increase waterharvesting yields. J. Soil & Water Conserv. 28: 118-121.

Schmidt, R.A., Jr. 1970. Locating snow fences in mountainous terrain, p. 220-225. *In:* Snow Removal and Ice Control Res., Highway. Res. Board Spec. Rep. 115. (Proc. of Int. Symp., April 8-10, Dartmouth College, Hanover, N.H.)

Schmidt, R.A., Jr. 1970. Sublimation of wind-transported snow—a model. U.S. Forest Serv. Res. Pap. RM-90. 24 p.

Schneider, T.R. 1962. Snowdrifts and water ice on roads. Natl. Res. Council of Canada Tech. Trans. 1038. 200 p.

Seely, E.H., and D.G. DeCoursey. 1975. Hydrologic impact of weather modification. Water Resources Res. Bull. 11: 365-369.

Slaughter, C.W. 1969. Snow albedo modification, a review of literature. U.S. Army Crops of Eng., Cold Regions Res. and Eng. Lab., Hanover, N.H., Tech. Rpt. 217: 25 p.

Slaughter, C.W. 1970. Evaporation from snow and evaporation retardation by monomolecular films, a review of literature. U.S. Army Crops of Eng., Cold Regions Res. and Eng. Lab., Hanover, N.H., Spec. Rep. 130. 31 p.

Solomon, R.H., P.F. Ffolliott, and D.B. Thorud. 1975. Characterization of snowmelt runoff efficiencies, p. 306-326. *In:* Watershed Management, Proc. Amer. Soc. Civil Eng. Symp., Utah State Univ., Logan.

Stanford Research Institute. 1974. The impacts of snow enhancement: technology assessment of winter orographic snowpack augmentation in the Upper Colorado River Basin. Univ. of Okla. Press, Norman. 624 p.

Stanton, C.R. 1966. Preliminary investigation of snow accumulation and melting in forested and cut-over areas of the Crowsnest Forest, p. 7-12. *In:* Proc. 34th Annu. Western Snow Conf. Seattle, Wash. April 19-21.

Steppuhn, H. 1976. Areal water equivalents for prairie snowcovers by centralized sampling, p. 63-68. *In:* Proc. Western Snow Conf., April 20-22. Calgary, Alberta.

Steppuhn, H., and G.E. Dyck. 1974. Estimating true basin snowcover, p. 314-328. *In:* Proc. Intern. Symp. on Adv. Concepts and Techniques in the Study of Snow and Ice Resources. Int. Hydrologic Decade, U.S. Nat. Comm. Nat. Acad. Sci.

Stoeckler, J.H. 1962. Shelterbelt influence on great plains field environment and crops. U.S. Dep. Agr. Prod. Res. Rep. 62, 26 p.

Stoeckler, J.H., and E.J. Dortignac. 1941. Snowdrifts as a factor in growth and longevity of shelterbelts in the Great Plains. Ecology 22: 117-124.

Sturges, D.L. 1975. Hydrologic relations on undisturbed and converted big sagebrush lands: the status of our knowledge. U.S. Forest Serv. Res. Pap. RM-140. 23 p.

Swank, G.W., and R.W. Booth. 1970. Snow fencing to redistribute snow accumulation. J. Soil & Water Conserv. 25.

Swanson, R.H., and D.R. Stevenson. 1971. Managing snow accumulation and melt under leafless aspen to enhance watershed value. Western Snow Conf. Proc., april 20-22, Billings, Montana. 7 p.

Tabler, R.D. 1968. Physical and economic design criteria for induced snow accumulation projects. Water Resources Res. 4: 513-519.

Tabler, R.D. 1971. Design of a watershed snow fence system, and first year snow accumulation, p. 50-55. *In:* Proc. 39th Annual Western Snow Conf., April 20-22, Billings, Mont.

Tabler, R.D. 1973. Evaporation losses of windblown snow, and the potential for recovery, p. 75-79. *In:* West. Snow Conf. Proc., Grand Junction, Colo., April 17-20.

Tabler, R.D. 1975a. Predicting profiles of snowdrifts in topographic catchments. p. 87-97. *In:* Proc. 43rd Western Snow Conf., Coronado, Calif., April 23-25.

Tabler, R.D. 1975b. Estimating the transport and evaporation of blowing snow, p. 85-104. *In:* Snow Management on the Great Plains, Great Plains Agr. Council Pub. 73, Univ. Nebraska, Lincoln.

Tabler, R.D., and K.L. Johnson. 1971. Snow fences for watershed management, p. 116-121. *In:* Proc. Symp. on Snow and Ice in Relation to Wildlife and Recreation, Ames, Iowa, Feb. 11-12.

Tabler, R.D., and R.A. Schmidt. 1973. Weather conditions that determine snow transport distances at a site in Wyoming, p. 118-126. *In:* The Role of Snow & Ice in Hydrology, Symp. on Properties and Processes, Banff, Alberta, Sept., 1972.

U.S. Army Crops of Engineers. 1956. Snow hydrology. North Pacific Div., Portland, Oreg. 437 p.

U.S. Army Crops of Engineers. 1971. Runoff evaluation and streamflow simulation by computer. North Pacific Div., Portland, Ore. 87 p.

U.S. Dept. of the Interior. 1973. Project skywater, an introduction to rivers in the sky. Bur. of Reclamation Unnumbered Pub. 31 p.

U.S. Dept. of the Interior. 1975. Augmentation potential through weather modification. Bur. of Reclamation, Div. Atmos. Water Resources Manage., Working Document, Unnumbered. 51 p.

U.S. Soil Conservation Service. 1965-7. Snow survey phase of Park Atmospheric Water Resources program of the U.S. Bureau of Reclamation. Annual Reports. U.S. Soil Conserv. Serv., Denver, Colorado.

Van Haveren, B.P. 1974. Soil water phenomena of a shortgrass prairie site. M.S. Thesis, Colorado State Univ., Ft Collins. 172 p.

Van Haveren, B.P. 1975. Airflow and gas exchange in snow—fact or fiction? p. 21-27. *In:* Proc. 43rd Western Snow Conf., Coronado, Calif., April 23-25.

Van Haveren, B.P., and W.D. Striffler. 1976. Snowmelt recharge on a shortgrass prairie site. p. 52-56. *In:* Proc. 44th Western Snow Conf., Calgary, Alberta, April 20-22.

Wight, J.R., E.L. Neff, and F.H. Siddoway. 1975. Snow management on eastern Montana rangeland, p. 138-143. *In:* Snow Management on the Great Plains, Great Plains Agr. Council Pub. 73, Univ. Nebraska, Lincoln.

Wilken, G.C. 1967. Snow accumulation in a manzanita brush field in the Sierra Nevada. Water Resources Res. 3: 409-422.

Willen, D.W., C.A. Shumway, and J.E. Reid. 1971. Simulation of daily snow water equivalent and melt, p. 1-8. *In:* Proc. Western Snow Conf., Billings, Mont., April 20-22.

Willis, W.O., C.W. Carlson, J. Alessi, and H.J. Haas. 1961. Depth of freezing and spring runoff as related to fall soil moisture level. Canad. J. Soil Sci. 41: 115-123.

Wilson, W.T. 1941. An outline on the thermodynamics of snowmelt, Trans. Amer. Geophys. Union. 23: 182-205.

Young, G.J. 1973. A data collection and reduction system for snow accumulations studies, p. 304-318. *In:* Rep. 319, Dep. of Environ., Ottawa, Canada.

Young, G.J. 1974. A stratified sampling design for snow surveys based on terrain shape, p. 14-22. *In:* Proc. Western Snow Conf. Anchorage, Alaska. April 16-20.

Zuzel, J.F., and W.T. Ondrechen. 1975. Comparing water supply forecast techniques, p. 327-336. *In:* Watershed Management, Proc. Amer. Ass. Civil Eng. Symposium, Utah St. Univ., Logan.

Chapter 9

Urban Impacts on Rangeland Hydrology

"Of all land-use changes affecting the hydrology of an area, urbanization is by far the most forceful" (Leopold 1968). The urbanization process is characterized by large areas of land rendered impervious by roads, footpaths, roofs, and parking areas. The area in which rainfall can infiltrate into the soil is reduced, depression and interception storage of precipitation may be reduced, and overland flow can take place readily on the relatively smooth impermeable surfaces. Hydrologic impacts of these changes need to be considered because, if projections for the United States based on historic growth and trends prove valid, the amount of urbanized land will double in the next 30 years (Pickard 1967).

Figure 9-1 shows projected major urbanized areas in the United States. Before the turn of the century, over 90% of the U.S. population is expected to be residing in these urban areas. The key descriptor for the future is intensification of people and of activity of many kinds. In addition to its vital roles of supporting human life and providing environmental amenities, the manner in which the urban water resources are managed may have a large influence on

Fig. 9-1. Anticipated major urbanized areas in the United States in 100 years (Amer. Ass. Civil Eng.

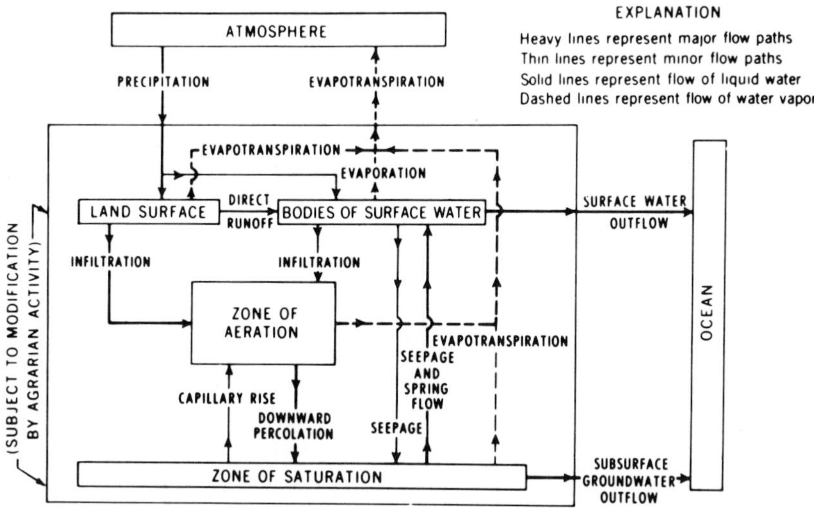

Fig. 9-2. Flow diagram showing preurban hydrologic system (after Cohen, Franke, and Foxworthy 1968).

our economic growth in the decades ahead. The water resource problems are especially acute in the more arid and semiarid areas such as are encountered in most rangeland areas. In addition to increasing water resource problems, urban expansions will reduce rangeland areas and increased recreational demands will further decrease forage resources alloted to livestock production. In the following discussions, comparisons will be made between urbanized versus "undeveloped" lands. In some instances the "undeveloped" areas are rangelands and in others they are cultivated or forest lands, but principles derived from all the research cited have some degree of application to effects of conversions of rangelands to urbanized lands.

Leopold (1968) indicates there are four interrelated but separable effects of urbanization on the hydrology of an area: changes in peak flow characteristics, changes in total runoff, changes in quality of water, and changes in the hydrologic amenities. The hydrologic amenities are what might be called the appearance or the impression which the river, its channel, and its valleys leave with the observer. In taking the pictorial representation of the hydrologic cycle in Figure 2-4 and presenting it in the form of a flow diagram, the preurban hydrologic system would resemble that shown in Figure 9-2. Urbanization results in many modifications to the flow system, and this is shown for water quantities in Figure 9-3. In general, the most dramatic hydrologic impact of urban development is that on peak flows in streams and storm drains. Also, the larger the urban complex the greater the hydrologic region that may be impacted in the acquisition of its water supply.

In general, there is a tendency of urbanization to produce higher peak discharges for a given frequency (Figure 9-4) although the effect would be expected to be smaller for large flows than for minor flows (Crippen 1965).

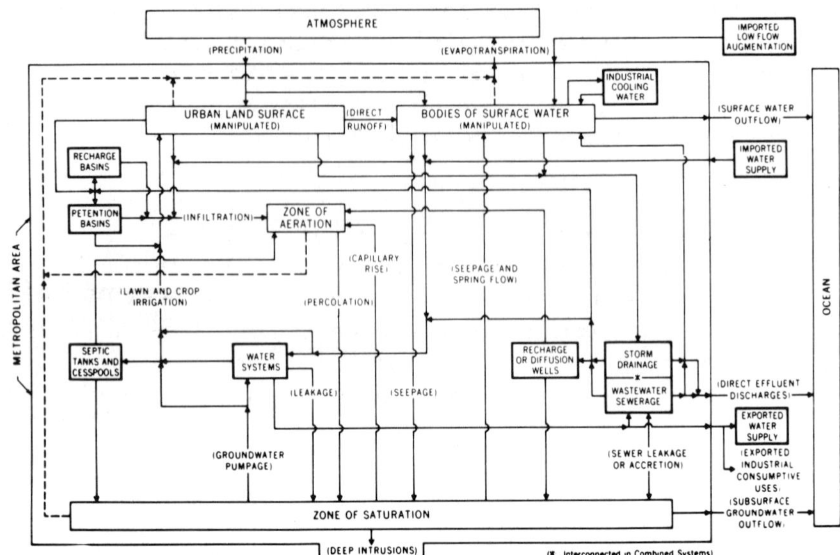

Fig. 9-3. Simplified summary of the urban hydrologic system. "Water systems" is intended to include treatment and distribution facilities of public systems and self-supplied industrial process systems. The term "manipulated" would include: (1) runoff management of the "urban land surface"; and (2) recreation, transportation, flood control measures, and property-value enhancement in connection with "bodies of surface water" (after Cohen, Franke, and Foxworthy 1968).

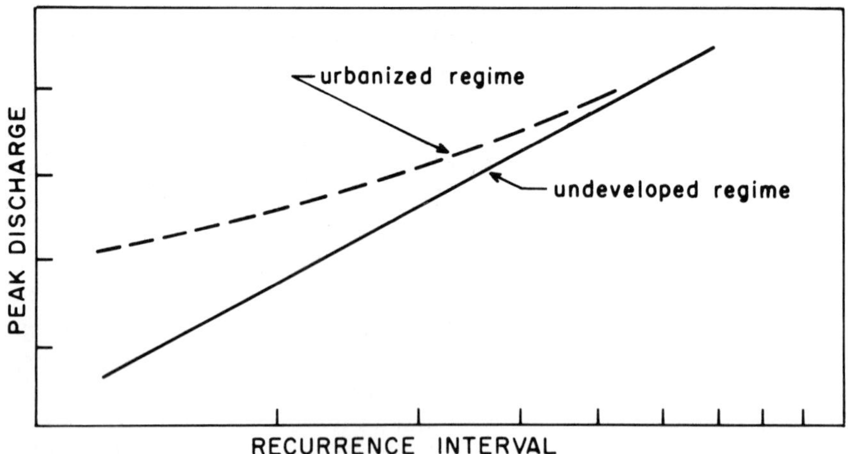

Fig. 9-4. Typical trends in peak flow frequency during urbanization showing greater effects on minor than on major flows (from Crawford 1968).

Table 9-1. Percentages of impervious land area within land-use categories with low, intermediate, and high estimate for each category.

Land-use category	Impervious land area: Low (%)	Intermediate (%)	High (%)
Single-family residential	12	25	40
Multiple-family residential	60	70	80
Commercial	80	90	100
Industrial	40	70	90
Public and quasi-public	50	60	75
Conservational, recreational and open	0	0	1

[1]Data from Stankowski (1974) who summarized several studies.

Indexes of impervious land surface areas may be derived on the basis of land use categories (Table 9-1) or population density (Figure 9-5). Data from Antonoine (1965) and Felton and Lull (1963) indicate the percentage of impervious surface area on residential properties to be as follows:

Lot size of residential area (m^2 [ft^2])	Impervious surface area (%)
560 (6,000)	80
560–1,395 (6,000–15,000)	40
1,395 (15,000)	25
7,200 (78,000)	8

As can be seen, the percentage of impervious surface area decreases markedly as the size of the lot increases.

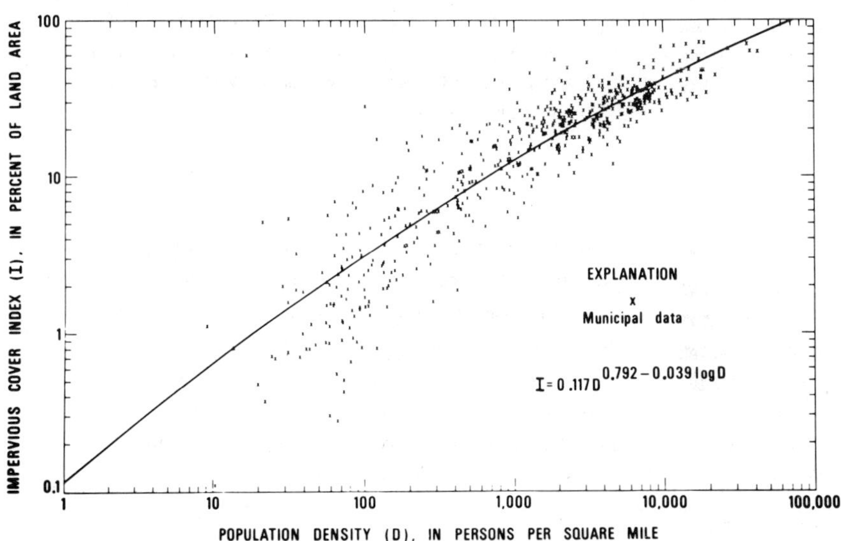

Fig. 9-5. Relation between manmade-impervious-cover index and population density for 567 municipalities in New Jersey (from Stankowski 1974). (Divide persons per sq. mile by 2.59 to obtain persons per km².)

Interception

The impact of urbanization on interception has not been evaluated in great detail. However, Lull and Sopper (1969) provide an example of the reasoning involved in trying to assess the impact on interception:

> Hardwood and conifer canopies in the Northwest intercept and evaporate about 15 to 20% of the growing-season rainfall before it reaches the forest floor. In winter, hardwoods intercept about 7% of the precipitation and conifers about 18 to 25%, depending upon how much precipitation occurs as snow. For example, of an average annual precipitation of 45 inches in central Massachusetts, hardwoods intercept about 5 inches and conifers about 9 inches, respectively 11 and 20% of the annual precipitation. . .
>
> City streets can intercept and store on their surfaces perhaps 0.004 to 0.01 inch of water. . . The amount of precipitation intercepted by roofs and sides of urban buildings, and evaporated from them, has probably never been estimated; but it is very likely less than 0.04 inch. In light of the great variety of construction that protrudes from city surfaces, any estimates would be highly speculative. For example, a city block of 1 acre occupied by a 10-story building would have about 2 acres in sidwalls, which would give an interception surface-area to ground-area ratio (including roof) of 3:1. In comparison, leaf areas, (both sides) of forest canopies have ratios that range from 2:1 to 9:1. . . If the sidewalls (only partially wetted) could intercept 0.01 inch of rainfall and the roof 0.04 inch, and if 120 rainfalls occurred during a year amounting to 45 inches precipitation, 16% would be intercepted—about the percentage intercepted by trees in summer.
>
> Residential areas, where the principal intercepting areas are lawns, would intercept somewhat less. From measurements of grass interception. . .we estimate a lawn 2 inches high could intercept about 0.01 inch of water. And if we add to this an estimated depression storage of 0.04 inch for the entire area, the total residential-area interception would be about 0.05 inch or, in relation to the 45-inch annual rainfall noted above, a total annual interception of about 13 percent

In another report, Crawford (1968) suggested:

> For impervious surfaces the only losses that will occur from rainfall are interception and depression storage. Interception storage is minor in the range from 0.01 to 0.03 inches.
>
> Depression storage on impervious surfaces may be considerable, perhaps as large as 0.2 inches in areas with low surface slopes. Water is lost from depression storage by evaporation and by seepage through joints and cracks in otherwise impervious surfaces.
>
> On pervious soil profiles, interception, depression storage and infiltration will occur. Interception losses remain relatively minor. Values from 0.03 to 0.15 inches have been reported for interception storage for vegetation ranging from grass to heavy forest. Interception parallels major losses due to depression storage and infiltration, and computer simulation runs checking the sensitivity of watershed response to interception processes show only minor influences.
>
> In summary, depression storage may cause significant losses on imprevious surfaces. On pervious surfaces depression storage and infiltration are of major importance.

Crawford then showed how depression storage and infiltration could be treated with the Stanford Watershed Model IV (Crawford and Linsley 1966).

Infiltration

The literature suggests that basin infiltration decreases with urbanization because of physical changes other than an increase in impervious area (e.g., change in vegetative cover, decrease in overland flow caused by gutters and storm drains, reduction of undrained surface depressions, and greater surface compaction). However, quantified measures of these relationships are not generally available. In general, infiltration may range from zero for paved areas and roofs to rates exceeding rainfall intensities in those uncommon, soil-covered areas not subject to traffic by vehicles or people. With the exception of one study (Felton and Lull 1963), all other studies reviewed on changes in infiltration characteristics of watersheds as urbanization progresses evaluate indirectly the changes in runoff quantities and patterns rather than through direct measurement of infiltration. The study by Felton and Lull (1963), in the northeastern U.S., indicated that lawns had an infiltration rate of 2.5 mm/min (0.10 inch/min) as compared with 15 mm/min (0.58 inch/min) into nearby forest soil. The fact that infiltration rates (measured with a ring infiltrometer) of lawns were only 1/6 that of forests was explained as the result of soil mixing, bulldozing into position, and compaction from mowing and trampling.

Evapotranspiration

Within many rangeland plant communities, evapotranspiration (ET) very nearly equals incoming precipitation or may even exceed the precipitation during many times of the year. Urbanization tends to produce a greater range in ET losses. For example, in Pasadena, California, annual ET was estimated at 10 cm (4 inches) for impervious areas; at 60 cm (24 inches) for residential areas; and at 90 cm (36 inches) for areas of lawns and trees (Gleason 1952). Annual precipitation is about 50 cm (20 inches) in this area. Urbanization may actually increase dry-weather flow of streams during dry years as a result of reduced ET losses or the discharges of sewage effluent (Muller 1967,

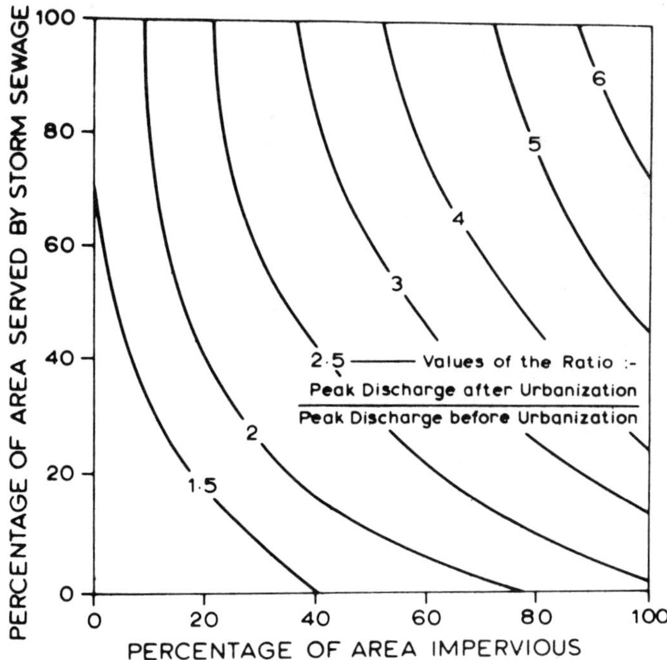

Fig. 9-6. Effect of urbanization on mean annual flood for a 259 ha (1-sq. mile) drainage area (from Leopold 1968).

 Seaburn 1970). Changes in evapotranspiration resulting from urbanization is an area where additional research is needed to insure efficient water use in the future.

Runoff and Peak Flows

The net effects of urbanization are that a higher proportion of rainfall becomes runoff, this runoff occurs more quickly, and flood flows are higher and "flashier", reaching stages more rapidly than was the case before urbanization. Several studies indicate that the effect of urbanization is greatest for small floods and as the size of the flood and its recurrence interval increase, the effects of urbanization diminish. This general statement holds true for all disturbance situations within rangeland ecosystems. As shown in Figure 9-4, peak discharges for large storms are nearly the same for urbanized and undeveloped regimes.

Figure 9-6 and 9-7 are from Leopold (1968) and each represents data from several studies. Figure 9-6 relates the percentage of the catchment paved and sewered to the posturbanization increase in the average peak discharge from a 259-ha (1-sq. mile) basin. The graph shows that mean annual floods increase

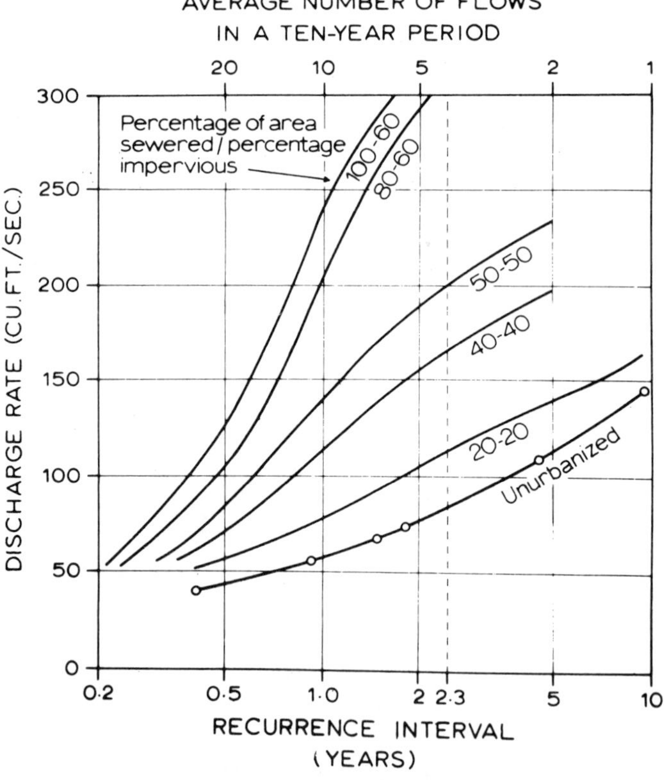

Fig. 9-7. Flood-frequency curves for a 259-ha (1 sq. mile) basin in various in various stages of urbanization (from Leopold 1968). (Multiply cfs by 0.028 to obtain cms or by 28.3 to obtain 1/sec.)

up to 6 times as areas become 100% impervious.

Figure 9-7 shows flood-frequency curves for a 259-ha (1-sq. mile) basin in various stages of urbanization. Though extreme floods are not included in the future, the effect of urbanization prevails much in the same way as depicted in Figure 9-6.

Hollis (1975) has also reviewed a number of relevant studies and attempted to synthesize the results into diagrams for practical application. In one instance, he plotted flood recurrence interval against percentage of basin paved for several values of peak discharge after urbanization to that before urbanization (Figure 9-8). The catchments varied in size, morphology, and geology, and are in widely differing climatic environments from the semiarid parts of the United States to Japan and western Britain. Despite some reservations Figure 9-8 suggests several things:

1. Floods with a return period of 1 year or more are not appreciably affected by a 5% paving of their catchment areas.
2. Small floods may be increased by a factor of 10 or more depending upon the degree of urbanization.
3. Floods with a return period of 100 years may be doubled in size by the complete urbanization of a catchment if that urbanization results in at least a 30% paving of the basin.
4. The effect of urbanization declines in relative terms as flood recurrence intervals increase.

Figure 9-9, also from Hollis (1975), is another visual representation of the idea that the relative increase, but not necessarily the absolute increase, in flood peaks brought about by urbanization declines as recurrence intervals lengthen.

The U.S. Soil Conservation Service (1975), has a design method for determining the effects of urbanization in a watershed on hydraulic and hydrologic parameters and presents methods for estimating peak discharge and runoff volumes. The method involves using the curve numbers developed

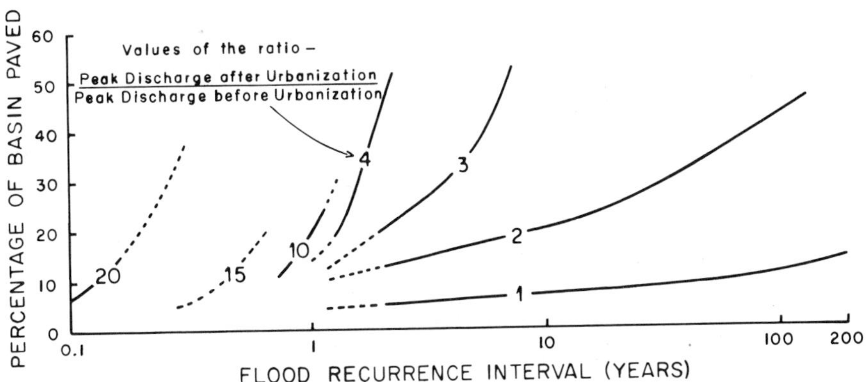

Fig. 9-8. Effect of urbanization on flood peaks (from Hollis 1975). Actual data points used to construct the eye-fitted curves are shown.

 for their national hydrologic model (see Chapter 6). Examples of curve numbers for different soils and land uses and the subsequent changes for suburban and urban land uses is shown in Table 9-2.

Water Quality

Of the many water quality parameters, impact of urbanization on sediment is probably of greatest importance to the rangeland hydrologist. It is well recognized that the average sediment yield from the landscape and the condition of the stream channels tend to change with the advancing forms of man's land-use activity. This is illustrated in Table 9-3. Guy (1970) indicates

Table 9-2. Runoff curve numbers[1] for selected agricultural, suburban, and urban land use. (Antecedent moisture condition II, and $I_a = 0.2S$).[2]

Land use description		Hydrologic soil group[2]			
		A	B	C	D
Cultivated land[3]: without conservation treatment		72	81	88	91
with conservation treatment[3]		67	71	78	81
Pasture or range land: poor condition		73	79	86	89
good condition		39	61	74	80
Meadow: good condition		30	58	71	78
Wood or forest land: thin stand, poor cover, no mulch good cover[4]		25	55	70	77
Open Spaces, lawns, parks, golf courses, cemeteries, etc. good condition: grass cover on 75% or more of the area		39 49	61 69	74 79	80 84
Commercial and business areas (85% impervious)		89	92	94	95
Industrial districts (72% impervious)		81	88	91	93
Residential[5]:					
Average lot size	Average % Impervious[6]				
1/8 acre or less	65	77	85	90	92
1/4 acre	38	61	75	83	87
1/3 acre	30	57	72	81	86
1/2 acre	25	54	70	80	85
1 acre	20	51	68	79	84
Paved parking lots, roofs, driveways, etc.[7]		98	98	98	98
Streets and roads:					
paved with curbs and storm sewers[8]		98	98	98	98
gravel		76	85	89	91
dirt		72	82	87	89

[1]For a more detailed description of agricultural land use curve numbers refer to National Engineering Handbook, Section 4, Hydrology, Chapter 9, Aug. 1972. (Soil Conserv. Serv. 1972).
[2]For detailed description of soil groups sec Chapter 11, page
[3]From U.S. Soil Conservation Service (1975).
[4]Treatments include terracing, grassed waterways, crop rotations, etc.
[5]Good cover is protected from grazing and litter and brush cover soil.
[6]Curve numbers are computed assuming the runoff from the house and driveway is directed towards the street with a minimum of roof water directed to lawns where additional infiltration could occur.
[7]The remaining pervious areas (lawn) are considered to be in good pasture condition for these curve numbers.
[8]In some warmer climates of the country, a curve number of 95 may be used. (Multiply acres by 0.405 to obtain ha.)

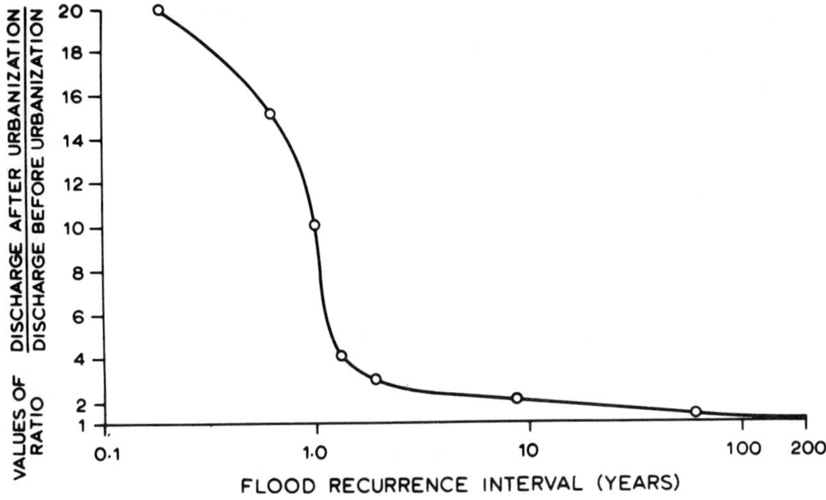

Fig. 9-9. Effect on flood magnitudes of paving 20% of a basin (from Hollis 1975).

the following:

> As in many other situations involving intensive use of resources and rapid growth, one can expect that sediment problems will be most serious during the urban construction period (E) (Table 9-3 from Guy). This is not to say that problems are not likely to occur during the stable period (G) because physical and esthetic values or quality standards with respect to both water and property are expected to increase with time. . .sediment problems can usually be classed into groups related to land and channel erosion, stream transport, and deposition processes. . . Land erosion, including the sheet, rill and gully forms, is likely to be most severe during the urban construction period (E), though it may be present to some degree regardless of land use. Channel erosion is most severe during the stabilization period (F), especially when channels have been realined, waterways have been constricted, and (or) the amount and intensity of runoff have been increased because of imperviousness and "improved" drainage. Sediment transport problems are usually associated with the pollution of water by sediment from either or both the esthetic or physical utilization viewpoints. . .

Table 9-3. Effect of land-use sequence on relative sediment yield and channel stability[1].

Land use	Sediment yield	Channel stability
A. Natural forest or grassland.	Low	Relatively stable with some bank erosion.
B. Heavily grazed areas.	Low to moderate	Some less stable than A.
C. Cropping.	Moderate to heavy	Some aggradation and increased bank erosion.
D. Retirement of land from cropping.	Low to moderate	Increasing stability.
E. Urban construction.	Very heavy	Rapid aggradation and some bank erosion.
F. Stabilization.	Moderate	Degradation and severe bank erosion.
G. Stable urban.	Low to moderate	Relatively stable.

[1]Data from Guy (1970) as modified from Wolman and Schick (1967).

An example of the effect of intensity of construction and drainage basin size on sediment yield is given in Figure 9-10. Wolman and Schick (1967) differentiated the following types of activity: natural or agricultural; building construction with sediment yields highly diluted before reaching channels; and undiluted sediment yields delivered to stream channels from construction sites. For very small areas in Maryland, Wolman (1964) states, "Because construction denudes the natural cover and exposes the soil beneath, the tonnage of sediment derived by erosion from an acre of ground under construction in developments and highways may exceed 20,000 to 40,000 times the amount eroded from farms and woodlands in an equivalent period of time."

Data are not available to quantify the effects of converting sparsely vegetated rangeland to urban uses although the effects might be expected to be less than those reported by Wolman (1964) in Maryland. Because the cover is often not providing the degree of protection afforded in humid climates and because the sediment concentrations may already be quite high, the construction period sediment discharge may not increase proportionately as much as in humid climates. The problem is still acute, however, because soil losses cannot be tolerated in most range areas because the soil profiles which are generally shallow already are at minimal levels for moisture storage and forage production.

Some relative magnitudes of sediment transport in terms of land use and sediment texture can be illustrated in Table 9-4. These values, derived by Knott (1973), are from Colma Creek Basin (San Mateo County), California.

In general it should be recognized that urbanization will often cause

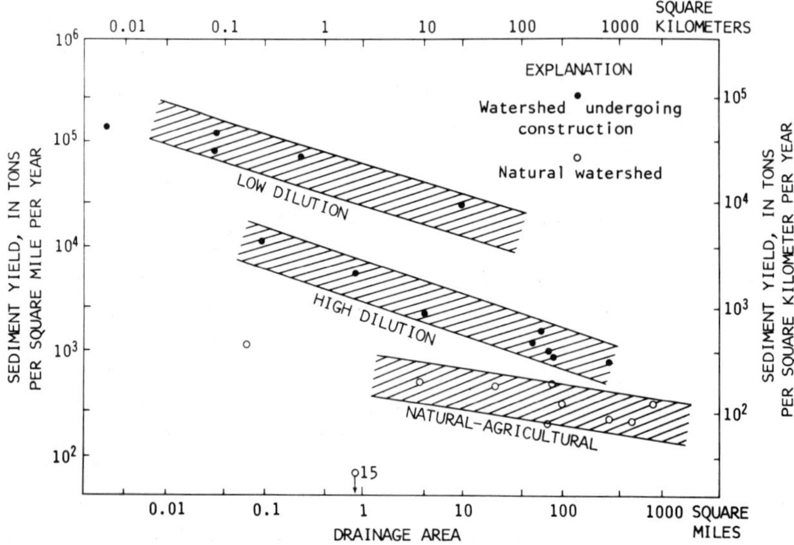

Fig. 9-10. Effect of construction intensity and drainage area on sediment yield. Most of the data are from the Baltimore and Washington, D.C., metropolitan areas. The term "dilution" refers to drainage from relatively stable construction areas (after Wolman and Schick 1967).

Table 9-4. Average sediment yield indexes as a function of land use and sediment texture for Colma Creek Basin, California[1].

Land use[2]	Sediment transported	Sediment yield index[3]
Open space	Sand	1.0
Urban	Sand	2.2
Agriculture	Sand	70.0
Construction	Sand	70.0
Open space	Silt-clay	1.0
Urban	Silt-clay	3.0
Agriculture	Silt-clay	60.0
Construction	Silt-clay	120.0
Open space	Total	1.0
Urban	Total	2.4
Agriculture	Total	65.0
Construction	Total	85.0

[1]Data from Knott (1973).

[2]1) Open space areas—soils were protected by native or reestablished vegetation, parks, and cemeteries.

2) Urban areas—include residential, commercial, and industrial communities.

3) Agricultural areas—soils have been under cultivation for many years (extent of conservation practices not indicated); areas of older exposed soil where vegetation has not been reestablished were included in this category.

4) Construction areas—freshly excavated areas and areas that had been exposed for as much as several years.

[3]Relative index values or relative erosion rates for specific types of land uses were estimated by determining the yield from several sub-basins with different land use. Values given are characteristic of moderate to large storms.

accelerated rates of erosion, especially during construction phases. Effects and equitable sediment control measures generally must be devised to suit specific conditions on specific sites. Guy and Jones (1972) have stated, "Effective control measures to prevent sediment damage should stem basically from the right of an owner of, or even a visitor to, a body of water such as a stream or lake, to view and use the water in its natural state. Thus, a serious problem is evident; the need to define the natural state of the water."

Leopold (1968) has noted that there are two principal effects of urbanization on water quality. The first is that the influx of waste materials tends to increase the dissolved-solids content and decrease the dissolved-oxygen content. Second, as flood peaks increase as a result of the increased area of imperviousness and decreased lag time, less water is available for groundwater recharge. In general, the stream becomes flashier in that flood peaks are higher and flows during nonstorm periods are lower. From the standpoint of the range hydrologist, water quality parameters other than sediment, as related to urbanization, are generally of minor importance. Suffice it to say here that urbanization will have an impact on other water quality parameters such as dissolved oxygen, biochemical oxygen demand, nutrients, dissolved solids, coliforms, thermal loading, and esthetics. In most cases the problem is very localized, and control measures (as with sediment) may have to be rather site specific. The Sanitary Engineering Journal, American Society of Civil Engineers, has had numerous articles describing some of the particular chemicals encountered in urban storm runoff. Efforts to relate the chemicals in storm runoff to activities of man are also receiving widespread interest (Viessman 1968).

Literature Cited

American Society of Civil Engineers. 1968. A systematic study and development of long range programs of urban water resources research. Amer. Soc. Civil Eng. Urban Hydrol. Proj. Rep., New York, N.Y. 125 p.

Antoine, L.H. 1964. Drainage and best use of urban land. Public Works (New York) 95: 88-90.

Cohen, P., O.L. Franke, and B.L. Foxworthy. 1968. An atlas of Long Island's water resources. New York Water Resources Comm. Bull. 62, Albany, N.Y. 117 p.

Crawford, N.H. 1968. Infiltration and losses in urban hydrology, p. A22-A31. *In:* Urban Water Resources Res., Amer. Soc. Civil Eng., Urban Hydrol. Proj. Rpt., Chapter 3.

Crawford, N.H., and R.K. Linsley. 1966. Digital Simulation in Hydrology: Stanford Watershed Model IV. 210p. Tech. Rep. 39, Dep. of Civil Eng., Stanford University.

Crippen, J.R. 1965. Changes in the character of unit hydrographs, Sharon Creek, California, after suburban development. U.S. Geol. Surv. Prof. Pap. 525-D:196-198.

Felton, P.M., and H.W. Lull. 1963. Suburban hydrology can improve watershed conditions. Public Works 94: 93-94.

Gleason, C.B. 1952. Consumptive use of water, municipal and industrial areas. Trans. Amer. Soc. Civil Eng. 117: 1004-1013.

Guy, H.P. 1970. Sediment problems in urban areas. U.S. Geol. Surv. Circ. 601-E 8 p.

Guy, H.P., and D.E. Jones, Jr. 1972. Urban sedimentation—in perspective. J. Hydraulics Div. Amer. Soc. Civil Eng. 98(HY12): 2099-2116.

Hollis, G.E. 1975. The effect of urbanization on floods of different recurrence interval. Water Resources Res. 11: 431-435.

Knott, J.M. 1973. Effects of urbanization on sedimentation and floodflows in Colma Creek Basin, California. U.S. Geol. Surv. Open-File Rep., Menlo Park, Calif. 54 p.

Leopold, L.B. 1968. Hydrology for urban land planning—a guidebook on the hydrologic effects of urban land use. U.S. Geol. Surv. Circ. 554. 18 p.

Lull, H.W., and W.E. Sopper. 1969. Hydrologic effects from urbanization of forested watersheds in the Northeast. U.S. Forest Serv. Res. Pap. NE-146. 31 p.

Muller, R.A. 1967. Some effects of urbanization on runoff as evaluated by Thornthwaite water balance models. p. 127-136. *In:* Proc. Third Annual Amer. Water Resour. Conf., Nov. 8-10, San Francisco.

Pickard, J.P. 1967. Dimensions of metropolitanism. Urban Land Inst. Res. monogr. 14k, Washington, D.C. 93 p.

Seaburn, G.E. 1970. Effects of urban development on direct runoff to East Meadow Brook, Nassau Country, Long Island, New York. U.S. Geol. Surv. Prof. Pap. 627-B. 14 p.

Stankowski, S.J. 1974. Magnitude and frequency of floods in New Jersey with effects of urbanization. U.S. Geol. Surv. Special Rep. 38. 46 p.

U.S. Soil Conservation Service. 1972. National engineering handbook, Sec. 4, Hydrology. U.S. Dep. Agr., Soil Conserv. Serv., Washington, D.C.

U.S. Soil Conservation Service. 1975. Urban hydrology for small watersheds. U.S. Dep. Agr., Soil Conserv. Serv., Washington, D.C., Tech. Release 55.

Viessman, W., Jr. 1968. Modeling of water quality inputs from urbanized areas, p. A79-A103. *In:* Urban Water Resources Res., Amer. Soc. Civil Eng. Urban Hydrol. Proj. Rep., Chapter 5.

Wolman, M.G. 1964. Problems posed by sediment derived from construction activities in Maryland. Rep. to the Maryland Water Pollution Control Comm., Annapolis. 125 p.

Wolman, M.G., and P.A. Schick. 1967. Effects of construction on fluvial sediment, urban and suburban areas of Maryland. Water Resources Res. 3: 451-462.

Chapter 10

Geomorphology

Geomorphology is, simply stated, the study of the landforms of the earth and the processes that shape them. Landforms can be described in a qualitative way for the purpose of general classification but this does not usually provide an insight to processes that shape the landscape. It is necessary to relate landforms to formative processes in order to predict future events and to understand how land use or land management will influence the landscape. Quantitative studies of geomorphic processes and their relation to drainage basin characteristics have provided hydrologists and earth scientists with many practical tools for evaluating the interrelations between landforms, plant cover, hydrology, and climate.

The Drainage Basin

The drainage basin is the logical unit of the earth's land surface to use in the study of relations between landform characteristics and processes. The basin is the natural integrator of such variables as precipitation, soil moisture, evapotranspiration, runoff, vegetation, lithology and structure, erosion, and sediment discharge as they relate to input and output in an open hydrologic system. Characteristics of discrete landforms, such as hillslopes or channels, can be measured and studied outside the context of the basin. However, the interaction of processes and the routing of water and sediment through the system is more meaningful if the basin is used. Also, relations can be developed in small headwater basins and extended cumulatively in a downstream direction to larger drainage areas.

Quantitative measurements of the topographic characteristics of drainage basins were being made in the nineteenth century but the first major advancements are attributed to Horton (1932). A little more than a decade later, Horton (1945) wrote his landmark paper that related drainage basin characteristics to hydrologic processes. Later work by Langbein et al. (1947) and Strahler (1952, 1957) serve as the foundation for quantitative geomorphic studies of the drainage basin.

Drainage basin characteristics include the topographic elements of relief, slope, stream channel system, and basin size and shape. In addition, a basin has certain soil, rock and vegetation characteristics, which strongly affect the development of topographic characteristics. If the topographic, lithologic, pedologic, and botanic characteristics are to be related to the hydrology of a basin, it is apparent that climate will also play an important role. In this chapter the quantitative expression of topographic and stream system characteristics as developed principally by Horton (1945) are considered and related to other drainage basin characteristics of soils and vegetation common to arid and semiarid rangelands.

Stream System Characteristics

The stream system in a drainage basin was expressed quantitatively by Horton (1945) in terms of stream order, stream numbers, stream lengths, drainage density, and bifurcation ratio.

Stream ordering as defined by Horton (1945) states that fingertip or unbranched tributaries are to be designated as first-order streams. Streams that receive first-order tributaries, but these only, are second-order streams; third-order streams receive second- or first- and second-order tributaries, and so on until the mouth of the basin is reached (Fig. 10-1.). When the order of the main stream is determined, it is projected headward through the basin along the stream that appears to be the main channel, which involves some subjective judgment. In order to alleviate some of the subjectiveness, Strahler (1952) modified Horton's methodology and made all unbranched tributaries first-order streams; two first-order streams join to make a second-order stream, and so on downstream (Fig. 10-1.). Either method permits a quantitative expression of stream order when comparing basins being studied. It should be noted, however, that investigators must indicate the method used to determine stream orders and source and scale of the maps or aerial photographs used. Determination of stream orders in a basin can produce vastly different results depending on the scale of the maps used (Leopold, Wolman, and Miller 1964).

The law of stream numbers expresses the relation between the number of streams of a given order and the stream order (Horton 1945). The number of streams of different orders in a basin closely approximate an inverse geometric series; as order increases, number decreases geometrically, in which the first term is unity and the ratio is the *bifurcation ratio.* Bifurcation means branching or dividing and the bifurcation ratio is simply the ratio of the

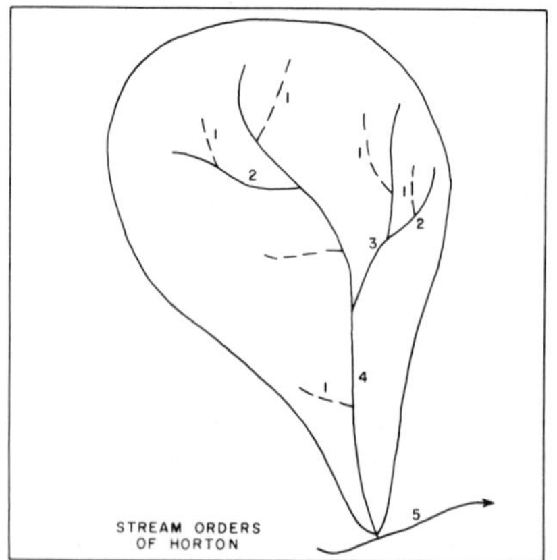

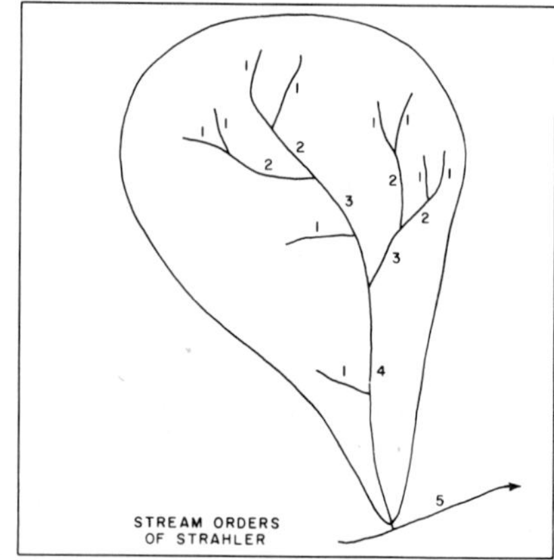

Fig. 10-1. Methods used for stream ordering by R.E. Horton and by A.N. Strahler.

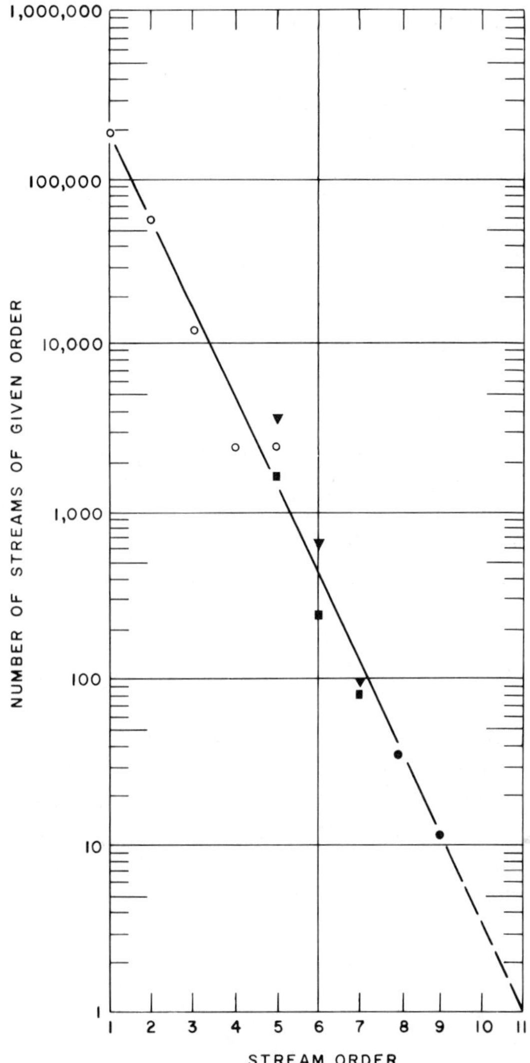

Fig. 10-2. Relation of number of streams of a given order to stream order (from Leopold and Miller 1956). Different symbols represent measurements on different streams.

number of streams of any given order to the number of streams in the next lower order.

Horton's (1945) law of stream lengths states that the average lengths of streams of each order in a basin closely approximate a direct geometric series in which the first term is the average length of first-order streams.

The three laws of stream-system composition as formulated by Horton demonstrate the adjustment of streams and their drainage basins in a quantitative way. Since the Horton paper was published, many hydrologists and geomorphologists have applied his methodology in studies of field problems related to a variety of objectives and a wide range of climatic environments.

An excellent example of the application of Horton's laws to a field study in northern New Mexico is described by Leopold and Miller (1956). Figure 10-2 shows the relation of stream order to number of streams in a basin having an eleventh-order stream at its mouth. Similarly in Figure 10-3, from the study by Leopold and Miller (1956), it is shown that the relation of stream order to stream length shows an orderly progression of drainage development. By carrying the Horton analysis further, Leopold and Miller (1956) show a regression of stream order to drainage area for the 670 square-mile (1,735 km^2) basin as shown in Figure 10-4. As the authors note, the maximum stream length is a function of drainage area and the regression plots as a straight line on semi-logarithmic paper as expected. In the paper by Leopold and Miller, they develop the relation between the stream system measurements of Horton and discharge in an ephemeral stream channel network that illustrates the application of quantitative geomorphology to hydrology.

The relation of drainage network measurements, both linear and areal, to streamflow and sediment yield will be discussed later in this chapter, as well as their application to land management problems.

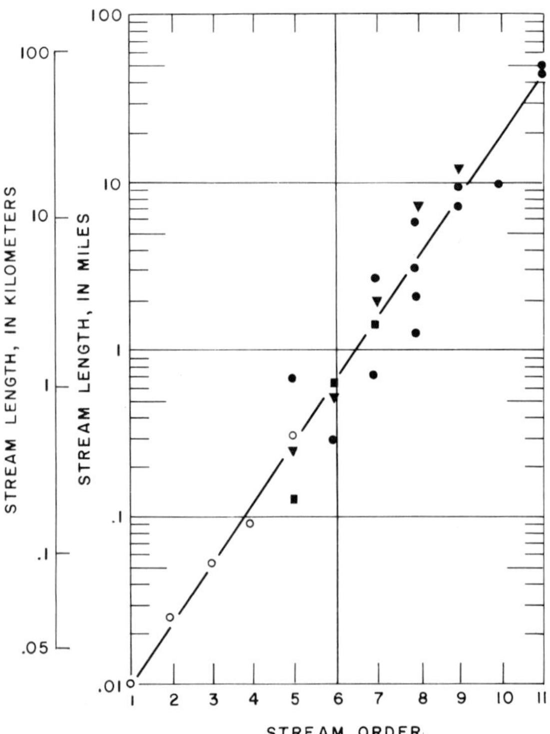

Fig. 10-3. Relation to stream length to stream order (from Leopold and Miller 1956). Different symbols represent measurements on different streams.

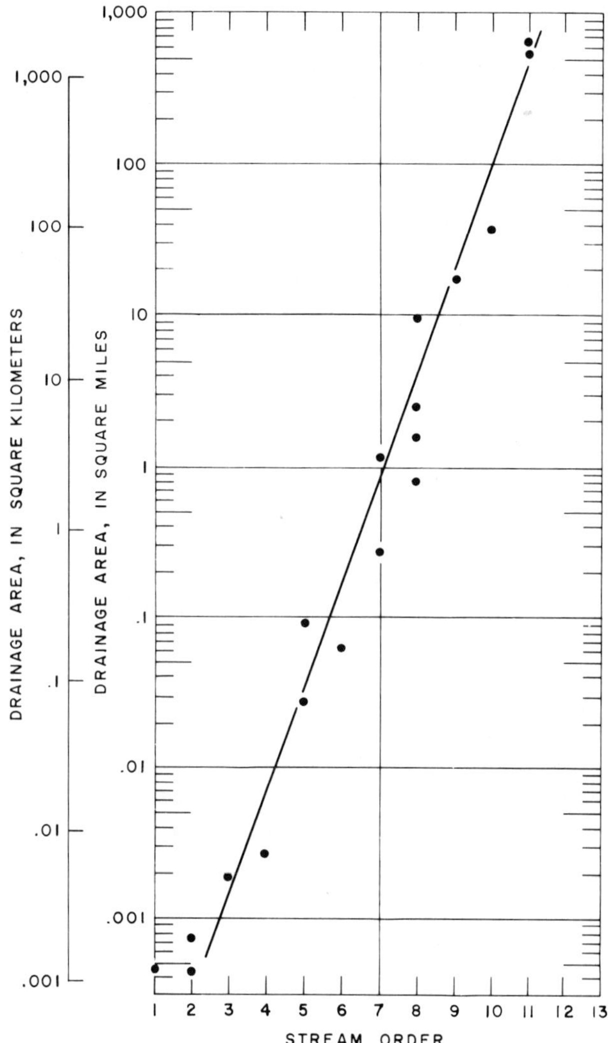

Fig. 10-4. Relation of drainage area to stream order for ephemeral streams in central New Mexico (from Leopold and Miller 1956).

Areal Characteristics of Drainage Basins

There are other measures of drainage basin characteristics that may be related to both stream discharge and sediment discharge. These include drainage density, basin shape, and drainage area.

The drainage density, as defined by Horton (1932), is the length of streams within a basin per unit of area, or miles of channel per square mile (km per km^2); it may be expressed

$$\text{Drainage density, Dd} = \frac{\Sigma L}{A} \qquad \textbf{(10-1)}$$

 where ΣL is the total length of streams and A is the drainage area. A basin with a high value for drainage density will have a fine-textured topography and short, generally steep, slopes. Conversely, a basin with a low drainage density will have longer slopes and more distance between streams. Drainage density is controlled to a large extent by the infiltration capacity of the soil or rock underlying the basin.

Basin shape, or geometry, has been expressed in several ways by investigators but it was found by Morisawa (1959) that only two, circularity ratio (R_c) and elongation ratio (R_e), showed a significant correlation with the runoff-rainfall ratio for 25 watersheds on the Appalachin Plateau (Strahler, 1964).

In a study of basin morphometry in Tennessee and Virginia (Miller 1953), basin circularity (C) was defined as the ratio of basin area (A_b) to the area of a circle having the same perimeter (A_c). In equation form the relationship may be expressed

$$C = \frac{A_b}{A_c} \tag{10-2}$$

The rationale for Miller's expression of basin shape is that a circle provides maximum area with a minimum perimeter while an oblong basin will have a large perimeter but a smaller drainage area. The circularity ratio serves as a quantitative index for comparison of basin forms. It has also been shown that basin shape is related to the hydrograph characteristics of flood flows. The time of water concentration for tributary flow in the main channel is less in circular or ovoid basins than it is in long, narrow basins.

Another expression of basin shape was described by Schumm (1956) and called the *elongation ratio.* It is the ratio of the diameter of a circle with the same area as the basin and the maximum length of the basin from the mouth to the headwater divide along the main channel. The elongation ratio, R_e, can be expressed by the equation

$$R_e = \frac{A_c}{L_m} \tag{10-3}$$

where A_c is the diameter of a circle of the same area as the basin, and L_m is the maximum basin length. Although shape does have some influence on the hydrologic response of a basin, such as streamflow and sediment discharge, other characteristics probably are more important (Strahler 1964).

Drainage area is generally the simplest areal characteristic of a basin to measure either from maps or aerial photographs. The relation between stream order and drainage area is similar to the relation of stream order to stream length (Leopold and Miller 1956).

Basin Relief

In addition to the linear and areal characteristics of a basin that can be related to hydrology and fluvial processes, the relief or gradient is also very important. Basin relief influences several basin parameters, such as channel slopes, hillslopes, and drainage density (Schumm 1956). A simple expression of relief, however, has proven to be very useful in geomorphic studies and is called *relief ratio*. It was first described by Schumm (1956) and is expressed as

$$R_r = \frac{R}{L} \tag{10-4}$$

where R_r is the relief ratio, R is the difference in altitude between the mouth of a basin and the headwater divide, and L is the maximum length of the basin measured in the same units as R along a line essentially parallel to the principal channel. Relief ratio correlates better with hydrologic characteristics in a basin if the relief measurements are made so that abnormally high, isolated points on the divide are not included. Also, basins with more or less uniform slopes (Hadley and Schumm 1961) give closer correlations of relief ratio and runoff or sediment yield.

Relation of Geomorphic to Hydrologic Characteristics

The quantitative description of drainage basin characteristics and some of the relationships that have been discussed can be applied to field problems with the objective of evaluating land use. It is true that the climate, vegetation, and hydrology are highly interdependent, thus masking complete knowledge of cause and effect. However, quantification of the physical variables brings us closer to understanding of the relations between geomorphic and hydrologic characteristics. For example, sediment yield is dependent on the gross erosion in the drainage basin, which is related to the vegetation cover, erodibility of soil, and the transport efficiency of the channel network. Similarly, streamflow is dependent on the amount and distribution in time and space of precipitation, infiltration characteristics of upland soils and streambeds, and basin size, shape, and slopes.

Hadley and Schumm (1961) made an intensive study of 99 small drainage basins in eastern Wyoming to delineate sediment sources in the drainage basin of the Cheyenne River. Reservoirs at the mouths of these basins were used to measure runoff and sediment yield, which, in turn, were related to geomorphic characteristics. One of the primary objectives of the study was to develop an erosion classification of rangelands that would be adaptable to reconnaissance investigations.

Drainage density was found to be an important geomorphic parameter in relation to the hydrology of the small basins. The texture of the topography is controlled by the rock type underlying a basin. The shale basins have high drainage densities and the sandstone basins have the lowest drainage densities. These differences in drainage density also reflect infiltration rates with the shale basins having the lowest rates. It was also shown that the infiltration rates were inversely related to runoff.

Drainage density was measured from aerial photographs for several basins underlain by different rock types having similar climatic conditions and for which runoff records were available. The regression of drainage density in miles per square mile (km/km^2) to mean annual runoff in acre-ft/sq. mile

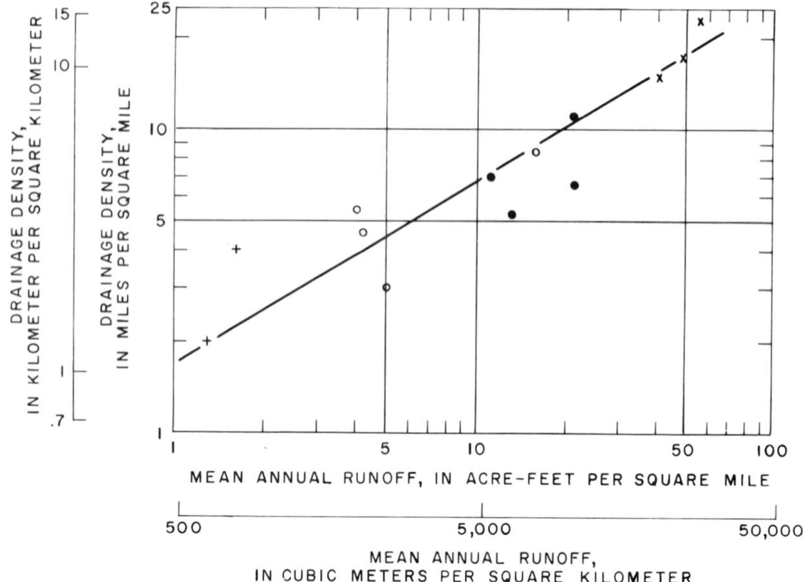

Fig. 10-5. Relation of mean annual runoff to drainage density for small basins in eastern Wyoming (from Hadley and Schumm 1961). Different symbols represent data from different drainages.

(m^3/km^2) is shown in Figure 10-5. The basins underlain by Pierre Shale with drainage densities of about 16 miles per square mile (10 km/km^2) yield a mean annual runoff of about 50-acre-per square mile (40,000 m^3/km^2). The basins underlain by sandstone with drainage densities of about 4 miles/per square mile (3 km/km^2) yield mean annual runoff of only 3 acre-ft per square mile (1,500 m^3/km^2).

Similarly, the relation between drainage density and sediment yield for small basins in eastern Wyoming (Hadley and Schumm 1961) show that the shale basins have the highest sediment yields and the sandstone basins have the lowest sediment yields (Table 10-1).

Relief ratios were measured by Schumm (1955) for several basins in Utah, New Mexico, and Arizona for which sediment yield data were available (Hains, Van Sickle, and Peterson 1952, King and Mace 1953). These data were combined with relief ratios and sediment yield from basins in eastern Wyoming (Hadley and Schumm 1961). Figure 10-6 shows the regression of relief ratio to sediment yield for the combined data segregated by rock type.

Table 10-1. Relation between drainage density and sediment yield by rock type.

Rock type	Drainage density (miles/sq. mile)	Mean annual sediment yield (acre-ft/sq mile)
Sandstone	5.4	0.13
Sandy shale	11.4	1.3
Shale	16.1	1.4

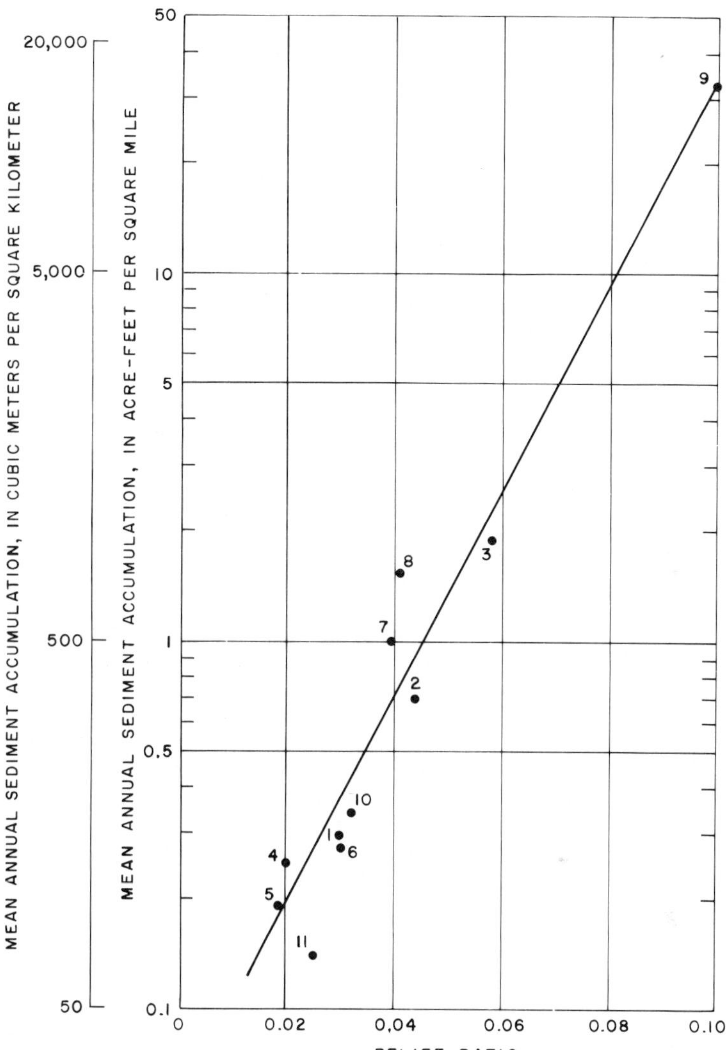

Fig. 10-6. Relation of mean annual sediment yield to relief ratio for basins underlain by different rock types. Numbers indicate stratigraphic units as follows: 1. Navajo sandstone, 2. Entrada sandstone, 3. Mancos shale, 4. Conglomerate and sandstone of Triassic and Permian age, 5. Shinarump conglomerate member of the Chinle formation, 6. Mesaverde sandstone, 7. Fort Union formation, 8. Pierre shale, 9. White River group, 10. Lance formation, 11. Wasatch formation (from Hadley and Schumm 1961).

The relation between relief ratio and sediment yield for a wide variety of rock types shows that relief ratio is a useful indicator of potential sediment yield. In a further test of the relation between relief ratio and sediment yield to reservoirs, Hadley and Schumm (1961) plotted data from 26 basins in eastern Wyoming underlain by a relatively uniform rock type, the sandy shale of the Fort Union Formation. The regression is shown in Figure 10-7; as expected,

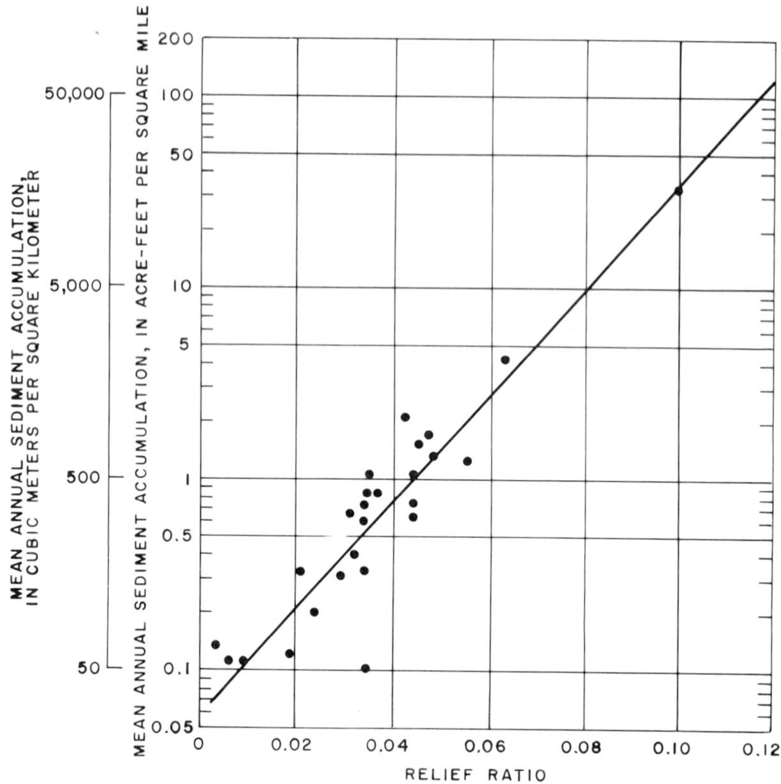

Fig. 10-7. Relation of mean annual sediment yield to relief ratio for basins underlain by the Fort Union formation (from Hadley and Schumm 1961).

the steeper basins produced higher sediment yields.

The correlation between mean annual sediment yield and relief ratio suggest that an evaluation of erosion conditions on rangelands may be approximated by a quantitative analysis of geomorphic characteristics. The geomorphic controls provide a common basis for consideration of the factors such as climate.

Effects of Climate on Geomorphic Processes

The relation of landforms to processes has been mentioned as well as the role of climate in process rates. Stoddard (1969) concludes that climate through process exerts some control on landforms although it may not be dominant. The interaction of climate with rock type and vegetation make it unrealistic to attempt to isolate climate as a geomorphic control. Nevertheless, we do find geomorphic characteristics in arid and semiarid regions, where most of our rangelands are located, that reflect a water-deficient environment. Peltier (1950) defined morphogenetic regions based on climate using the parameters of mean annual precipitation and mean annual temperature. Using these two parameters, he developed hypothetical effects of frost action, chemical weathering, mechanical weathering, wind erosion, and fluvial erosion. As pointed out by Stoddart (1968), however, Peltier's classification may be too simplistic because it does not consider the magnitude and frequency of precipitation and flood events.

The Semiarid Cycle of Erosion

Most of the streams are ephemeral in a semiarid environment, flowing only in response to summer thunderstorms or spring snowmelt, and the distribution and areal extent of individual thunderstorms often is limited to only a few square miles. Therefore, as the runoff and transported sediment moves downstream away from the storm, flow losses by adsorption in sandy channels are often high and deposition of sediment occurs. It is not unusual to observe flows of 300 cfs or more in the channels of ephemeral streams during a summer thunderstorm and that these flows are completely lost by infiltration into the streambed in a reach less than 10 miles (16 km). As the flow diminishes in volume, the sediment concentration generally increases and deposition occurs (Hadley and Schumm 1961). With these observations as background, Schumm and Hadley (1957) described a semiarid cycle of erosion that may explain some of the sediment transport processes that are active between the point of detachment on an upland slope and deposition in the channel.

Field examination of drainage basins in eastern Wyoming and northwestern New Mexico in semiarid environments (Schumm and Hadley 1957) showed that two general cases frequently are found: (1) basins with actively eroding headwater channels and rapidly aggrading lower valley reaches, or (2) moderate erosion in headwater channels with deep gullies trenching downstream valley reaches. These two distinctly different conditions are commonly found in adjacent drainage basins with similar environmental conditions. When a gully occurs at the downstream end of a basin and is actively eroding headward, it is an efficient conveyance channel for sediment transport. However, as the gully head migrates upstream, previously ungullied tributary channels become graded to it. The rejuvenation of tributary channels will steepen side slopes and greatly increase the sediment being transported in the basin. When the sediment being delivered to the main channel exceeds the capacity of the limited streamflow to transport it and aggradation begins in the middle and lower reaches of the stream, the cycle is repeated with the beginning of renewed aggradation. The relict features of this process are often preserved in the alluvial deposits of valleys in western United States.

Effects of Plant Cover on Processes

The deficiency of soil moisture resulting from low annual precipitation and high evapotranspiration rates in arid and semiarid regions is responsible for sparse plant cover on upland slopes. This lack of ground cover makes the large area of exposed soil surface on upland slopes highly susceptible to erosion. Most measurements of erosional processes on hillslopes have been made at seasonal or annual intervals, or over a period of several years. Such observations produce average rates of erosion and mask the effect of individual storm events.

In a study by Hadley and Lusby (1967), the results of measuring hillslope erosion from a single thunderstorm in western Colorado are reported. The maximum intensity of rainfall for a 10-minute period was 1.98 inches/hr (50.2 mm/hr) and the total rainfall was 0.90 inch (22.8 mm). Data from measurement of erosion pins along six hillslope profiles before and after the storm showed that the average depth of erosion was 0.009 foot (2.74 mm). The volume of eroded material transported out of the small basin studied was about 0.11 acre-foot (136 m^3), assuming all slopes were eroded uniformly. For comparison with unit rates of sediment yield from small basins, the sediment yield for the storm was equivalent to 5.9 acre-feet per square mile (2,790 m^3/km^2).

In another study in the Badger Wash basin of western Colorado (Branson and Owen 1970), the relationships between vegetation and hydrologic characteristics for 17 watersheds were analyzed. It was found that highly significant correlations exist for the relation between percent bare soil and runoff. Bare soil measurements may provide rapid and inexpensive estimates of runoff for drainage basins similar to those studied by Branson and Owen.

The relation between bare soil and sediment concentration was developed for four small drainage basins near Eagle, Colorado, (Lusby, 1979). The percent bare soil was altered artificially on two of the basins in an experimental study to evaluate the effect on hydrology of removing sagebrush cover and replacing it with grass. All of the basins were dominated by sagebrush during the pretreatment years and there was very little understory of grasses; thus averaging about 60% bare ground. The sediment concentration with nearly 60% bare ground was 0.058 as shown in Figure 10-8. After the sagebrush was removed from two basins and the sown grass became established, the percent bare ground was reduced to about 38% and the sediment concentration was only 0.015, a threefold reduction.

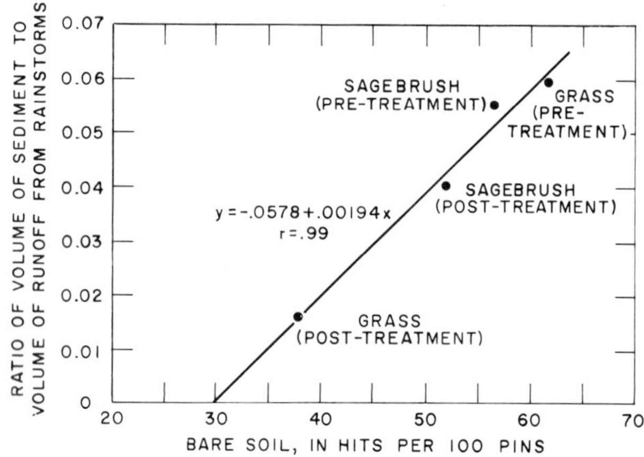

Fig. 10-8. Relation of sediment concentration by volume to percent bare soil for four small basins in central Colorado (data from Lusby 1979).

The exposure of a hillslope to solar radiation has a marked effect on microclimate and the rate at which geomorphic processes operate. Northerly-facing slopes generally are steeper, less dissected, and support a denser growth of plant cover than southerly-facing slopes. These contrasts in plant cover are accompanied by similar differences in surface runoff and erosion. Because of the differences in erosion there is a tendency for the thalweg of the major stream channel in a basin that is oriented east and west to be shifted to the south side of the valley floor by the debris fans and colluvial aprons of eroded material derived from south-facing slopes. Erosional debris tends to accumulate at the base of south-facing slopes in arid and semiarid climates where most streams are ephemeral and infrequent flows are not able to transport all of the sediment delivered to the channels out of the basin. The resulting migration of the axial channel to the south side of the valley causes an asymmetrical development of many drainage basins.

Microclimatic differences that cause valley asymmetry are manifested in slope gradients and drainage density. Slope gradients are compared with exposure on the rose diagram in Figure 10-9 for an area of 0.80 km^2 (0.5 sq. mile) in eastern Wyoming. The slope of the land surface and the direction of exposure were measued at 50 points equally spaced in a grid. The diagram shows that the steepest slopes face north, northeast, and northwest; whereas the gentlest slopes face south, southeast, and south-southwest.

Vegetation counts made on several slopes using step-point transects show that plant cover on southerly-facing slopes is only 28% of that occurring on

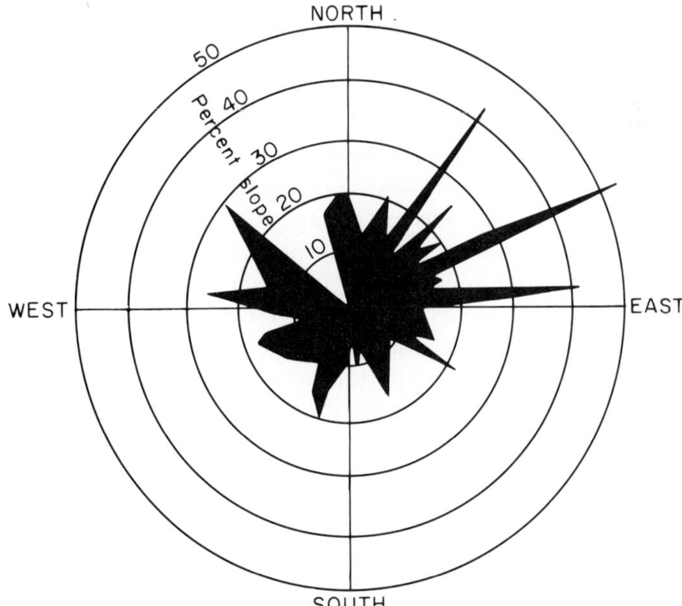

Fig. 10-9. Diagram showing relation of slope exposure to degree of slope at 50 measurement sites in eastern Wyoming (from Hadley 1961).

northerly-facing slopes. The sparse cover on southerly-facing slopes is the result of a soil-moisture deficiency due to higher evapotranspiration rates.

Drainage density, expressed as the total length of channels per unit drainage area (km/km^2), is greater for south-facing slopes than for north-facing slopes. This is a further reflection of higher erosion rates and sediment yields accompanied by more surface runoff on south-facing slopes. Table 10-2 presents the results of drainage density measurements in six small drainage basins in eastern Wyoming. The data show that the drainage on the south-facing sides of the basins is more than twice that on the north-facing sides.

Table 10-2. Drainage density, for six basins in east-central Wyoming.

		Drainage density	
Drainage basin	Area (km^2)	North-facing side (km/km^2)	South-facing side (km/km^2)
1	0.41	8.1	24.4
2	.26	5.3	19.5
3	.23	10.0	27.4
4	.36	10.0	13.7
5	1.97	8.1	13.7
6	.60	9.0	10.8
Average		8.4	18.3

Drainage-basin asymmetry has been expressed as the difference in valley-side slope angles on the north- and south-facing slopes within a single basin (Emery 1947, Melton 1960). Hadley (1961) has expressed basin asymmetry simply as a measure of the deviation of the main channel from a position along the central axis of the basin. Several measurements are made of the distance from the main channel to both north and south drainage divides. The lines of measurement to the divides are perpendicular to the axis of the basin along the main channel (Fig. 10-10.) The ratio of the mean value of all measurements from the channel to the northern divide to the mean value of all measurements to the southern divide is termed the *index of symmetry, (I)* (Hadley 1961). An index of 1.0 represents perfect symmetry. The basin shown in Figure 10-9 has an index of 1.97, indicating an appreciable displacement of the main channel to the south side of the basin.

The significant differences in vegetation, soil moisture, and erosion found within drainage basins with a large percentage of hillslopes oriented either to the south or north can be an important consideration in land-use planning. Land-use and land-treatment practices, such as contour furrowing, that will conserve soil moisture and control erosion on south-facing slopes can improve plant cover and reduce sediment yields appreciably.

Sediment Sources and Geomorphic Processes

Sheet, rill, and gully erosion are separate phases of the degradation process on rangelands. Most striking is the formation of rills, gullies, and arroyos. Factors contributing to gully formation are often ill-defined, although grazing, intense summer storms, and general climatic changes have been noted as

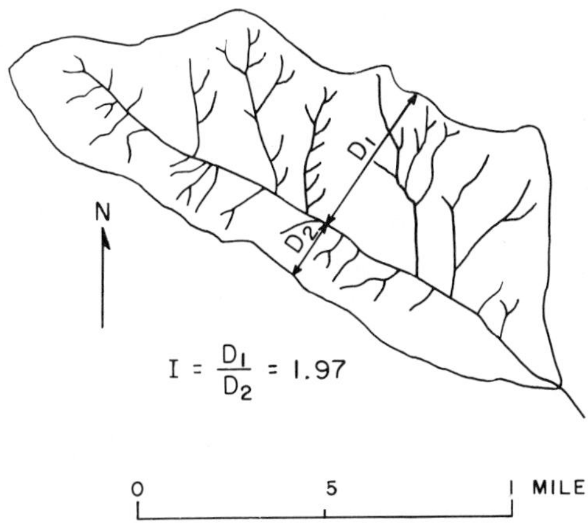

Fig. 10-10. Map of drainage basin showing drainage network and asymmetrical development. I is the index of symmetry (from Hadley 1961).

possible contributors. Some of these aspects of the western United States have been discussed by Bryan (1925, 1940), Bailey (1937, 1941), Peterson (1950), Antevs (1952), Leopold and Miller (1954), Schumm and Hadley (1957), Hadley (1960), and Bull (1964). Though sheet erosion, defined by Baur (1952) as removal of a fairly uniform layer of soil or material from the land surface by the action of rainfall and runoff, is less obvious than formation of rills and gullies, it cannot be ignored as a source of sediment (Glymph

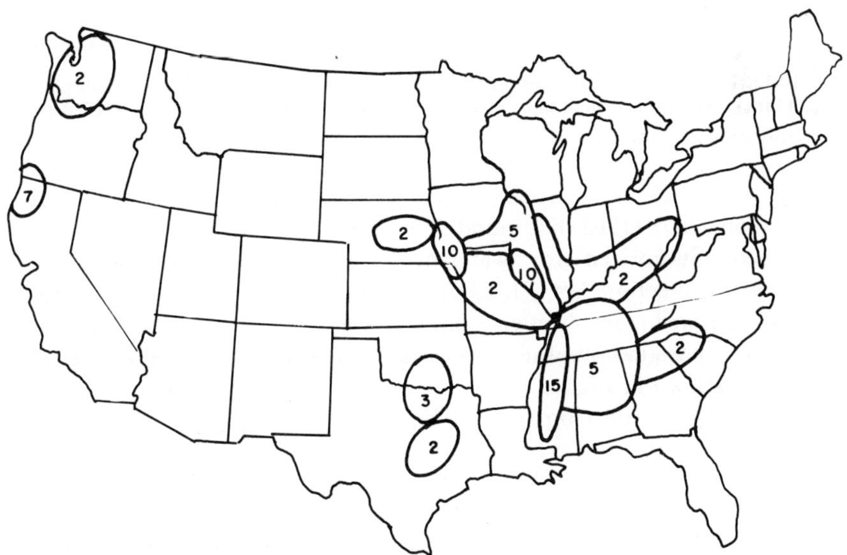

Fig. 10-11. Generalized regions of the amount of sediment in streams, expressed in tons/ha/yr (from Stall 1972, used by permission).

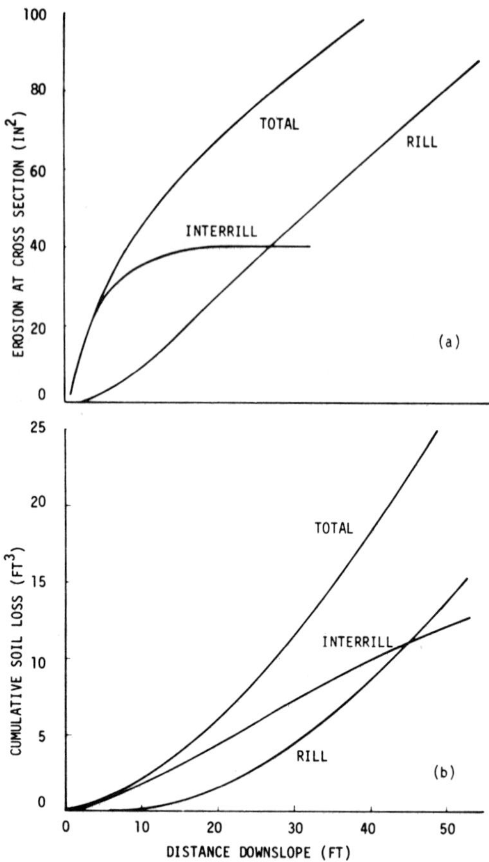

Fig. 10-12. Eroded soil apportioned between rill and interrill sources for rill-susceptible plot (a) and rill-resistant plot (b) in Indiana (Meyer, Foster and Romkens 1972). (sq. inches $\times$ 6.452 = cm^2, $ft^3 \times 0.028 = m^3$)

1957). Gottschalk (1962) has found that estimates of sediment sources in 157 watersheds authorized for watershed protection measures under the Watershed Protection Act (PL 83-566) indicated that 73% of their sediment yield was derived from sheet erosion, and 10% from gully erosion and flood plain scour. The amount of sediment in streams has been generalized in some regions as shown in Figure 10-11.

Meyer, Foster, and Romkens (1972) have shown the geomorphic factors that affect the source of eroded soil is related to rill suceptibility and slope shape (Figs. 10-12 and 10-13). Besides the influence of rilling, the length, steepness, shape of the slope, and the vegetation canopy, they showed a breakdown in erosion resistance. The location where the eroded soil originates and the erosive agents that are dominant in detaching and transporting the soil can significantly affect the particle size characteristics of the sediment.

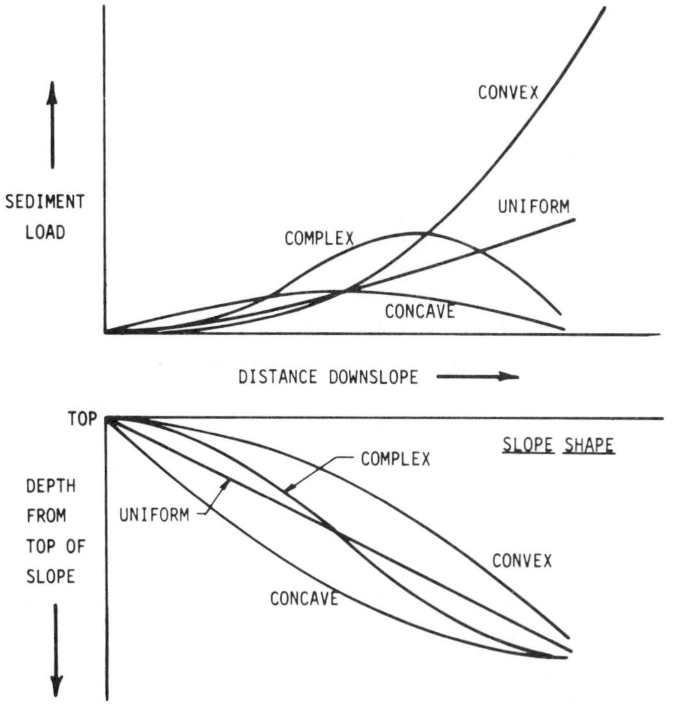

Fig. 10-13. Influence of slope shape on sediment load and rate of erosion. Steepness of sediment-load curves indicates rate of erosion and negative slope indicates deposition (Meyer, Foster, and Romkens 1972).

Fig. 10-14. Map of aerial extent and severity of landslide problems in the United States (from Swanston 1971).

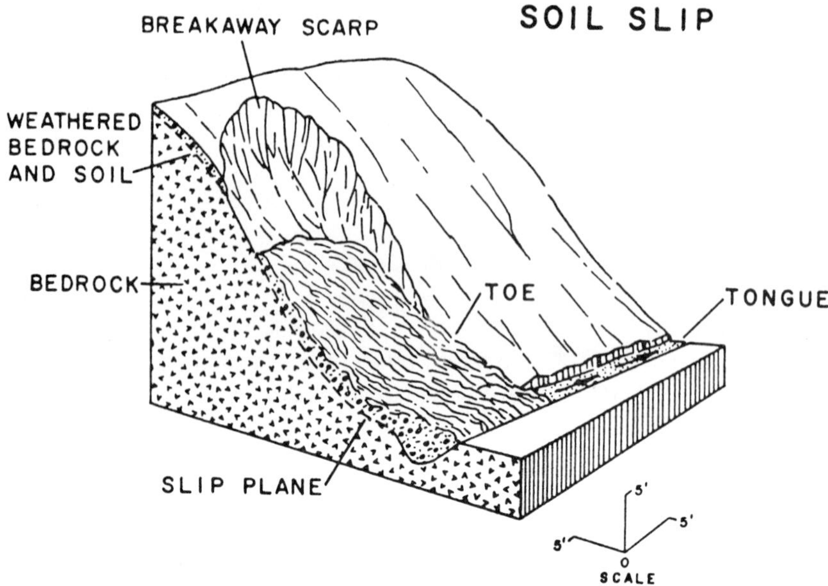

Fig. 10-15. Idealized diagram of a soil slip involving saturated soil and rock debris. Periodic flushing of the channel by stormflow allows material covering lower portion of slip scar to creep down the slope and into the channel. This cycle contributes large quantities of debris to flood waters (from Bailey and Rice 1969).

Fig. 10-16. Piping erosion along a tributary of the Rio Pureco 15 miles (24 km) west of Albuquerque, New Mexico.

Obviously sediment, like runoff, is not generated uniformly throughout the watershed.

Soil slips (small landslides) constitute another type of erosional agent. These phenomena are common in the mountainous regions in southern California (Fig. 10-14). They also occur elsewhere where they are associated with intense, high-volume storms. Bailey and Rice (1969) note that slips often exhibit four distinct features: 1) a slip plane, formed where the presence of accumulated stresses exceeds the resistance to shear; 2) a breakaway scrap, formed as sliding and nonsliding materials separate; 3) a toe, where detached materials come to rest; and 4) a tongue, semifluid mud which extends from the base of some slips down the drainage channel (Fig. 10-15). Results of studies by Anderson, Coleman, and Zinke (1959), Corbett and Rice (1966), Bailey and Rice (1969), Rice, Corbett, and Bailey (1969), and Scott and Williams (1974) in the San Gabriel Mountains of California show that soil slips are indicators of unstable slope conditions and are directly related to shallow soils (2 ft [0.6 m] or less), oversteeped slopes (80% or greater), heavy and prolonged rainfall, and insufficient vegetational protection (rooting habits and density being most important). Land use in such areas should be minimized to ensure maximum vegetative protection and the additional stability afforded by deep vegetation roots.

Another type of erosion that is common in arid and semiarid rangelands is termed "piping." This phenomenon appears when the subsoil erodes from under the surface leaving tunnels (Fig. 10-16). In order for piping to take place, there must be a source of water, the surface infiltration rate must exceed the permeability rate of some subsurface layer, there must be an erodible layer (immediately above the retarding layer, the water above the retarding layer must be under a hydraulic gradient, and there must be an outlet for the lateral flow (Fletcher et al. 1954). Rodent activity and dessication cracks near existing channels can often be triggering agents for "piping." Erosion from these intensive sources is often difficult to predict and is not accounted for in most prediction equations.

Literature Cited

Antevs, E. 1952. Arroyo-cutting and filling. J. Geol. 60: 375-385.

Anderson, H.W., G.B. Coleman, and P.W. Zinke. 1959. Summer slides and winter scour-dry-wet erosion in southern California mountains. U.S. Forest Serv., Pacific Southwest Forest and Range Exp. Sta. Tech. Pap. 36. 12 p.

Bailey, R.W. 1937. A new epicycle of erosion. J. Forest. 35: 997-1005.

Bailey, R.W. 1941. Land erosion—normal and accelerated—in the semiarid west.Trans. Amer. Geophys. Union 22: 240-250.

Bailey, R.G., and R.M. Rice. 1969. Soil slippage: an indicator of slope instability on Chaparral watersheds of southern California. Prof. Geogr. 21(3): 172-177.

Baur, A.J. 1952. Soil and water conservation glossary. J. Soil and Water Conserv. 7: 41-52, 93-104, 144-156.

Branson, F.A., and J.R. Owen. 1970. Plant cover, runoff, and sediment yield relationships on Mancos Shale in Western Colorado. Water Resources Res. 6: 783-790.

Bryan, K. 1925. Date of channel trenching (arroyo cutting) in the arid Southwest. Science 62: 338-344.

Bryan, K. 1940. Erosion in the valleys of the Southwest. New Mexico Quart. 10: 227-232.

Bull, W.B. 1964. History and causes of channel trenching in western Fresno County, California. Amer. J. Sci. 262: 249-258.

Corbett, E.S., and R.M. Rice. 1966. Soil slippage increased by brush conversion.USDA Forest Serv. Res. Note PSW-128. 8 p.;

Emery, K.O. 1947. Asymmetrical valleys of San Diego County, California. Bull. So. Calif. Acad. Sci. 46(2): 61-70.

Fletcher, J.E., K. Harris, H.B. Peterson, and V.N. Chandler. 1954. Piping. Trans. Amer. Geophys. Union 35: 258-262.

Gottschalk, L.C. 1962. Effects of watershed protection measures on reduction of erosion and sediment damages in the United States. Int. Ass. Sci. Hydrol. Pub. 59: 426-447.

Glymph, L.M., Jr. 1957. Importance of sheet erosion as a source of sediment. Trans. Amer. Geophys. Union 34: 419-426.

Hadley, R.F. 1960. Recent sedimentation and erosional history of Fivemile Creek, Fremont County, Wyoming. U.S. Geol. Survey Prof. Pap. 352-A: 1-16.

Hadley, R.F. 1961. Some effects of microclimate on slope morphology and drainage basin development. U.S. Geol. Surv. Res., 1961. Pap. No. 16: B-32—B-34.

Hadley, R.F., and S.A. Schumm. 1961. Sediment sources and drainage basin characteristics in upper Cheyenne River basin. U.S. Geol. Surv. Water-Supply Paper 1531-B: 137-197.

Hadley, R.F., and G.C. Lusby. 1967. Runoff and hillslope erosion resulting from a high-intensity thunderstorm near Mack, western Colorado. Water Res. 3(1): 139-143.

Hains, C.F., D.M. Van Sickle, and H.V. Peterson. 1952. Sedimentation rates in small reservoirs in the Little Colorado River basin. U.S. Geol. Surv. Water-Supply Pap. 1110-D: 129-155.

Horton, R.E. 1932. Drainage basin characteristics. Trans. Amer. Geophys. Union 350-361.

Horton, R.E. 1945. Erosional development of streams and their drainage basins—hydrophysical approach to quantitative morphology. Bull. Geol. Soc. Amer. 56: 275-370.

King, N.J., and M.M. Mace. 1953. Sedimentation in small reservoirs on the San Rafael Swell, Utah. U.S. Geol. Surv. Circ. 256. 21 p.

Langbein, W.B., and others. 1947. Topographic characteristics of drainage basins. U.S. Geol. Surv. Water-Supply Pap. 968-C. p. 99-114.

Leopold, L.B., and J.P. Miller. 1954. A postglacial chronology for some alluvial valleys in Wyoming. U.S. Geol. Surv. Water-Supply Pap. 1261. 90 p.

Leopold, L.B., and J.P. Miller. 1956. Ephemeral streams—hydraulic factors and their relation to the drainage net. U.S. Geol. Surv. Prof. Pap. 282-A. p. 1-36.

Leopold, L.B., and M.G. Wolman, and J.P. Miller. 1964. Fluvial processes in Geomorphology. W.H. Freeman and Co., San Francisco and London. 522 p.

Lusby, G.C. 1979. Effects of converting sagebrush cover to grass on the hydrology of small watersheds at Boco Mountain, Colorado. U.S. Geol. Surv. Water-Supply Pap. 1532: J. 22 p.

Melton, M.A. 1960. Intravalley variation in slope angles related to microclimate and erosional environment. Bull. Geol. Soc. Amer. 71: 133-144.

Meyer, L.D., G.R. Foster, and M.J.M. Romkens. 1972. Source of soil eroded by water from upland slopes, p. 177-189. *In:* Present and Prospective Technology for Predicting Sediment Yields and Sources. U.S. Dep. Agr. Res. Serv. ARS-S-40.

Miller, V.C. 1953. A quantitative geomorphic study of drainage basin characteristics in the Clinch Mountain area, Virginia and Tennessee. Columbia Univ., Dep. of Geol., Tech. Rep. 3. 30 p.

Morisawa, M. 1959. Relation of quantitative geomorphology to streamflow in representative watersheds of the Appalachin Plateau province, p. 11-13. Columbia Univ., Dep. of Geol., Off. Naval Res. Geog. Branch, New York. *In:* Project NR 389-042, Tech. Rep. 20.

Peltier, L. 1950. The geographic cycle in periglacial regions as it is related to climatic geomorphology. Ann. Ass. Amer. Geogr. 40: 214-236.

Peterson, H.V. 1950. The problem of gullying in western valleys, p. 407-434. *In:* Applied Sedimentation. John Wiley & Sons, Inc. New York.

Rice, R.M., E.S. Corbett, and R.G. Bailey. 1969. Soil slips related to vegetation, topography, and soil in Southern California. Water Resources Res. 5: 647-659.

Schumm, S.A. 1955. The relation of drainage basin relief to sediment loss. International Union Geodesy and Geophys. 10th General Assembly (Rome). Trans. 1: 216-219.

Schumm, S.A. 1956. Evolution of drainage systems and slopes in badlands at Perth Amboy, New Jersey. Bull. Geol. Soc. Amer. 67: 597-646.

Schumm, S.A., and R.F. Hadley. 1957. Arroyos and the semiarid cycle of erosion. Amer. J. Sci. 255: 161-174.

Scott, K.M., and R.P. Williams. 1974. Erosion and sediment yields in mountain watersheds of the Transverse Ranges, Ventura and Los Angeles Counties, California—analysis of rates and processes. U.S. Geol. Surv. Water-Resources Invest. 47-73: 66 p.

Stall, J.B. 1972. Effects of sediment on water quality. J. Environ. Quality 1: 353-360.

Stoddart, D.R. 1969. Climatic geomorphology. R.J. Chorley (ed.) Chapter 10, p. 189-200. *In:* Introduction to Fluvial processes, Metheun and Co. Ltd. London.

Strahler, A.N. 1952. Dynamic basis of geomorphology. Bull. Geol. Soc. Amer. 63: 923-938.

Strahler, A.N. 1957. Quantitative analysis of watershed geomorphology. Trans. Amer. Geophys. Union 38: 913-920.

Strahler, A.N. 1964. Quantitative geomorphology of drainage basins and channel networks, p. 40-74. *In:* Handbook of Applied Hydrology, Ven Te Chow, editor. McGraw-Hill Book., New York. Section 4-II.

Swanston, D.N. 1971. Judging impact and damage of timber harvesting forest soils in mountainous regions of Western North America, p. 1-6. *In:* Proc. Western Reforestation Coordinating Comm., Western Forest. & Conserv. Ass., Portland, Ore. Nov. 30, 1971.

Chapter 11

Rangeland Hydrologic Models: Theory to Practice

Introduction to Modeling

Hydrologic models provide a way to transfer knowledge from a measured or study area to an unmeasured area where management decisions are needed. They provide a quantitative expression of that which is being observed, analyzed, or predicted. To address the problems of hydrologic processes on rangeland watersheds, scientists have increasingly sought the insights provided through mathematical modeling. The insights are sought via investigation of properties of simple mathematical models constructed to mimic the operation of natural systems with respect to the characteristics of interest. If the model is examined and insight to its operation is gained, then the attempt is to infer similar properties, for the characteristics of interest, in the natural system. Thus, a model is generally intended to reproduce the behavior of variables (previously studied) which are determined to be important.

Dooge (1973) stated:

> is any structure, device, scheme, or procedure, real or abstract, that interrelates in a given time reference, an input, cause or stimulus, of matter, energy, or information an output, effect, or response, of information, energy, or matter.... The emphasis is on the function of the systems—that it interrelates, in some time reference, an input and an output. In mechanics, we tend to talk about inputs and outputs; physicists and philosophers often speak of causes and effects; workers in biological science talk of stimuli and responses. Reference to an input does not restrict the concept of a single input. The input could consist of a whole group of inputs so that we would have an input vector rather than a variable. In some cases, the input could be completely distributed in space and thus represented by a function of both space and time. The definition refers to inputs (and outputs) as consisting of matter, energy, or information. In some systems, both the input and output would consist of material of some sort; in others, attention would be concentrated on the input and output of energy; while in other systems, the concern would be with the input and output of information. There is no need, however, for the input and output to be alike. It is perfectly possible to have a system in which an input of matter will produce an output of information or vice versa. That there is no necessity for the natures of the input and output to be the same has been emphasized in the definition by using the reverse order to describe the natures of the input and the output. The essence of the systems—which can be real or abstract—is that it interrelates two things.

In the systems approach, attention is directed to the horizontal relationship of Figure 11-1; i.e., the concern is how the system converts an input into an output. If we can describe this transformation with a mathematical operation, we are not then concerned with the nature of the system—the vertical elements of the figure describing the components of the system, their connection

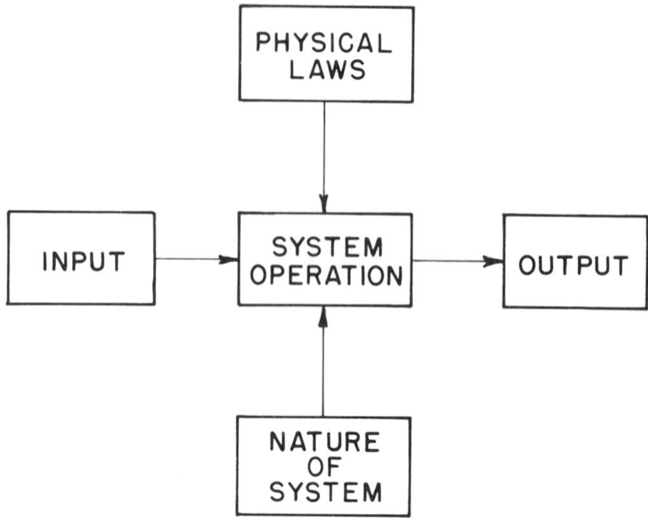

Fig. 11-1. The concept of system operation (from Dooge 1973).

with one another, or with the physical laws which are involved. A classical example of the systems approach is the use of the unit hydrograph to predict storm runoff.

In the systems approach, the details of how the system operates are gener- generally ignored because they are too complex to be easily understood or because there is no need to accurately describe the complete physical process. Thus, the model employs a conceptual analog system that is designed to perform in analogy to the system as a whole and in certain parts. This is different than an intent to emulate the actual hydraulic processes of the water paths.

If, on the other hand, the problem requires synthesis or simulation rather than analysis, it is then necessary to consider the vertical elements in Figure 11-1. In such an instance, if input and output data (e.g., precipitation and runoff) are not available for the watershed in question, one must resort to synthetic procedures (e.g., use of synthetic unit hydrograph procedures) relating parameters of the synthesis procedure with watershed characteristics. Thus, the effect of the structure of the system is considered because the physics would not be expected to change between watersheds.

Systems Classification

There are many ways to classify systems. A *physical system* is a system in the real world. A *sequential system* is a physical system which consists of input, output, and some working medium (matter, energy, or information) known as *throughput* passing through the system. A *dynamic system* is a physical system which receives certain quantitative inputs and accordingly acts concertedly under given constraints to produce certain quantitative outputs. The hydrologic system is there a physical, sequential, dynamic system. In a watershed system, for example, the throughput consists essentially of water and energy of various forms. The input-output relationship of the system may be represented mathematically by

$$y(t) = \phi.x(t) \qquad \textbf{(11-1)}$$

where x(t) and y(t) are respectively time functions of input and output, and ϕ is the transfer function which represents the operation performed by the system on the input to transform it into output. As previously mentioned, the well-known unit hydrograph is a transfer function of the watershed system.

When the operation of the system is independent of the output obtained, the system is an *open-loop system*. If the operation is dependent on the output of a feed back action, the system is a *closed-loop system*. Most local or regional hydrologic systems are open-looped because their output does not significantly exert any control of the system. On the other hand, the groundwater system on an island and the global hydrological system may be considered as closed-looped because the output of such systems will affect to a large extent the balance of the system. For such systems, additional boundary conditions, due to feed-back systems, should be considered in the system analysis.

Furthermore, if the behavior of the system can be altered by modification of the system, the steps taken are known as *exerting control*. The exerting control on hydrologic systems may be either natural or artificial. In a watershed system, for example, the climatic trend and cycles are the natural exerting control that may alter the characteristics of the watershed. The man-made dams and reservoirs on river basins represent artificial exerting controls. The exact modification of the hydrologic system and the true nature of the exerting control are often not clearly understood.

If we concentrate on the relationship between the three elements involved—input, system, and output—then problems which arise can be conveniently classified. Such a classification is shown in Figure 11-2. Where

PROBLEMS ARISING WITH SYSTEMS

TYPE OF PROBLEM		INPUT	SYSTEM	OUTPUT
ANALYSIS	PREDICTION	√	√	?
	IDENTIFICATION	√	?	√
	DETECTION	?	√	√
SYNTHESIS (SIMULATION)		√	??	√

Fig. 11-2. Classification of systems problems (from Dooge 1973).

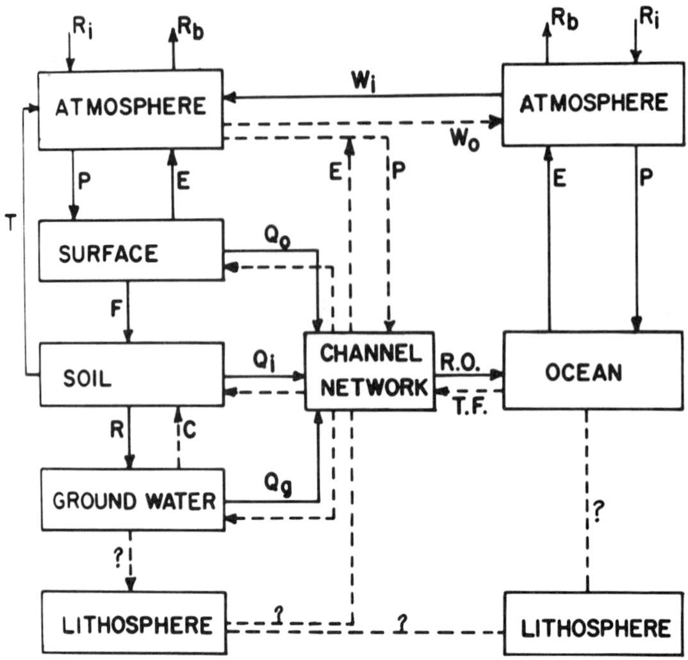

Fig. 11-3. Block diagram of the hydrologic cycle (from Dooge 1973). Symbols used are:
C = capacity rise, E =, F = infiltration, P = precipitation
Q_g = groundwater, Q_i interflow, Q_o overflow,
R = recharge of groundwater; R_b = back radiation,
R = incoming radiation, R.O. = runoff, T = transpiration, tidal flow, W_i = incoming atmosperhic moisture, and W_o = outgoing atmospheric moisture.

two factors are known and the third is required, we have a problem in analysis. Such problems, as noted by Dooge (1973), can be classified according to the unknown output, the problem of identifying an unknown system operation, and the problem of detecting an unknown input signal. Problems of synthesis are less frequently encountered in watershed analysis work—yet they may arise. In the study of our hydrological systems, we are frequently required to solve the problem of synthesis involved in the simulation of a system (abstract mathematical system or a real model) for which records of input and output are available.

Hydrologic Systems

Textbooks on hydrology commonly represent the hydrologic cycle in pictorial form. Such a picture represents a high degree of abstraction from the actual hydrologic cycle. The cycle has been lumped insofar as all the land surface and ocean surface have been taken as single elements and the complexities of local variation ignored. It is on the basis of such abstractions that problems of hydrological analysis and synthesis have been presented in the past.

However, using a systems approach, we would represent the hydrologic

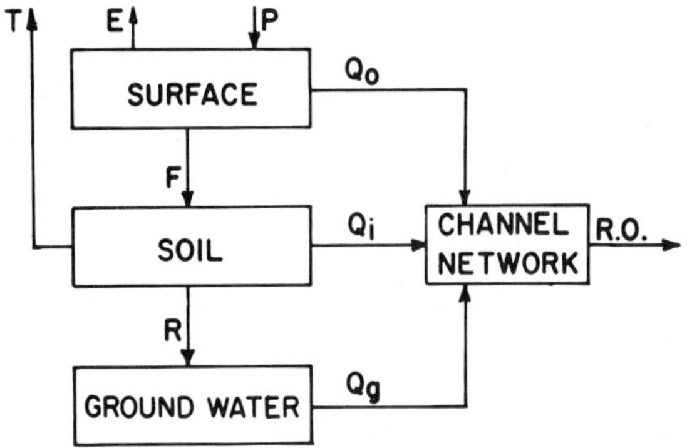

Fig. 11-4. The catchment as a system (Dooge 1973). See Fig. 11-3 for definition of symbols.

cycle in the form shown in Figure 11-3. This diagram actually contains more information and reflects a lesser degree of abstraction than does Figure 2-4 (Chapter 2). The cycle operates as a result of the surplus of incoming radiation (r_i) over back radiation (R_b).

In rangeland hydrology we are concerned primarily with individual basins or catchment areas. Following the discussion by Dooge (1973), let's narrow our concern to the particular subsystem of the total hydrologic cycle shown in Figure 11-4. In this isolation process, it is necessary to cut across certain lines of transport of moisture from one part of the cycle to the other. The broken lines (Fig. 11-3) thus produced represent either inputs or outputs from the subsystem representing the catchment area.

Though classical hydrology describes the hydrologic cycle in terms of *surface runoff, interflow,* and *ground-water flow,* in practice quantitative hydrology usually ignores this three-fold division and considers the hydrograph being made up of a *direct storm response* and a *base flow.* Thus, in the analysis of the relationship between rainfall and runoff, the system

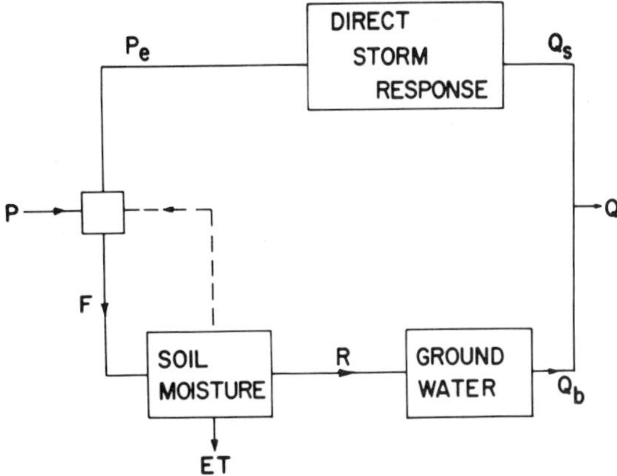

Fig. 11-5. The simplified catchment model (from Dooge 1973). Pe = precipitation excess, Q^b = flow, and Q^a = quantity of runoff. See Fig. 11-3. for definitions of other symbols.

 analyzed by the practical rangeland hydrologist corresponds closely to that indicated in Figure 11-5. The system shown is seen to consist of three subsystems: (1) the subsystem involving direct storm response, (2) the subsystems involving ground water response, and (3) the subsystem involving the soil phase which has a feedback loop to the separation of precipitation excess and infiltration.

It is worthwhile noting that a systems approach has thus far been limited to analysis of the subsystem involving direct storm response, which has been studied by unit hydrograph procedures. The other two subsystems represent difficult unknowns which must be approached in other ways.

Methods of System Investigation

The methods of system investigation fall into two principal categories:

1. Parametric hydrology.
2. Stochastic hydrology.

Parametric Hydrology

As defined by Amorocho and Hart (1964), parametric hydrology is the development of relationships to generate, or synthesize, nonrecorded hydrologic sequences. From the earliest days in hydrology, simple parametric models of surface runoff have been used. The first models usually related runoff to rainfall and drainage area and took the form either of purely empirical equations or a design procedure such as the rational method. As understanding was gained, the models used became more complex, although the necessity for hand computation limited the complexity. With increasing complexity a division in the direction of model development evolved. This division was into component modeling on the one hand and integrated system modeling on the other.

a. *Component modeling.*—The land phase of the hydrologic cycle can be broken into several components. The major components might be considered infiltration, evapotranspiration, aquifer response, and the routing of both overland flow and channel flow. Each of these components has attracted specialists who have developed empirical approximations to the processes controlling the component. The development has been toward an understanding of the physical laws governing the component and an attempt to make the empirical models approximate more closely the underlying physical situation.

b. *Integrated system modeling.*—One of the purposes of developing better conceptual models for the individual components (infiltration, evapotranspiration, routing of overland and channel flows, etc.) was to improve the overall modeling of the total system. Integrated system modeling involves all aspects of the land phase of the hydrologic cycle, and it borrows on knowledge gained from all aspects of component modeling. However, the development added complexity, which, to an extent, limited the ability to use the better models. The advent of electronic digital computers removed most limitations on the development of integrated system models.

Uses of Parametric Models

Parametric models usually have been developed to help in the solution of very practical operation problems. They have demonstrated their utility in many different types of situations. Examples include the design of the capacity of a small reservoir, the determination of the effect of channel improvements upon the flood-frequency characteristics of catchment, the effect of urbanization on flood peaks, and the extension of streamflow records for small basins on the basis of rainfall records, as well as for the solution of problems concerning aquifer response. The well-known and widely used flood-runoff routing equations are further examples of practical use of parametric modeling. Parametric models have developed into a powerful tool with many applications in water resource investigations.

Stochastic and/or Probabilistic Hydrology

As defined by Amorocho and Hart (1964), stochastic hydrology is the use of statistical characteristics of variables to solve problems. The word stochastic comes from the Greek word, *stochastikos,* meaning skillful in aiming. Assuming that this refers to a person's ability to shoot a bow and arrow, then the indicator of his ability would be the nearness of the arrows to the bullseye in the target he was shooting at. Upon examination of the target, one would find the density of arrows greatest near the center and least around the edge. In addition, one would find that the location of each arrow as it occurred would be random. Thus, the word, stochastic, has come to refer to the random nature of a variate with respect to time. In watershed modeling, it generally referes to the random nature of hydrologic phenomena, such as river flow, precipitation, wind velocity, etc.

The stochastic approach to modeling basically means that the distribution characteristics of a variate are determined from sample data and used with a random number generator to produce synthetic sequences of the variate. The distribution characteristics considered in such processes are the mean, standard deviation, skew, and serial correlation. If there is more than one variate, the cross correlation among variates is considered.

A stochastic system is similar to a probabilistic system except the former is *time-dependent* (sequence of variables is observed and the variables may or may not vary with time) and the latter is *time-independent* (sequence of variable occurrence is ignored and the chance of their occurrence is assumed to follow a definite probability distribution in which variables are considered random).

Although the various methods of systems investigations have certain basic differences, they share two characteristics of prime importance:

1. Their dependence on historical records of the values of the parameters.
2. The assumption of stationariness or time invariance of the hydrologic systems.

As pointed out by Amorocho and Hart (1964), if we consider precipitation and runoff to be the principal system inputs and outputs, respectively, then the first characteristic means that, to the extent that historical records of the input and output are affected by systematic and random errors of measurement, by inhomogeneity, and by lack of completeness, the results or parametric of stochastic hydrology are affected also. The second characteristic requires that hydrological systems must not change with time, relative to their behavior during the recorded year.

Stochastic modeling is a relative newcomer to the science of hydrology. The approach in stochastic modeling is to give a statistical simulation of the measured response of the system. The statistical simulation is based upon the stochastic model, for which the parameters must be derived. With the statistical model, many "equally likely" time series can be simulated and used in planning and design of water resource systems.

a. *Methods of stochastic simulation*—Three major problems face the stochastic modeler. First is the choice of a proper model for describing hydrologic phenomena; second is the choice of a proper probability distribution for the input and the estimation of parameters for simulation; third is the prediction of probability distributions for the system output when the system and the distribution of inputs are known.

In order to use a stochastic model, some probability distribution for the input must be assumed. Therefore, there has been a considerable effort to determine the statistical properties of hydrologic data, particularly of streamflow and rainfall. Both major cyclic phenomena and daily flow sequences for streamflow have been studied. Rainfall has been studied mainly to determine a probability distribution for storm rainfall, for broad regional values of rainfall intensity-frequency relations, or for modeling rainfall for simulation purposes. Some attention has also been directed at water-quality data. One of the problems of stochastic hydrology is that an operational assumption must be made concerning the distribution of flows. However, statistical theory can be used to place levels of accuracy upon predictions which include the errors introduced by the assumption of both distribution and parameter values.

b. *Uses of stochastic models.*—Stochastic models have been widely used for system analysis and synthesis, particularly for the simulation of inputs (e.g., the temporal and spatial variability of precipitation) to complex systems.

Representation of System Processes

Following the identification of a prototype system, the various system processes may be represented by physical or mathematical models. Figure 11-6 from Riley and Hawkins (1976) indicates that the two general categories of mathematical modeling are simulation and mathematical programming. Riley and Hawkins state:

> Mathematical programming is an optimizing procedure whereby a solution is sought in terms of a specific objective function. Frequently, this procedure requires considerable simplification of the real system. Simulation is an attempt to represent as realistically as possible (or necessary) the processes of the real world. Simulation by physical models has found application to many practical problems, such as the design of highway bridges and hydraulic structures. However, for complex systems such as those encountered in water

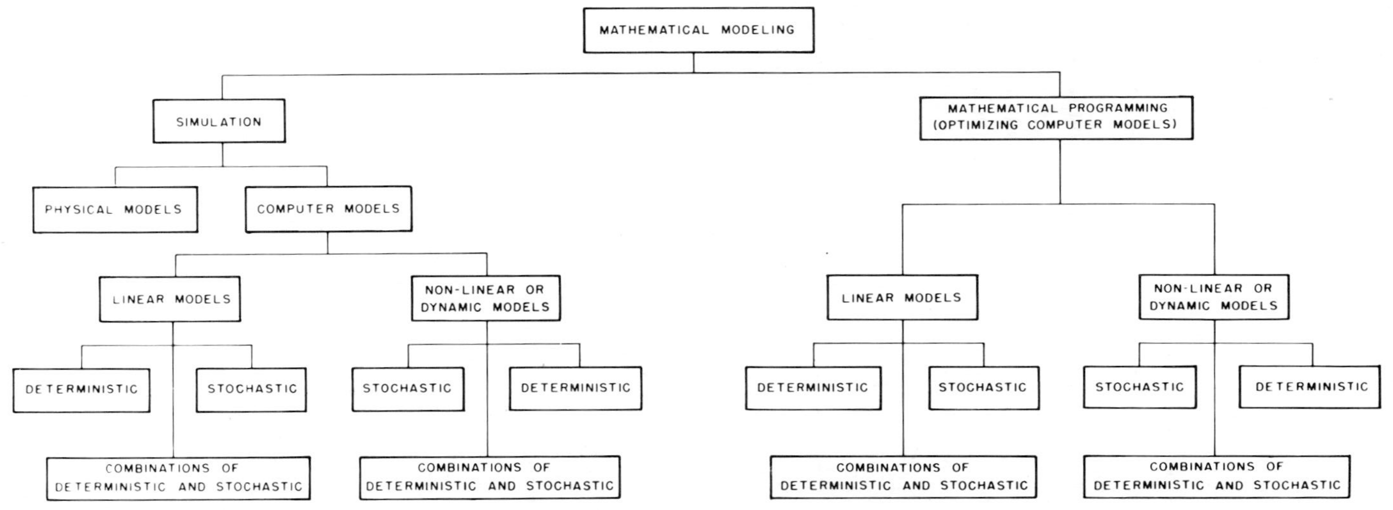

Fig. 11-6. A classification diagram for mathematical modeling (from Riley and Hawkins 1976).

resource management, mathematical simulation often proves to be the only feasible tool for predicting the system behavior.

Mathematical simulation is achieved by using algebraic relationships to represent the various processes and functions of the prototype systems, and by linking these equations into a systems model. Hopefully, simulation models have three basic properties: realism, precision, and generality.

Thus, computer simulation is basically a technique for analysis whereby a model is developed for investigating the behavior or performance of a dynamic prototype system subject to particular constraints and input functions. The model behaves like the prototype system with regard to certain selected variables, and can be used to predict probable responses when some of the system parameters or input functions are altered.

As illustrated by Figure 11-6, it is possible to employ either stochastic or deterministic techniques, or various combinations of both, in the representation of a system. The approach adopted is dependent upon a number of conditions, including availability of information about the system and the kinds of problems which the model is required to solve. The predictive power of the model within the system response space usually will vary with the degree to which the model is stochastic or deterministic.

Table 11-1 lists some advantages of computer simulation models as seen by Riley and Hawkins. There are disadvantages as well but they are generally much fewer than the advantages. The models require special expertise to operate. They can be very expensive and can develop a false sense of security

Table 11-1. Some advantages of simulation modeling.[1]

Basically, computer simulation models are advantageous because of:

1. The answers they give:
   a. Some answers otherwise unattainable.
   b. Evaluation of a wide array of alternatives.
   c. Non-destructive testing.
   d. Distribution of errors and judgment variations among several coefficients.
   e. Allows assembly of many processes, etc., into an integrated package.
2. The questions they ask:
   a. Indications are provided in quantitative terms of progress toward system definition and conceptual understanding.
   b. The relative importance of various system processes and input functions is suggested.
   c. Priorities are suggested in terms of planning objectives and data acquisition.
   d. A clear identification is required of problems and objectives associated with the system being studied.
3. The insights they provide:
   a. A basis for coordinating information and efforts of personnel across a broad spectrum of scientific disciplines
   b. Models are a very effective teaching device.

[1]From Riley and Hawkins (1976).

about the accuracy of the input when used "blindly."

The development of a working mathematical model of a hydrologic system requires two major steps: creation of the conceptual model and development of a working computer model.

The first step, conceptualization, requires representing the various elements and systems of the real world (Fig. 11-7). The conceptualization includes using the known information to postulate hypotheses of the system elements and their interrelationship. Efforts are made to use the available data to create the conceptual model which are subsequently revised or improved as new information becomes available.

In the second step, the working computer model is made to quantify the various processes and relationships identified in the conceptual model. This step generally involves further simplification with the result that the model becomes a rather gross approximation of reality.

The important steps involved in the model development process are shown in Figure 11-8. Detailed explanations of this process are given by Riley and Hawkins (1976).

Examples of Hydrologic Models for Rangeland Watersheds

There are presently numerous hydrologic models in use, each having unique features making it efficient in problem solving for the objectives for which it was developed. One might ask why we need so many models, and the answer seems obvious. Each model and the approximations used in develop-

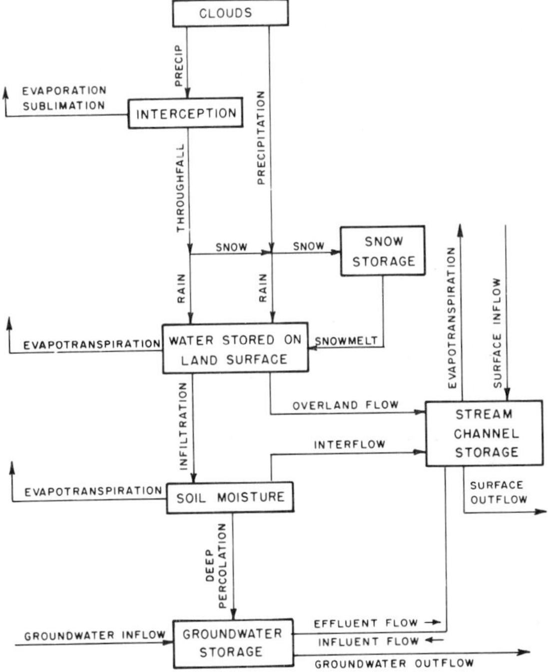

Fig. 11-7. A flow diagram of the hydrologic system within a typical drainage basin (from Riley and Hawkins 1976).

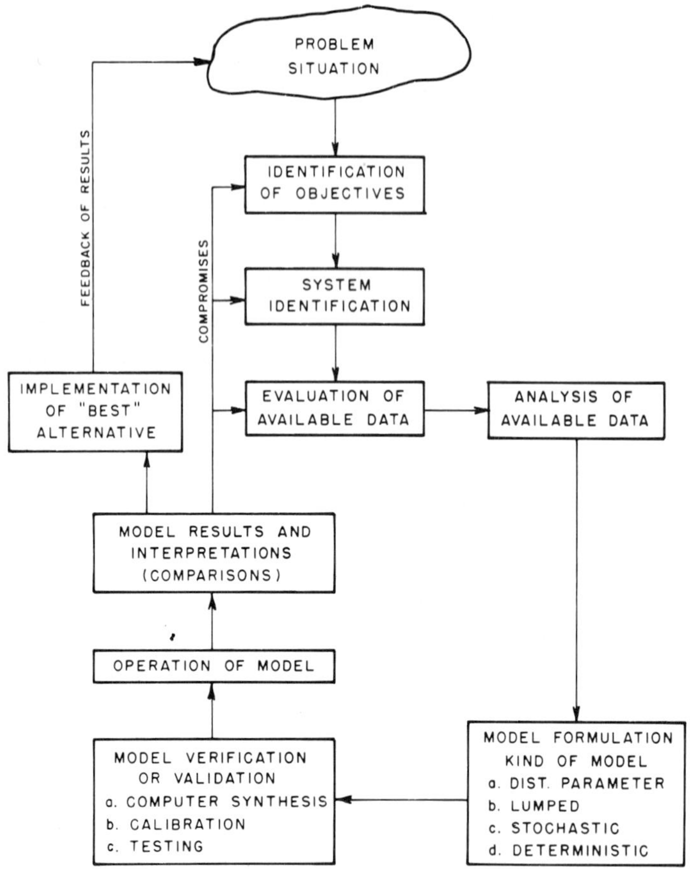

Fig. 11-8. Steps in the development and application of a simulation model (from Riley and Hawkins 1976).

ing it affect the application of the model for new objectives or a different climatic and physiographic set of conditions. An analogy to the transportation industry might be made where one would try to develop one truck to handle all needs. Such a truck would be undoubtedly very large and complicated and in most instances would work efficiently only a small part of the time. In other instances, we would have to make numerous trips with pieces of the product to be transported. The same situations would arise with hydrologic models wherein they would be horribly complicated and inefficient in some instances, and require multiple operations on pieces of the problem in other instances.

Parametric Methods

In correlation and regression analysis procedures, various combinations of variables are tested to explore their effects on the hydrologic system. The combination that yields a relationship which most closely approximates the

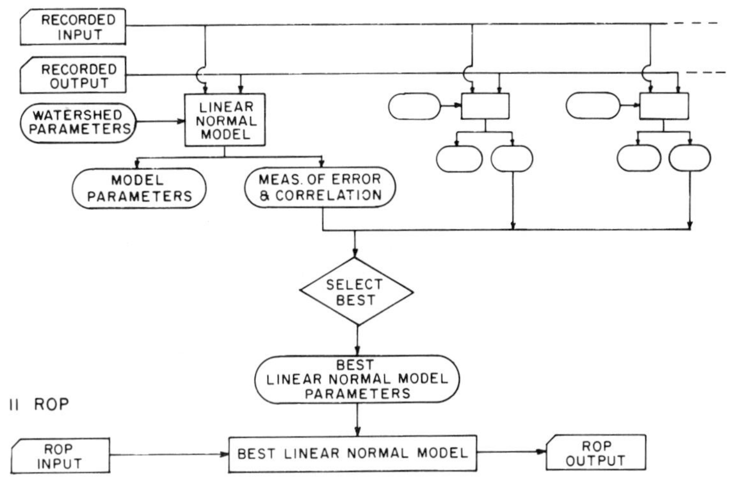

Fig. 11-9. Flow chart of the method of correlation analysis (from Amorocho and Hart 1964). ROP refers to reconstruction or prediction (and formulation of the best linear model).

recorded output function in terms of the recorded input function and other arbitrary parameters is adopted as the best production equation. Figure 11-9 illustrates a typical sequence of operations.

Precipitation and runoff are usually the primary input and output parameters. Amorocho and Hart (1964) describe the procedure as follows:

> The input is operated on by a system that has the form of a linear normal model involving additional arbitrary watershed parameters. These parameters are chosen by judgment for a number of models on the basis of a general knowledge of the hydrology of the watershed. Each individual model yields both the series of regession coefficients as determined by methods of regression analysis, and a set of error estimates and measures of correlation, as determined by analysis of variance. By examining the error estimates of all models and the measures of correlation, the best model is selected. A great deal of subjectivity underlies the entire process. Not only is it possible to obtain almost equally good prediction equations on the basis of sets of different parameters, but judgment must be exercised to avoid using physically irrelevant parameters or parameters which possess strong independence.

Figure 11-10 illustrates one type of regression model where the peak annual flood is operated on or determined by a number of watershed parameters. In Figure 11-10 Q_T = annual peak discharge in cfs for a recurrence interval of T years; A is the drainage area in square miles; S is the main channel slope in feet per mile; S_t is a measure of the surface storage area; I is the 24-hour rainfall in inches for a recurrence period of T years; t is a measure of freezing conditions in midwinter; and 0 is an orographical factor.

An example of the use of such a model is Benson's (1962) study of 164 station records in New England. Interestingly, in this work, inclusion of the precipitation factor (I) does not improve the multiple correlation coefficient

$$Q_T = a(A)^b (S)^c (S_t)^d (I)^e (t)^f (O)^g \qquad (1a)$$

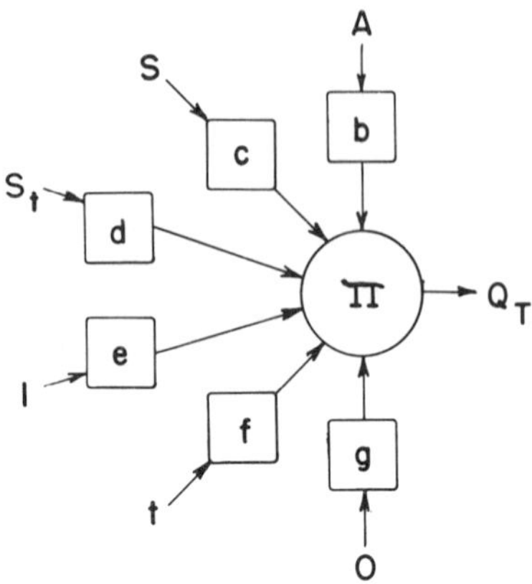

$$\log Q_T = \log a + b(\log A) + c(\log S) + d(\log S_t) + e(\log I) + f(\log t) + g(\log O) \qquad (1b)$$

Fig. 11-10. Regression models (from Dooge 1973). 1b represents a standard regression model and 1b represents a by transformation of a standard regression model. See text for definition of symbols.

(R) beyond the 0.959 shown for equation 11-2

$$Q_T = 4.52 \frac{AS^{0.4}t^{0.4}O^{1.1}}{S_t^{0.3}} \qquad \textbf{(11-2)}$$

which is the final regression equation with simplified exponents. There was no appreciable loss of accuracy with the simplification. The fact that the rainfall parameter did not improve the prediction accuracy suggests that it is highly correlated with the other parameters of the model.

Examples of regression models as used in rangeland areas are found in Schreiber and Kincaid (1967), Osborn and Lane (1969), Fogel and Duckstein (1970), and elsewhere.

Characteristics of the procedure include:

1. There is no assurance that among the models investigated the optimum has really been included.
2. The prediction equation cannot be generalized to other similar systems.
3. Inverse correlations with some of the variables may result when such relationships are illogical from physical considerations.

4. Correlation analyses are powerful tools when applied to the test of well-defined hypotheses in many fields of physical hydrology or in experimental study of systems that are well defined by functional relationships obtained independently.

Soil Conservation Service Curve Number Method

Perhaps the most widely used method of calculating direct runoff volumes from rainstorms is the technology first pioneered by and still strongly identified with the U.S. Soil Conservation Service (SCS). In the simplest sense it may be considered a model: deterministic, nonlinear, and parametric (*one* parameter). Because of its widespread acceptance and its ability to reflect changes in land condition, it is described in some detail here.

Much of what follows is taken from the SCS National Engineering Handbook, Section 4 (Hydrology), hereafter simply called, "NEH-4" (U.S. Soil Conservation Service 1972). This is the ultimate reference for much of the "Curve Number Method," although the presentation is authoritative rather than developmental. Background for the method is sketchy and scientific references to it are rare. Nonetheless, it is commonly applied and enjoys widespread popularity (if not understanding), and will likely be in use for the foreseeable future.

NEH-4 presents two separable parts of rainstorm hydrology: 1) the rainfall-runoff procedures described here; and 2) a unit hydrograph synthesis technique which will not be described here. Numerous alternate methods are available for the construction of composite hydrographs given the runoff volume and distribution: the unique contribution is the rainfall-runoff model.

Runoff model:

The runoff equation used is, in an elementary form

$$Q = \frac{(P - I_a)^2}{(P - I_a + S)} \qquad \textbf{(11-3)}$$

for all $P \geq I_a$; where P = storm rainfall in inches; I_a = the initial abstration or rainfall before runoff is initiated (inches); S is a watershed storage in inches, envisioned as a measure of the maximum possible difference between P and Q; and Q is the direct storm runoff in inches.

The two-parameter equation (11-3) is simplified by a relationship between I_a and S:

$$I_a = 0.2\ S \qquad \textbf{(11-4)}$$

which was found to be an average condition for many small watersheds studied by the SCS during the development of the method. Actual values as shown in NEH-4 vary between about $I_a = 0.01$ S to $I_a = 2$ S, making the substitution of (11-4) into (11-3) and simplifying yields

$$Q = \frac{(P - 0.2\ S)^2}{(P + 0.8\ S)} \qquad \textbf{(11-5)}$$

 for all $P \geq 0.2$ S. This is the SCS runoff equation as currently used.

Thus, runoff (Q) is taken as a function of storm rainfall (P) for a family of functions on the parameter S. Because S may vary from 0 to ∞, it is transformed to a more intuitively pleasing parameter called *Curve Number* (CN) by the identity

$$CN = \frac{1000}{10 + S} \tag{11-6}$$

CN may vary from 0 at $S = \infty$ (a completely pervious watershed with no possible runoff) to 100 at $S = 0$ (a completely impervious watershed with runoff equalling rainfall). Dimensional considerations suggest that the 1000 and 10 in Equation 11-6 are in inches, and thus CN is dimensionless. Sobhani (1975) proposed a metric form of

$$CN = \frac{2540}{25.4 + S} \tag{11-7}$$

Table 11-2. Descriptions of hydrologic soil groups.[1]

Group A:	Soils having high infiltration rates even when thoroughly wetted, consisted chiefly of deep (3 – 6ft +) (1 – 2m), well- to excessively drained sands (loamy sands, sandy loam and sands) and/or gravel. These soils have a high rate of water transmission and would result in a low runoff potential. 1. Paralithic soils—Granitic soils more than 20 inches deep with a deep decomposed contact zone.
Group B:	Soils having moderate infiltration rates consisting chiefly of moderately deep (20 inches +) (50 m), moderately well to well-drained soils with moderately fine to moderately coarse textures. These soils have a moderate rate of water transmission. 1. Paralithic soils—Granitic soils less than 20 inches with a decomposed contact zone.
Group C:	Soils having slow infiltration rates consisting chiefly of (1) soils with a layer that impedes the downward movement of water, and (2) soils with moderately fine to fine texture and a slow infiltration rate. These soils have a slow rate of water transmission. 1. Change of one or two percolation classes within 10 (25 cm) of the surface, depending on roots and structures. 2. Moderately fractured limestone: less than 20 inches. 3. Change in permeability: less than 20 inches 4. Soils showing moderate compaction in the upper 8 inches (20 cm) of profile.
Group D:	Soils having very slow infiltration rates when consisting chiefly of (1) clay soils with a high swelling potential; (2) soils with a high permanent water table; (3) soils with clay pan or clay layer at or near the surface; and (4) shallow soils over nearly impervious materials. These soils have a slow rate of water transmission. 1. Less than 12 inches (30 cm) of soil over flat-lying sandstones, limestones, etc. 2. Lithosols—generally averaging less than 1 inch per hour infiltration. 3. Clay layer or shallow over shale. 4. Change of two or more percolation classes. 5. Soils showing heavy compaction in the upper 8 inches of profile. 6. Non-wetting soils, dry silt sand, etc.

[1]Most major soils (to the soil series level) in the United States are classified according to their hydrologic grouping in the 1972 version of the Soil Conservation Service National Engineering Handbook.

where S is in centimeters. Equation 11-5 will then calculate runoffs in centimeters if the input P is in centimeters.

Use of the equation requires only an input storm rainfall and the Curve Number parameter for the subject watershed. Note that storm intensity plays no role in this model.

Curve numbers:

Appropriate values of CN must be selected to apply the technique. As currently outlined in NEH-4, the choice depends on four considerations: 1) soil type, 2) vegetative type, 3) cover, and 4) soil moisture as expressed through antecedent precipitation. These will be discussed individually in the following, although, as will be seen, all must be considered together when making the selection.

Soil Type: Soils may be classified according to their hydrologic behavior into four SCS classes: A, B, C, and D, with A being the most pervious and D the least. Detailed criteria are available in NEH-4 and other references for selecting the appropriate soil group for a given watershed. A short summary of these is given in Table 11-2. Table 11-3 gives an even more abbreviated form which may be used to group soils which have been described but not categorized.

Infiltration rates (f_c) may be associated with soil groupings, although such is not a complete and clear definition. Table 11- 4 outlines these rates. Furthermore, more detailed values may be assumed by breaking the soil groups into ± categories as shown in Table 11-5. The user should be aware of the approximate nature of these associations and the lack of reinforcing field data.

Vegetation Type: Adjectival descriptions, such as "sage-grass," forests," etc., define the watershed vegetal type. The user must select the most appropriate cover to fit the case under consideration. Unfortunately, information for only a limited number of types has been published; judgment should be used to select a hydrologically equivalent type for situations not covered. Table 11-6 gives catalog values and examples for pasture and range circumstances.

Cover: Curve Number is influenced by the extent of protective cover on the receiving watershed. The causative mechanism assumed is protection of the

Table 11-3. Decision table for hydrologic soil groups on the basis of drainage and depth.[1]

	Drainage class					
Depth	Excessive	Well	Moderately well	Somewhat poorly	Poorly	Very poorly
Deep	A	[2]	B	[3]	D	D–
Moderately deep	A–	B	[4]	C	D	D–
Shallow	B+	B	C	C–	D	D–
Very shallow	B	C	D	C	C	C–

[1]From Gifford, Hawkins, and Williams (1975).
[2]B+ if Aridisols, Inceptasols, and Entosols; A– otherwise.
[3]D+ on flood plains, C– on alluvial fans.
[4]C+ if Ardisols, Interceptisols, and Entosols; B– otherwise.

Table 11-4. Infiltration rates associated with hydrologic soil group.[1]

Soil Group	Infiltration Rate f_c (inches/hr)
A	> 3.00
B	1.25 – 3.00
C	0.50 – 1.25
D	< 0.50

[1]From Gifford, Hawkins, and Williams (1975).

Table 11-5. Subgroupsings of hydrologic soil groups and infiltration rates.[1]

Soil Group	Infiltration rate f_c (inches/hr)
A	3.50
A–	2.35
B+	2.71
B	2.13
B–	1.54
C+	1.13
C	0.88
C–	0.63
D+	0.42
D	0.25
D–	0.08

[1]From Gifford, Hawkins, and Williams (1975).

Table 11-6. Runoff curve numbers for hydrologic soil-cover complexes for pasture or range, in watershed condition II and I^a = 0.2S.[1,2]

Treatment or or practice	Hydrologic condition	Hydrologic soil group			
		A	B	C	D
None	Poor	68	79	86	89
None	Fair	49	69	79	84
None	Good	39	61	74	80
Contoured	Poor	47	67	81	88
Contoured	Fair	25	59	75	83
Contoured	Good	6	35	70	79

[1]Note: I_a = Initial abstraction. S = Potential maximum retention (inches). Watershed condition II defined in Table 11-8.
[2]From U.S. Soil Conserv. Serv. 1972.

ground surface from raindrop impacts and consequent erosion from puddling and surface sealing by eroded fine soil particles. Furthermore, there is undoubtedly an intrinsic associative relationship between soil properties and plant cover which enhances the cover-CN function.

The precise definition of the cover term is occasionally ambiguous. It is is usually defined simply as the fraction (percent) of the ground surface covered by live vegetation and litter. Variations in range field surveys may include standing dead vegetation, tree canopies, and small or large rocks. Most of the NEH-4 information is thought to refer to a no-rock situation.

Often simple adjective descriptions of cover are given, i.e. either "Good," "Fair," or "Poor". The relationships as shown in Table 11-7 seem to be consistent with NEH-4 and may be used.

Table 11-7. Cover adjective and present cover.[1]

Percent cover[2]			
From	To	Midpoint	Descriptive adjectives
0	30	15	Poor
30	70	50	Fair
70	100	85	Good

[1]From Enderlin and Markowitz (1962).
[2]Note: Cover without rocks.

Soil Moisture: The effects of soil moisture on storm runoff are dealt with through the index of 5-day antecedent rainfall, dependent upon the time of year. Thus, antecedent moisture conditions (I, II, or III) are defined as shown in Table 11-8. Antecedent Moisture Condition (AMC) II is taken as the reference status and CNs are adjusted up or down in accordance with categories and the design conditions as will be subsequently explained.

Table 11-8. Antecedent moisture conditions for use with curve numbers.[1]

	Rainfall (inches) in last 5 days	
AMC	Dormant season	Growing season
I	< 0.5	< 1.4
II	0.5 – 1.1	1.4 – 2.1
III	>1.1	>2.1

[1]From U.S. Soil Conserv. Serv. 1972.

To carry out the adjustment, several means are available. The ultimate reference is Table 10.1 in NEH-4, which is presented in shortened form here as Table 11-9. An approximate algebraic equivalence is given by Sobhani (1975); the following expressions are accurate (on the average) to within about one CN.

$$CN_{I} = \frac{CN_{II}}{2.334 - 0.01334\ CN_{II}} \qquad \textbf{(11-8)}$$

$$CN_{III} = \frac{CN_{II}}{0.4036 + 0.0059\ CN_{II}} \qquad \textbf{(11-9)}$$

As previously stated, AMC II is taken as the reference status; that is, unless otherwise noted, the CN moisture condition is AMC II. For example, a CN of 75 at AMC II would be 57 at AMC I, and 88 at AMC III. Calculated runoff would be likewise affected.

The origin of the CN-AMC relationships is not known, nor is its derivation

Table 11-9. Relationships between Curve Numbers (CN) and anticedent moisture condition (AMC) classes.[1]

	Classes	
I	II	III
100	100	100
87	95	98
78	90	96
70	85	94
63	80	91
57	75	88
51	70	85
45	65	82
40	60	78
35	55	74
31	50	70
22	40	60
15	30	50
9	20	37
4	10	22
0	0	0

[1]From U.S. Soil Conserv. Serv. 1972.

documented. The abrupt thresholds discourage strong faith and are perhaps an engineering convenience. As a matter of practical usage, most calculations for peak flow are done with ACM II, although most hydrologic events on rangelands occur at AMC I.

Algebraic expressions of curve numbers:

NEH-4 gives a number of tables and figures for use in Curve Number selection. They cover a variety of soil and vegetal types and covers. By utilizing these relationships and those in Tables 11-4 and 11-7, equivalent algebraic approximations for CN as a function of cover and infiltration rates can be derived for different vegetative types. These are given in Tables 11-10 and 11-11. Such are useful when dealing with computer applications of the methodology, especially when dealing with changes in cover and infiltration rates, as might result from alterations in land conditions.

Sensitivity:

In general, and for most conditions, the methodology is more sensitive to the selection of an accurate Curve Number than an equally accurate design rainfall. That is, a given relative error in the selected Curve Number will cause more error in the calculated runoff Q (from Equation 11-5) than to equally poor estimate of the input rainfall (Hawkins 1975). This situation is exacerbated by the fact that extensive information on rainfall duration and frequency is usually available, while very few detailed studies of the validity of Curve Numbers have been made.

Simanton, Renard, and Sutter (1973) showed that the Curve Numbers for watersheds in southeastern Arizona decrease with increasing watershed size (Table 11-12). This decrease reflects greater infiltration losses in the stream beds with increasing watershed size. The work was restricted to watersheds in

Table 11-10. Values of a and b for CN = a - bX where X is percent cover for various vegetation types and soil groups.

Vegetation type	Soil group	a	b	Notes
Juniper-grass	C	88	0.32	1
	B	82	0.42	1
Sage-grass	C	86.5	0.46	1
	B	73.5	0.415	1
Herbaceous	D	95	0.115	1
	C	90	0.19	1
	B	84	0.25	1
Oak-aspen	C	79	0.44	1
	B	74	0.51	1
Desert brush	D	93	0.06	2 3
	C	80	0.06	2 3
	B	84	0.06	2 3
Ponderosa pine	C	83	0.14	2 4
	B	73	0.31	2 4
Pasture or rangeland	A	77	0.56	3
	A	63	0.28	6
	B	83	0.28	
	C	89	0.18	
	D	91	0.13	5
Annual grass	A	75	0.44	3
	A	60	0.13	6
	B	83	0.26	
	C	89	0.18	
	D	91	0.13	7
Forests (P=25 inches)	A	50.5	0.286	
	B	71.5	0.229	
	C	81.5	0.229	
	D	87	0.21	7
Roads	A	73	0	
	B	83	0	
	C	88.5	0	
	D	90.5	0	
Bare rock		96	0	8
Water surfaces		100	0	

All above curve numbers for AMC II, and $I_a = 0.2S$, Cover without rocks. [1]From Enderlin and Markowitz (1962). [2]From Simanton, Renard, and Sutter (1973). [3]For X≤50%. [4]For 10%<X<80%. Nonlinear relationship to Y = 83 and Y = 73, respectively at X = 0. [5]From NEH-4 (in table form), reduced and converted to above coefficients. [6]For X>50%. Note similarity between annual grass and rangeland coefficients except for Soil Group "A". [7]From unpublished tables from U.S. Forest Service personnel (personal communication). [8]Assumes Initial Abstraction = 0.08 inch.

the hydrologic soil groups B and C (Table 11-2) having contributing areas less than 560 acres (227 ha). On larger areas, the precipitation variability and channel losses would require using a distributed model with weighted Curve Numbers. A single Curve Number value may be meaningful when dealing with large storms on homogeneous agricultural watersheds, as found in the eastern United States, but small storms, trivial runoffs, and great land variety are common in many western wildland situations. The difference between the results using an average Curve Number and weighted curve numbers as previously shown may be the contrast between *no* runoff and some runoff, which may, indeed, be important.

In order to deal with this, a distributed form of Formula 11-5 may be used, as given in the following

$$Q = \sum_{i=1}^{n} a_i \frac{(P - 0.2S_i)^2}{P + 0.8S_i} \quad \textbf{(11-10)}$$

which may used for all $P \geq S_i$ and where the watershed is broken into n sub-units each of a separate CN_i (and thus S_i) and fractional area a_i. This strategy should give more realistic results with smaller storms and will also create the CN-storm size behavior pattern previously discussed. Examples of the application of this model to grazing hydrology can be found in Gifford and Hawkins (1973, 1979).

In summary, Hawkins[1] points out some important considerations in use of the SCS model.

1. Background for the technique is obscure and undocumented.
2. The system was originally intended for agricultural lands of the mid-west, east, and southeast. It has been extrapolated to western wildland situations without sufficient examination or trial development.
3. The system is more sensitive to errors in Curve Number than to errors in precipitation up to about 9 inches (23 cm) of precipitation. This is unfortunate, as there are usually well-defined precipitation frequency curves, etc., readily available, but little information on local Curve Numbers.
4. Much of the current use of the SCS technique is more a matter of "what shall we agree to say is going to happen" (for comparative purposes) than "what really will happen."
5. For most rangeland situations, the following choices are available:

A. For homogeneous small watersheds with no live channels or constant source reas, use a constant Curve Number and Equation 11-5.

$$Q = (P - 0.2S)^2/(P + 0.8S), \; P < 0.2S$$

B. For heterogeneous small watersheds, use a distributed or arrayed Curve Number, and thus

$$Q = \Sigma a_i \frac{(P - 0.2S_i)^2}{P + 0.8S_i} \quad \text{for all } P > 0.2S_i \quad \textbf{(11-11)}$$

The a_i is the fraction of the watershed in the ith element with $S = S_i$ ($\Sigma a_i = 1.0$). This reproduces the decreasing Curve Number situation and eventually levels out to a constant Curve Number.

C. For watersheds with live channels or constant impervious areas, and for storms which do not cause overland flow, use $Q = CP$, where C is relative area of channels and impervious areas ($P \leq C \leq 1.0$).

D. For watersheds with live channels and storm rainfall such that overland flow also occurs, this reduces to the same as "B" above. The impervious areas and channels have Curve Number (CN) = 100 or S = 0.

[1] R.H. Hawkins, personal communication.

Partial System Synthesis with Linear Analysis

As background for this and other methods, it is necessary to define the meaning of system *analysis* and system *synthesis*. Amorocho and Hart (1964) define the terms as follows:

System Analysis—The relationship between input and output is established by a mathematical process involving the use of measured input and output data only, without any attempt to describe the interal mechanisms of the system in explicit form. This relationship has the form of a unique function, which is made to operate on the input in order to produce the output. In general, this function is not required to have a physical meaning or to possess parameters fulfilling conditions of dimensional consistency.

System Synthesis—The investigator attempts to describe the operation of the system by a linkage or combination of components, whose presence is presumed to exist in the system and whose functions are known and predictable. The linkage of components must be made in such a manner that the correct output is produced whenever a specified input is applied. In general, the process of synthesis does not yield a unique model of the unknown system.

Pure synthesis or analysis can be performed on a system independently, or a combination of both can be employed. A method of partial synthesis with linear analysis as described by Amorocho and Hart (1964) is given first because it is the basis of the classical unit hydrograph procedure which precedes chronologically the process of general synthesis and general analysis in hydrologic practice.

The basic operations involved in the unit hydrograph procedure are represented in Figure 11-11. In essence, it is assumed that total precipitation input is first modified by storage and infiltration over the surface of the watershed to yield a "rainfall excess" function. The surface elements of the watershed operate then on the rainfall excess function to produce a surface runoff

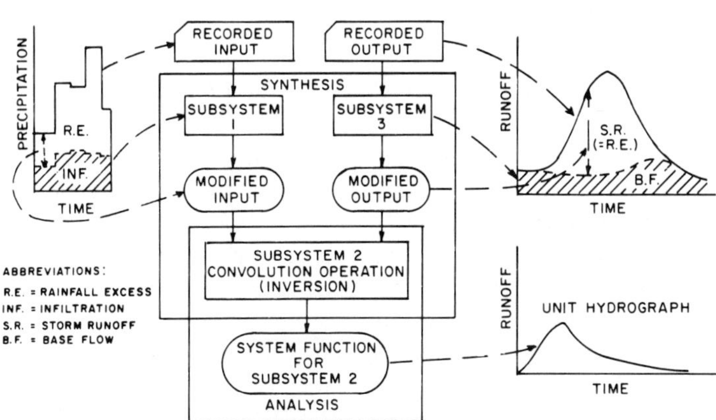

Fig. 11-11. Flow chart of a partial system synthesis with linear analysis (from Amorocho and Hart 1964).

 function, which is modified in turn by the base flow in order to yield the flood runoff, recorded as output at a gaging station. This sequence represents a synthesis operation involving three subsystems, designated in Figure 11-11 by numbers 1, 2, and 3, whose combined effect is assumed to be equivalent to the operation of the watershed.

Subsystem 1 performs the operation of subtracting the values of an infiltration function with the recorded input. This is determined by empirical procedures such as the "antecedent precipitation index," by judgment or by iteration. Subsystem 3 separates the so-called "hydrograph components" from the recorded output function. This is usually done by judgment and by semisubjective procedures involving the determination of the "normal regression curve" for the watershed. Subsystem 2 is a linear convolution. It is analyzed by one of the various numerical methods of inversion that has been described in the literature (Chow 1964) in order to determine the "unit hydrograph," which is assumed to be the invariant system function of subsystem 2. The convolution operation merely lags and sums the unit hydrograph or instantaneous unit hydrograph while multiplying it by the associated rainfall excess intensity.

In the reconstruction or prediction process, the input whose output is to be reconstructed or forecast is entered into the synthetic system which is a simple cascade (see Fig. 11-13). It should be noted that since the analysis operation has been performed on the basis of modified inputs and outputs, the system function of subsystem 2 (the unit hydrograph) depends on the assumptions made for these modifications. In other words, the unit hydrograph obtained fits specifically the input and output so modified; if the assumptions vary, the unit hydrograph will also.

Since hydrologic systems in general are nonlinear, for the validity of the method under discussion, the following interpretation is commonly made: that the operation of the watershed on the rainfall excess is truly linear, and that the over-all nonlinearity of the system is preserved by proper nonlinear modifications of the gross input and output (subsystems 1 and 3). It can only

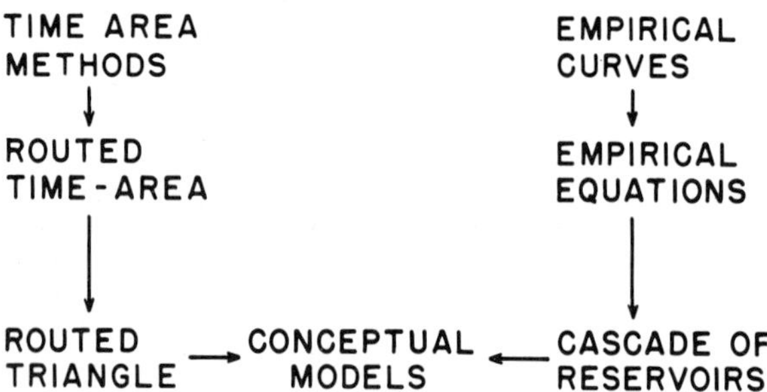

Fig. 11-12. Regression models (from Dooge 1973). Methods for constructing synthetic unit hydrographs as shown on the left side make the general assumption that each catchment had a unique hydrograph, and those on the right side make the general assumption that all unit hydrographs might be represented by a single curve, a family of curves, or a single equation.

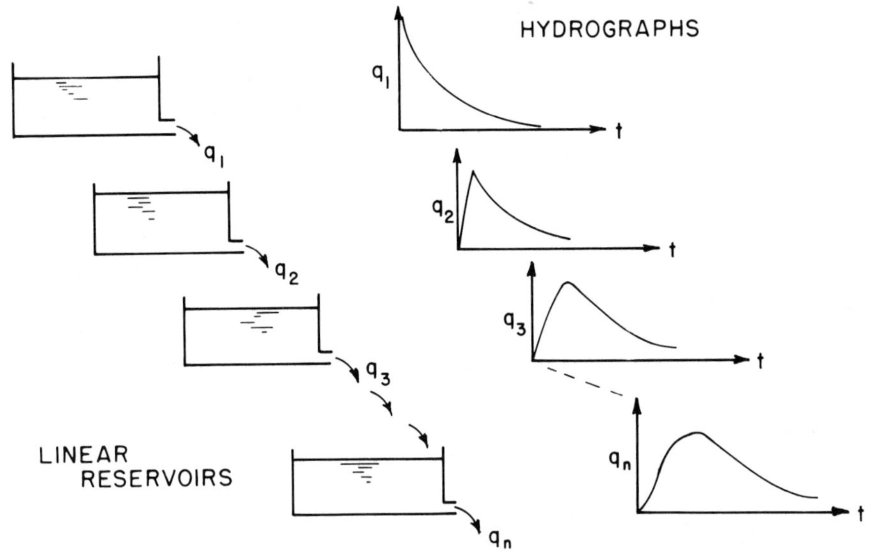

Fig. 11-13. Routing of instantaneous inflow through a series of linear storage reservoirs, a cascade.

be accepted as a rough approximation because no physical element of a watershed is truly linear.

It should be pointed out that for a relatively long period of time the topography of the watershed and its significant geomorphological characteristics do remain relatively constant. Also, while component processes are nonlinear, their combination can produce a linear effect. Consider, for example, $y_1 = c_1x^3$ and $y_2 = c_2x^2$. While each is nonlinear, if the total response is $y = y_1/y_2$, a linear response $y = cx$ is obtained. Finally, from a watershed manage-

Table 11-11. Curve Numbers as a function of cover (CD) and infiltration rate (f.)

Vegetative cover	Limits of application CD (%)	F (inches/hr)	Equation CN =
Juniper-grass	none	none	92.4–4.8f–(.25+08f)CD
Sage-grass	none	none	96.0–10.5f–(.49–035f)CD
Herbaceous	none	none	92.5–5.Of–(.12–.072f)CD
Oak-aspen	none	none	83.4–4.25f–(.39+.055f)CD
Ponderosa pine	10–80		90.0–0.0f–(.02+13f)CD
Pasture or	none	<2.13	92.06–4.26f–(.11–.08f)CD
rangeland	<50	>2.13	92.06–4.26+(.154–.204f)CD
	>50	<2.13	113.91–14.55f–55f–.28CD
Annual grass	none	<2.13	92.5–4.47f–(.12+.065f)CD
	<50	>2.13	95.37–5.82f+(.019–131f)CD
	>50	>2.13	118.53–16.72f–(.461–.095f)CD
Forests	none	<1.13	89.0–8.24f–(.20+.02f)CD
	none	>1.13	102.41–14.55f–(.20+.02f)CD
Roads	0	<2.13	92. – 4.24f
	0	>2.13	98.45 – 7.27 f

[1]From Gifford, Hawkins, and Williams (1975)

Table 11-12. Predicting equations showing the effect of drainage area on Curve Numbers (CN) for two different methods of analysis.[1]

Set of equations and type watershed	Prediction equation[2]	Standard error (CN)
First set:		
Shrub-covered	$CN = 91x^{-0.0088}$	±10.6
Grass-covered	$CN = 93x^{-0.0078}$	± 8.7
Combined grass and shrub	$CN = 92x^{-0.0086}$	±14.3
Second set:		
Shrub-covered	$CN = 85.75x^{-0.0087}$	± 1.6
Grass-covered	$CN = 88.00x^{-0.0085}$	± 1.1
Combined grass and shrub	$CN = 86.74x$	± 1.1

[1]From Simanton, Renard, and Sutter (1973).
[2]Where x = drainage area in acres with the largest watershed tested = 560 acres.

ment standpoint the question to be posed is, what is the error involved in assuming a linear system?

The degree of approximation which the linearity assumption may yield depends then on the degree of actual nonlinearity of the system. There are many instances in system analysis and design where linear approximations are available. In general, since linear analysis is much simpler and much better known than nonlinear analysis, it offers more appeal to the investigator. When it is used to approximate nonlinear cases, however, great caution should be exercised in interpreting the results.

As described by Amorocho and Hart (1964), the basis for the construction of synthetic models in hydrology is a statement of continuity, which can be expressed as an equation of state of the form

$$I = Q + \Delta s \quad t_1 \rightarrow t_2 \qquad \textbf{(11-12)}$$

where I = total inflow,
Q = total outflow, and
Δs = change in internal storage

all referred to the time interal t_1 to t_2. In differential form, the equation becomes

$$i(t) = q(t) + (ds/dt) \qquad \textbf{(11-13)}$$

where i(t) = rate of total inflow to system,
q(t) = rate of total outflow to system, and
ds/dt = rate of change in internal storage.

A system of this type, defined by continuity of matter, is said to be "closed."

The process of general synthesis ordinarily begins with the postulation of a more-or-less complex model, whose structure is based on qualitative and semi-quantitative knowledge of the phenomena involved in the hydrologic cycle. This model contains elements defined by explicit functions, which describe operations effected on various portions of the input and the storage. The recorded input is processed through the model, and the resulting output is compared with the recorded output of the natural system. If an acceptable agreement is not found, one or more of the functions of the component subsystems are modified and adjusted, and the process is repeated in a systematic way until there is adequate correspondence between the synthetic and the recorded outputs.

Synthetic Unit Hydrograph Models

The development of unit hydrograph theory from the classical hydrology point to the present state as modified with the emergence of the systems approach is a topic much too detailed to cover in this material. There are numerous references which will assist the reader in this effort including Chow (1964), Dooge (1973), and Chapman and Dunin (1975), to mention a few. Every applied hydrologist dreams of being able to forecast direct storm runoff from a watershed map where no records are available for the deriviation of a unit hydrograph. Dooge (1973) stated:

> In classical hydrology, synthetic unit hydrographs developed along two main lines, both of which converged at the time of the emergence of parametric hydrology. These two lines of development are shown in Figure 11-12. The methods at the left-hand side made the general assumption that each catchment had a unique unit hydrograph, and those at the right-hand side made the general assumption that all unit hydrographs might be represented by a single curve, or a family of curves, or a single equation.

As studies progressed on synthetic unit hydrographs, it was soon apparent that a one-parameter method was not sufficiently flexible for adequate representation, and independent developments led to the use of the two-parameter gamma distribution or the Pearson Type III empirical distribution (see Soil Conservation Service [1966] for explanation of gamma and Pearson Type III distributions).

Nash (1958, 1959) suggested the two-parameter gamma distribution had the shape required for an instantaneous unit hydrograph (IUH) and pointed out that the distribution could be considered as the impulse response for a cascade of equal linear reservoirs. Thus, the cascade model had two parameters—the number of reservoirs (N) and the storage delay time (K)—and generated a specific family of shapes (Fig. 11-13).

In mathematical form, the impulse response or instantaneous unit hydrograph (IUH) of the Nash model is:

$$h_o(t) = \frac{(t/k)^{n-1} e^{(t/k)}}{KTn} \quad \textbf{(11-14)}$$

where $h_o(t)$ is the ordinate of the IUH, n is the number of reservoirs, t is discharge time, T is duration of inflow, k is the storage delay of each of the reservoirs, and e means exponential. Other investigators such as Gray (1961), Wu (1963), and Diskin and McCarthy (1972), have used this function to describe derived unit hydrographs and to synthesize further hydrographs. Parameters of such an instantaneous unit hydrograph (n and K) can be determined by the method of moments using the rainfall excess hyetograph and the direct runoff hydrograph (Chow 1964), or by optimization (Singh 1974). These parameters can be related to watershed parameters to facilitate the runoff simulation on watersheds without records.

Figure 11-14 shows a flow chart of the Stanford Watershed Model Mark II as it might be used to simulate the land runoff phase of the hydrologic cycle. This figure shows the main features of the Stanford model, although the model has since been developed to the Mark IV (Crawford and Linsley 1966) (Fig. 11-15) and Mark V stages in which the model performance is improved at a cost of additional complexity.

The Kentucky Watershed Model is basically a FORTRAN version of the Stanford Watershed Model Mark III with a few of the Mark IV modifications made by James (1970) for use in eastern watersheds. For more detailed discussions of Stanford/Kentucky model structure and application, the reader is referred to reports by Anderson and Crawford (1964), Crawford and Linsley (1966), Liou (1970), Ross (1970), James (1970), Ricca (1972), and Striffler (1973).

Snowmelt Models

Snowmelt is an important element in many rangeland areas where the opportunity to manage the snow can be very important for forage production, water harvesting, mining reclamation, etc. Other work on snowmelt is discussed in Chapter 9. Figure 11-16 illustrates the snowmelt routine in the Kentucky version of the Stanford model. Inclusion of this routine in the Kentucky model enables runoff simulation from snow as well as rain, or both.

Snowmelt routines have also been successfully developed by Amorocho and Espildora (1966). The flow chart for their simulation is shown in Figure 11-17.

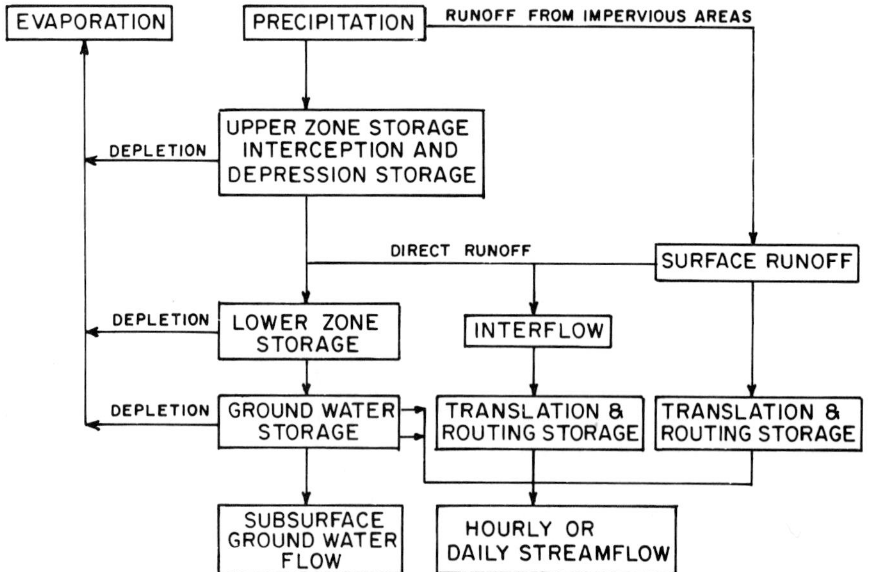

Fig. 11-14. Stanford Watershed Model Mark II.

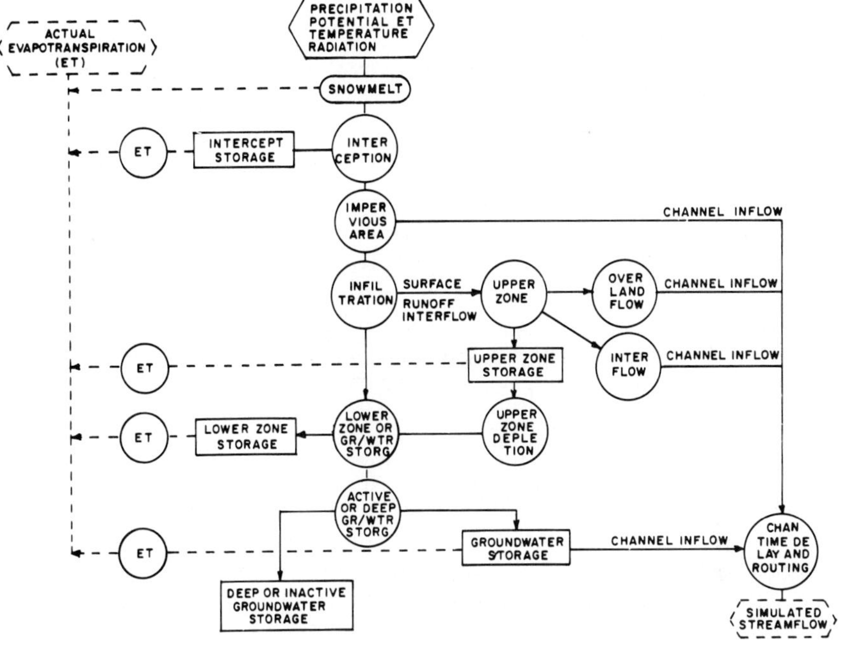

Fig. 11-15. Stanford Watershed Model Mark IV.

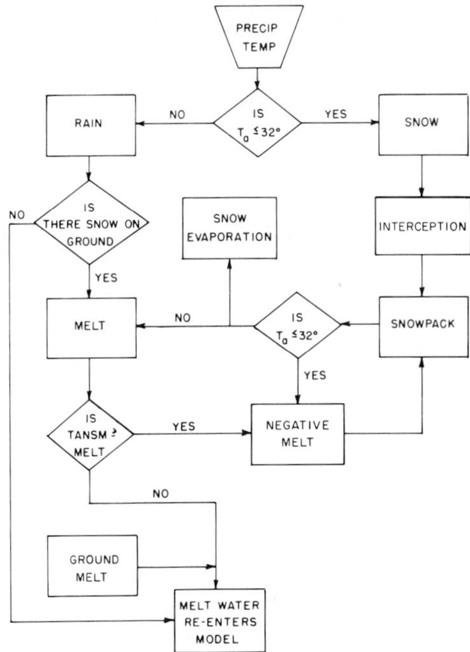

Fig. 11-16. Snowmelt flow chart for the Kentucky Watershed Models. Symbols used are: T_a = temperature of air and TANSM = refreezing of liquid water in the snow pack and also the liquid water = holding capacity of the snow pack.

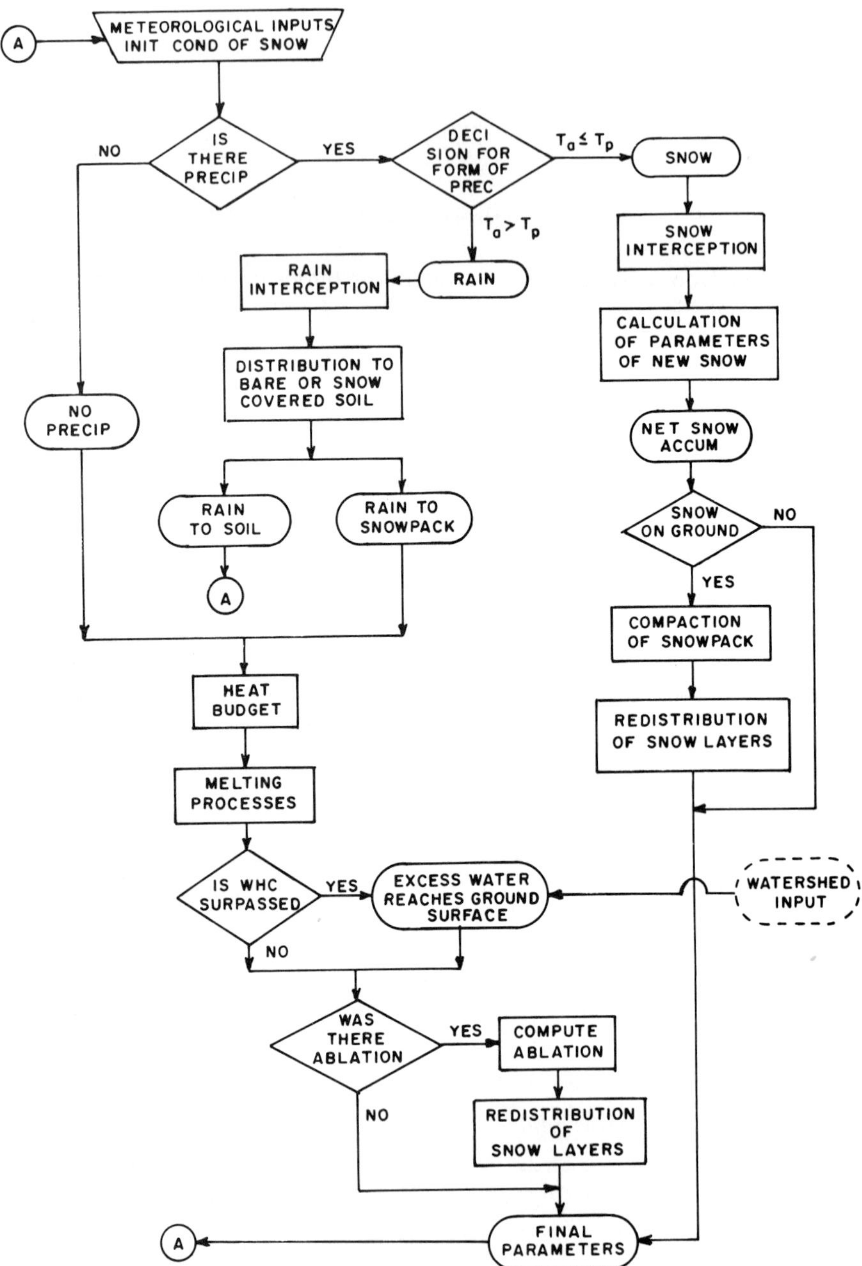

Fig. 11-17. Simulation of snowmelt (from Dooge 1973). Symbols are T_a = temperature of the air, T_p temperature of the precipitation, and WHC = water holding capacity.

U.S. Department of Agriculture Hydrograph Laboratory Model

The hydrologic model developed by Holton et al. (1975) expresses the watershed hydrology as a continuum and would be classified as a parametric model (Fig. 11-3), having some elements based on empirical approximations of physical hydrology. As is the case with the Stanford model, it has had only limited application to rangeland or wildland watersheds. This model is designed to serve the purposes of agricultural watershed engineering with emphasis on separating out the details of what actually happens during the runoff process as a basis for planning the engineering structures and procedures that will control the times, routes, and amounts of waterflow. The model consists of a series of empiricisms selected to provide a mathematical continuum from a ridgetop to a watershed outlet in terms of input data readily available to an analyst. A flow chart for the mainline program is shown in Figure 11-18. As shown in this figure, the model considers the precipitation, hydrologic grouping of soils and land use, evapotranspiration, infiltration, coefficients for routing the flow, and the hydrogeology of the watershed with subroutines for each phase. The model routines are divided so that, as better subroutines are developed for each part of the process, they can be inserted without interfering with other routines in the sequences.

Figure 11-19 illustrates the results of simulations from the model for four watersheds in the United States. The results as shown are encouraging, although only the Hastings, Nebraska, watershed with its 24 inch (61 cm) average precipitation (approximately 75% cultivated) is near to conditions encountered in rangelands.

This model was revised to include chemical and erosion routines and has been used recently to predict the concentration and amount of chemical in the runoff water and on the sediment at the watershed outlet (Frere, Onstad, and Holtan 1975). It was also used to predict the location and concentration of chemicals that are leached moved spatially through the soil of the watershed. As refined, the model (called ACTMO) includes such management options as the time, rate, and type of chemical applied, changes in crop pattern, and tillage practices. Although the model (ACTMO) has been used for very limited conditions, it does demonstrate what can be done and a direction for future research in meeting the requirements of the Federal Water Pollution Control Act Amendments of 1972 (PL 92-500) regarding nonpoint pollution control. Figure 11-20 is a flow diagram of the input-output files of ACTMO. Table 11-13 lists the types of information transmitted in ACTMO.

Kinematic Cascade Models

The application of the kinematic wave formulation to watershed modeling by Henderson and Wooding (1964) and Wooding (1965a, 1965b, and 1966) and the subsequent demonstration of its general applicability to hydrologic problems by Woolhiser and Ligget (1967) has led to its increasing utilization in watershed modeling (Brakensiek 1967; Woolhiser 1969 and 1973; Eagleson 1972; Smith 1976; Singh 1975; Lane Woolhiser, and Yevjevich 1975).

In this modeling approach, simplified watershed geometrics are repres-

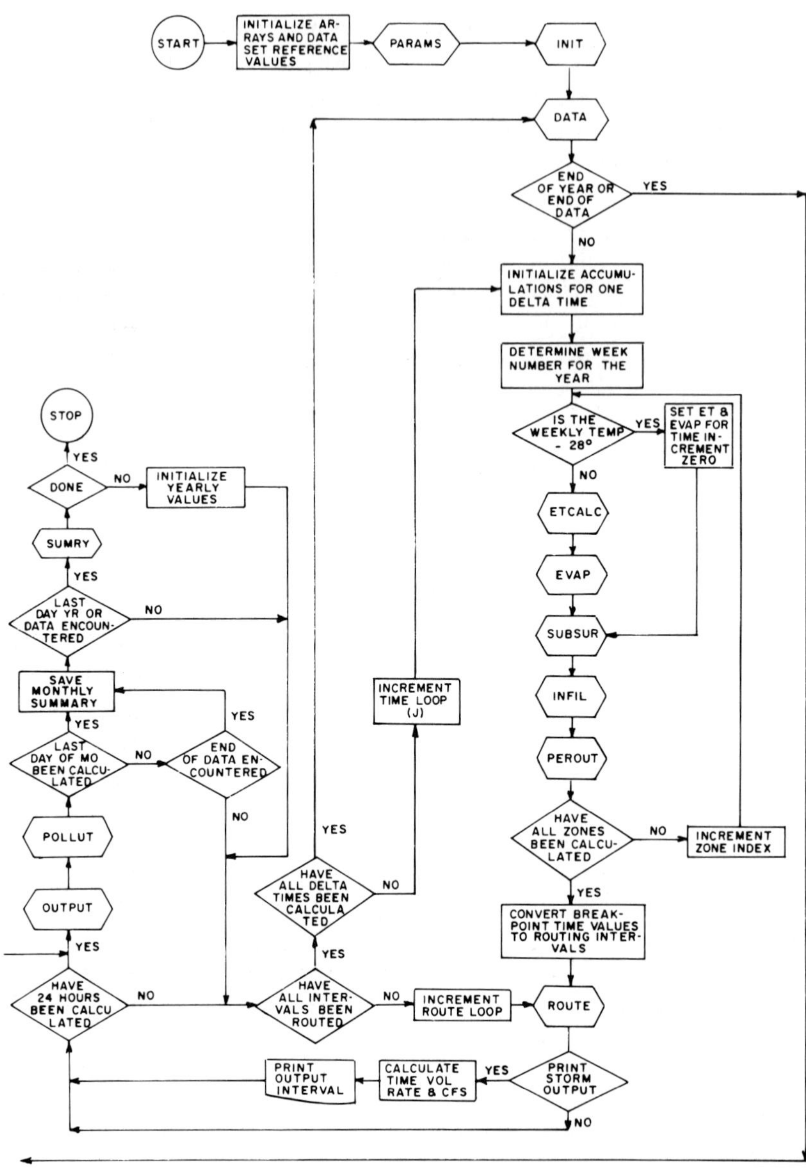

Fig. 11-18. Flow chart of USDAHL-74 mainline program (Holtan, et al. 1975).

Table 11-13. Description of the information transmitted in the input-output files of ACTMO.

File	Information
Weather	Rainfall, temperature, pan evaporation.
Watershed	Zone sizes and cascading, soil hydrologic properties, crop pattern and characteristics, tillage operations.
Storm	Length, slope, and cascading of zones, runoff infiltration, and rainfall intensity.
Soil	Soil texture.
Erosion	Compartment areas, enrichment coefficient, soil texture, infiltration, runoff, erosion, and deposition.
Hydrological	Cultivations, soil properties, time-averaged temperature, soil moisture, and evapotranspiration, lateral and vertical flow, soil water when the storm starts.
Chemical	Date, amount, and type of chemical application; treated area; soil chemical interactions; and critical levels.
Output of chemical behavior ...	Chemical lost in runoff and erosion. Concentrations in runoff. Chemical left in each compartment. Location and concentration of peak in each compartment.

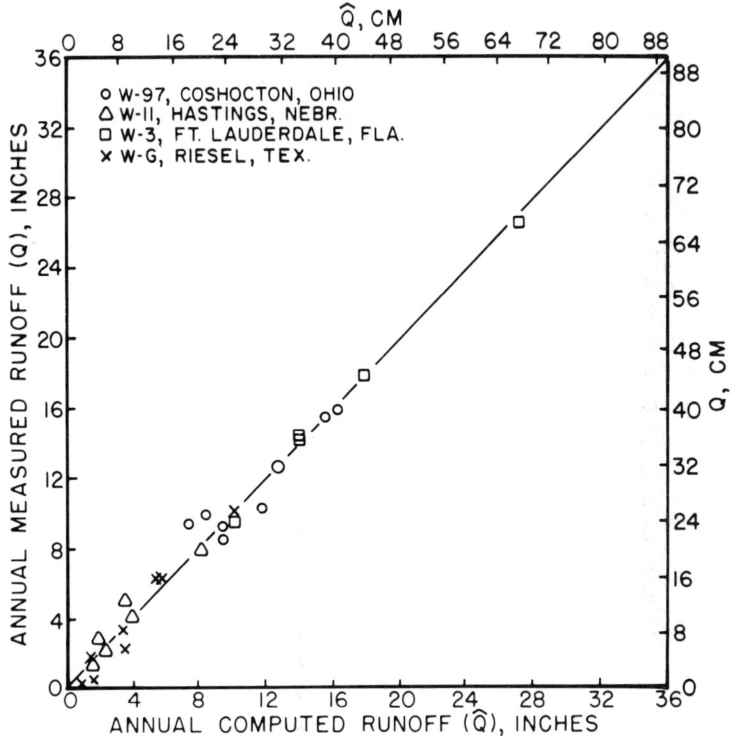

Fig. 11-19. Chart showing accuracy of USDAHL-74 model for estimating annual runoff as compared with annual measured runoff at four watersheds (Holtan et al. 1975).

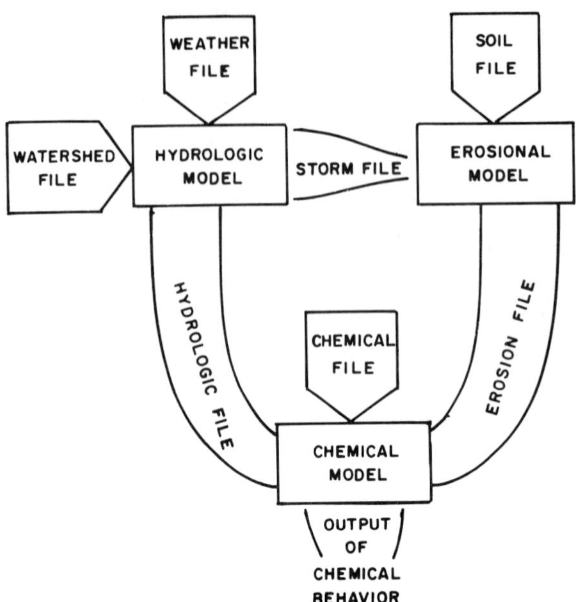

Fig. 11-20. Flow diagram for the input-output files in ACTMO (from Frere, Onstad, and Holtan 1975).

ented using planes, converging sections, planes in cascade, channels and combinations of these (Fig. 11-21). The mathematical surface configurations represent a prototype watershed to various degrees of correspondence. An objective evaluation of these approximations is given by Lane, Woolhiser and Yevjevich (1975).

The kinematic equations for a plane section are, the continuity equation:

$$\frac{\alpha h}{\alpha t} + u\,\frac{\alpha h}{\alpha x} + h\,\frac{\alpha u}{\alpha x} = q\,(x\ t) \qquad \textbf{(11-15)}$$

and the kinematic-momentum equation:

$$A = uh = \alpha h_n \qquad \textbf{(11-16)}$$

where,

h =local flow depth,
u =local average velocity,
q =rate of lateral inflow=rainfall excess,
t = time coordinate,
x =space coordinate, and
α and n are kinematic wave parameters of the friction relationship.

The solution to this system of equations characterizes the overland flow of the plane. For rainfall of pulse type, analytical solutions have been obtained (Wooding 1956b, Kibler and Woolhiser 1970, Singh 1974). For complex rainfall, however, analytical solutions are less feasible and the hybrid approach of Singh (1975) was developed. This hybrid solution is part analytical and part numerical.

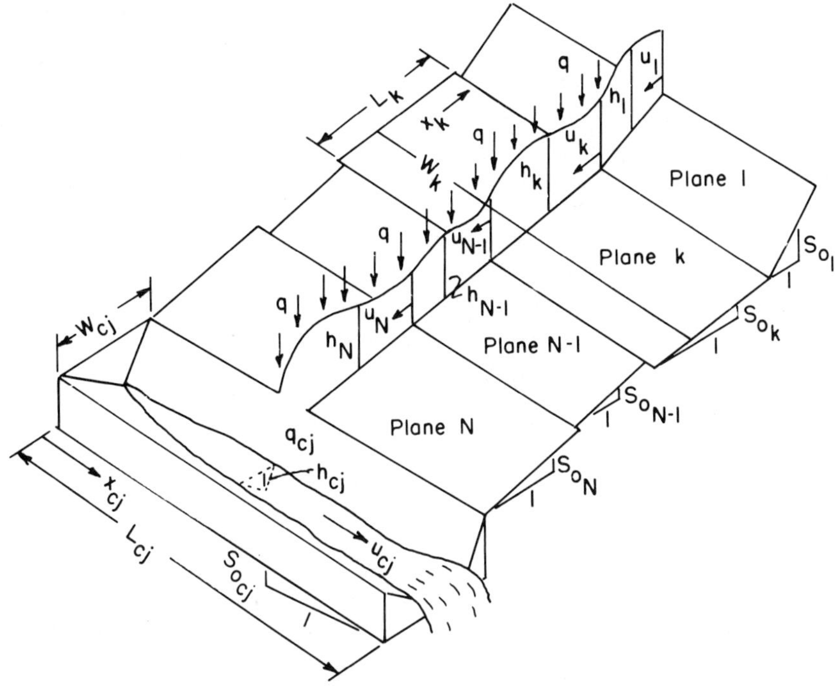

Fig. 11-21. Cascade of N planes discharging into the Cj[th] channel (from Singh 1975).

Modeling of rangeland watersheds has been successfully demonstrated using the kinematic wave equations on watersheds in Nebraska, Arizona, Texas, and South Dakota. Computer-use times are quite large with this model but the improvements such as Singh (1975) described led to appreciable cost reductions. This method has great promise where management alternatives are involved. The method which nearly represents the physical principles involved is now being used with erosion estimating routines to simulate water quality problems from very small watersheds (Smith 1976; Frere, Onstad, and Holtan 1975).

To illustrate the types of information that can be obtained with models, L.J. Lane, Hydrologist at the Southwest Watershed Research Center, applied a kinematic model to an assumed watershed shape shown in Figure 11-22. (For similar application to an actual rangeland watershed, see the gully advance example in the chapter on erosion and sedimentation.)

The precipitation time distributions of Figure 2-29 were then used as input to the simplified model assuming a storm duration of one hour and a storm depth of two inches. The hyetographs produced are shown in Figure 11-23.

The simplified infiltration function (Philip 1957) shown in Figure 4-8 was assumed, and in equation form is:

$$A + 1/2\ st^{-1/2} \qquad \textbf{(11-17)}$$

 where f(t) is infiltration rate, A and S are parameters, and t is time. For this example, $A = 0.50$ inch/hr and $S = 1.00$ inch/hr$^{3/2}$. The point to note is that each storm quartile has the same total depth, duration, and infiltration function, with the only difference being the rainfall time distributions which were developed from Figure 2-29 as is explained later. The estimated rainfall rates are shown in Figure 11-23 together with the infiltration rate and the resulting runoff.

For the illustration, a simplified watershed model was selected, composed of two lateral planes contributing to a single channel (Fig. 11-22). This model, referred to as the Wooding model, is a simple model but does preserve selected properties of some rangeland watersheds, particularly the hydro-

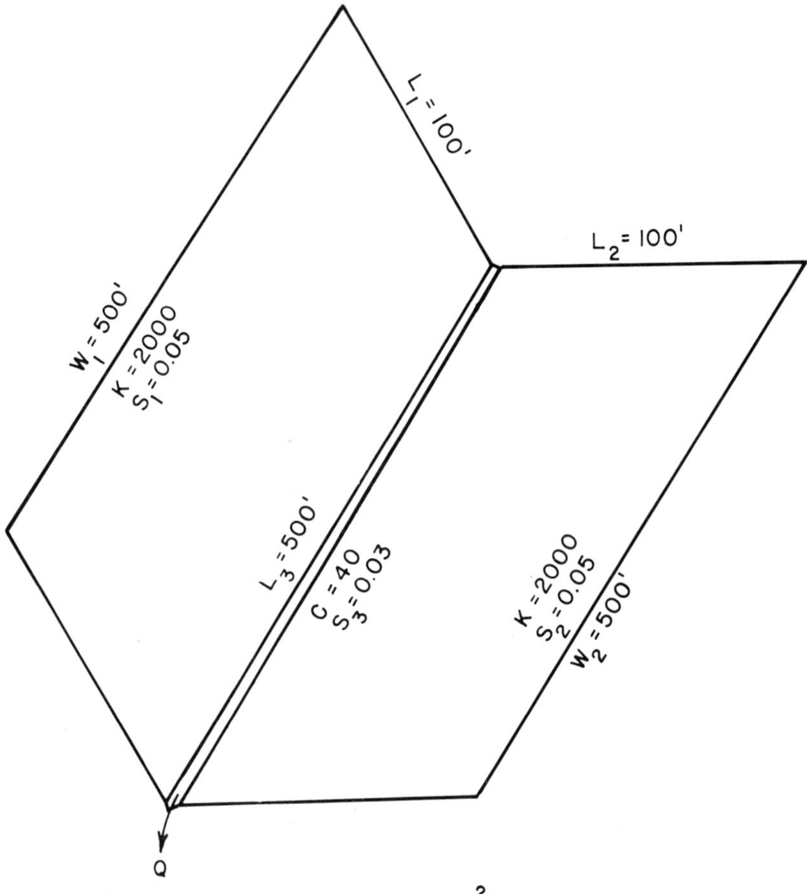

Fig. 11-22. An assumed watershed for kinematic simulation, a simplified model consisting of two planes and a channel (from Wooding 1965a). Symbols used are:
L = length of overland slope, W = width of plane, S_1 and $_2$ = slope of plane,
S = channel slope, C = Chezy roughness coefficient, H = hydralic roughness coefficient, and Q = runoff.
(Multiply ft by 0.305 to obtain m, multiply acres by 0.405 to obtain ha.)

graph of surface runoff. Each plane is characterized by a length of overland flow (100 ft, 30.5 m), a width 500 ft, 152 m), a slope (0.05), and a hydraulic roughness coefficient (K = 2000). The channel is assumed trapezoidal with a bottom width of 1.0 ft (0.3 m) and side slopes of 0.5. The downstream channel slope is 0.03 and its Chezy roughness coefficient is 40.

For laminar flow, the Darcy-Weisbach friction factor f was defined as

$$f = K/R_e \tag{11-18}$$

where R_e is the Reynolds number and K is the roughness coefficient. For turbulent flow, the Chezy form of the friction factor was used, defined as

$$f = 8g/C^2 \tag{11-19}$$

where g is the acceleration due to gravity and C is the Chezy coefficient. For the transition from laminar to turbulent flow at a transitional Reynolds number R_c, the friction factors from equations 11-18 and 11-19 are matched by setting

$$C = \sqrt{8gR_c/K} \tag{11-20}$$

Therefore, for the planes, the roughness is described by a coefficient K and a transition number R_c.

For laminar flow, N = 3.0 and

$$\alpha = 8gS/K\upsilon \tag{11-21}$$

where N is the kinematic wave parameter in equation 11-16, υ is the kinematic viscosity, and the other variables are as described. For turbulent flow (Chezy form), N = 1.5 and

$$\alpha = C\sqrt{S} \tag{11-22}$$

In the simulation, K was chosen as 2000 corresponding to sparse vegetation (Woolhiser 1974; Lane[2]) and R_c was assumed to be equal to 500. Thus, the Chezy C value for overland flow on the planes is 8.02.

Equations 11-15 and 11-16 were solved for the 2.3 acre (0.9 ha) watershed for the rainfall patterns shown in Figure 11-23. That is, rainfall excess as a difference between rainfall rate and infiltration rate from Figure 11-23 was routed over the watershed surface represented by Figure 11-22. These hydrographs of surface runoff are shown in Figure 11-23 to illustrate the influence of rainfall distribution patterns on surface runoff.

The cumulative distribution curves for each of the four quartiles (Fig. 2-29) were differentiated to obtain the hyetographs shown in the Figure 11-23. Some of the differences in the hydrograph peak are associated with the higher rainfall intensity, but major differences in the hydrograph shapes are associated with the differing rainfall patterns. In the example, the rainfall and infiltration were assumed to be time varying but uniform in space. Spatial variability, as is always encountered in nature, while compounding the results, will also affect the hydrograph shape.

A cautionary note is that the foregoing results are for simplified models—

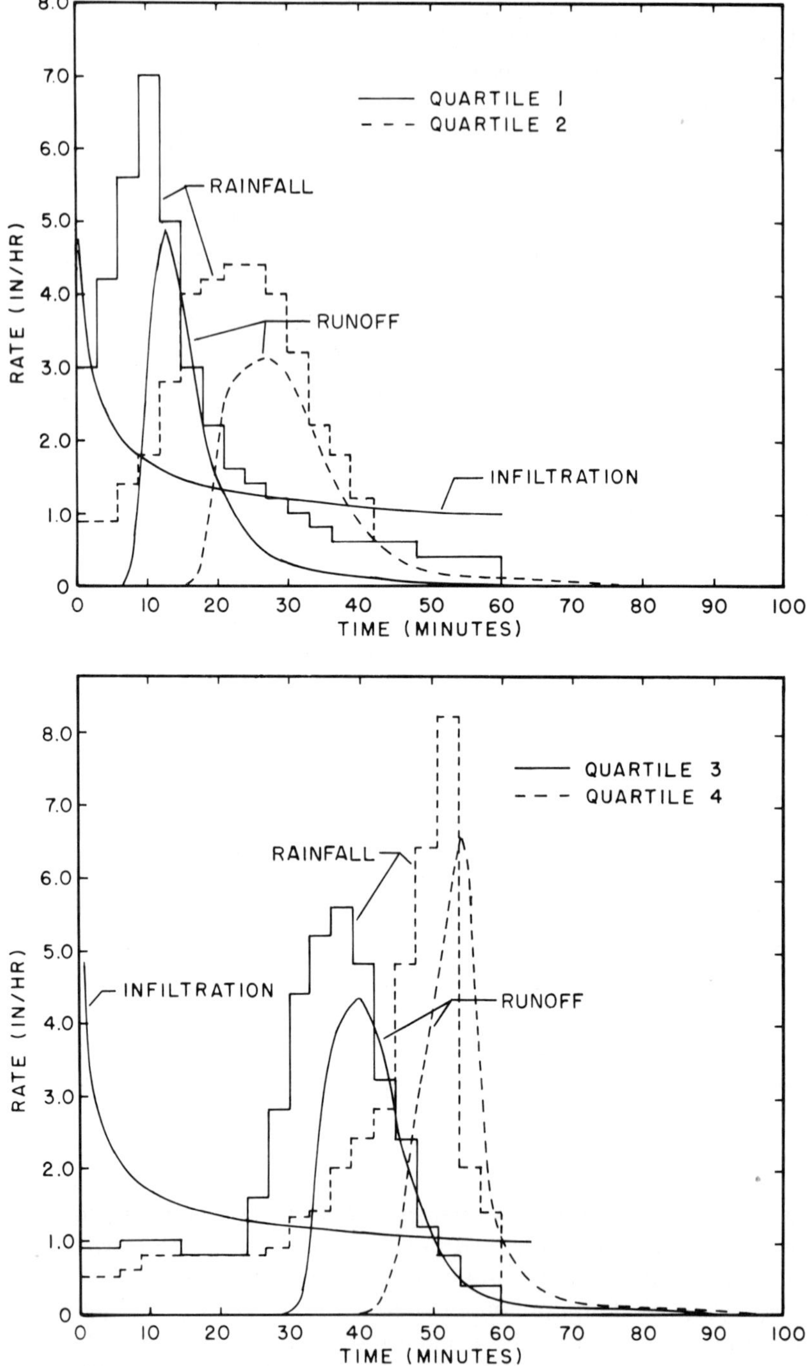

Fig. 11-23. Rainfall hyetograph, infiltration, and simulated runoff for each of four quartiles of storm duration as shown in Fig. 2-29., for a 2.3 acre simulated watershed (Lane 1975, personal communication).

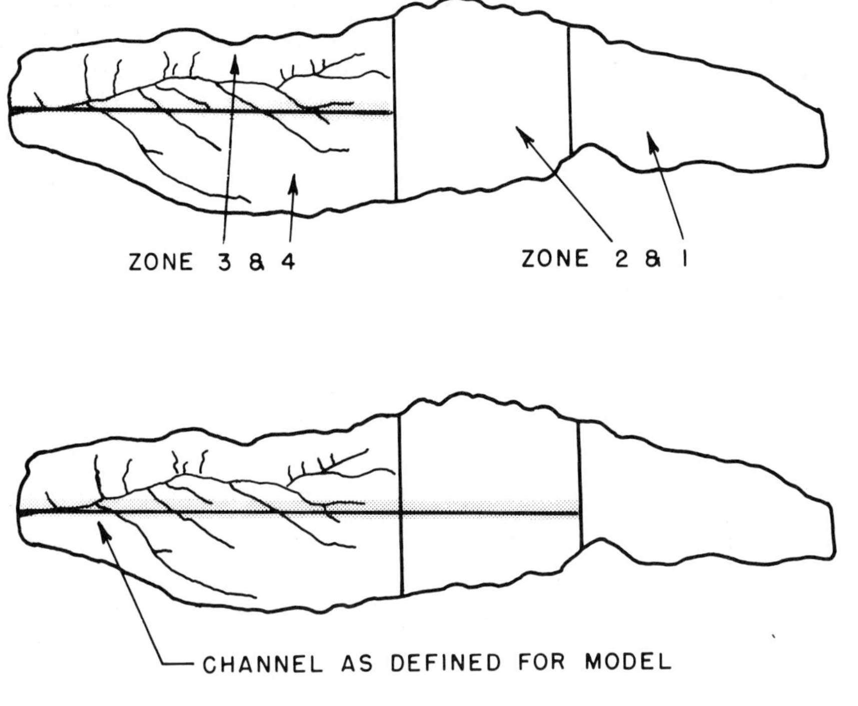

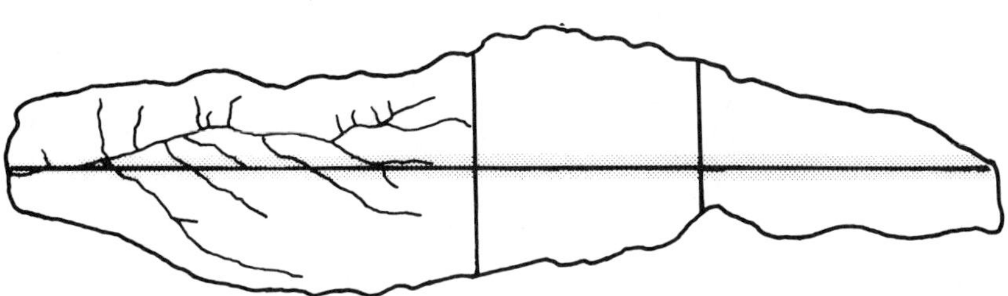

Fig. 11-24. Map of the 4.02-acre Watershed on the Santa Rita Experimental Range illustrating the four zones of the watershed and channel lengths used for modeling purposes (from Wallace and Lane 1976).

both for rainfall excess and watershed geometry. However, the simulation results appear reasonable and provide a basis for further research. That is, hydrograph features illustrated in Figure 11-23 provide a working hypothesis on the influence of rainfall patterns upon surface runoff. This is a useful result from a simplified model applicable to rangeland hydrology.

Kinematic cascade models may also be helpful to illustrate what might happen to the rainfall runoff relationship as a result of a geomorphic change in watershed feature. Wallace and Lane (1976) used a 4.02-acre (1.62 ha) watershed on the Santa Rita Experimental Range in southeastern Arizona to calibrate their kinematic cascade model. The watershed with incised drainage in the lower portion has an upland area with poorly defined drainage. They found it convenient to divide the watershed into four zones with a single, main channel (Fig. 11-24).

Although they used the Φ index (an average rate of infiltration derived from a time-intensity graph of rainfall) for obtaining rainfall excess, they then used the same kinematic equations (Equations 11-15 and 11-16) as in the previous example. The main channel was assumed to be 50%, 100%, 150%,

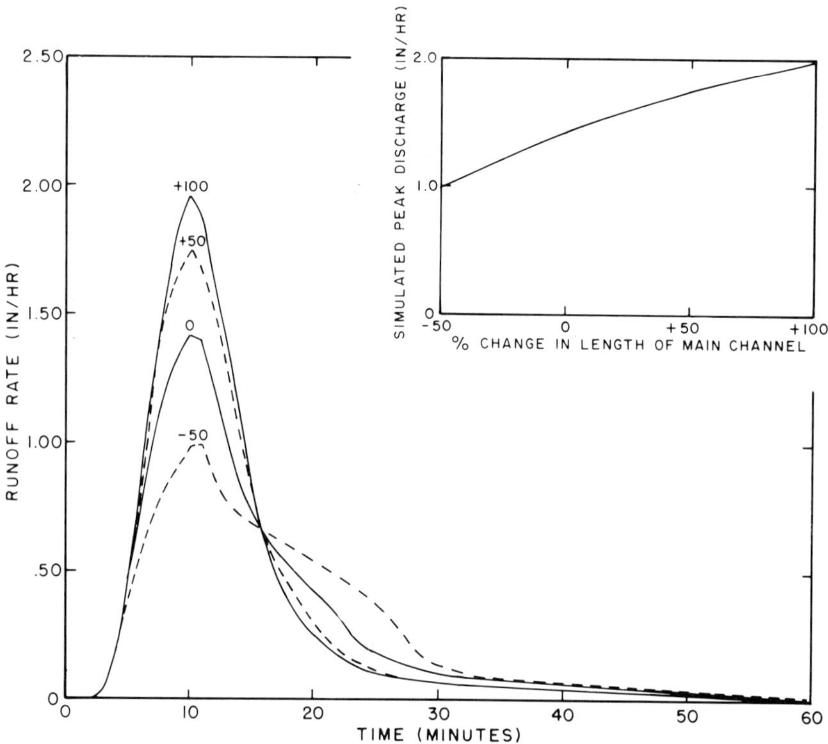

Fig. 11-25. Simulated hydrographs for Watershed 76.001, uniform rainfall excess. Simulation of gully advance of 50%, 100%, 150%, and 200% of the present length of the main channel (from Wallace and Lane 1976).

and 200% of its present length for illustrative purposes. Recognizing this oversimplification may not be truly realistic, they assumed all other watershed features were constant, i.e., only the main channel length changes. The resulting runoff hydrographs are shown in Figure 11-25 with the upper right inset showing the change in peak discharge with changing main channel length.

Because of the simplifications in their example, Wallace and Lane (1976) state that the results represent only a qualitative estimate of the influence of a gully advance on the surface runoff hydrograph but they do allow forming a hypothesis for more comprehensive research into the hydrologic importance of gully advances in small semiarid watersheds. For additional reading, see Simons, Li, and Stevens (1975).

Such a model might be used to illustrate or predict changes resulting from grazing a rangeland watershed. The grazing might lead to decreasing surface roughness (e.g., d and n in equation 11-16) and to changes in the coefficients in the infiltration relationship (Equation 11-17) with resulting changes in the volume and time distribution of runoff. Additional geometric watershed complexities (e.g., other representative of Fig. 11-22) can be used to mimic a prototype watershed with only additional computer costs restricting the correlation between the model and prototype conditions.

Dawdy-O'Donnell Model

Dawdy and O'Donnell (1965) pioneered in a systematic optimization of model parameters. They reported their model (Fig. 11-26) was deliberately kept simple so emphasis could be directed to parameter sensitivitiy and optimization. The model was restricted to four storage elements having simple hydrologic significance analogous to surface, channel, soil moisture, and groundwater storage. Inputs to the model are precipitation (P) and evapotranspiration. The later consists of surface storage evaporation (E_r) and transpiration from soil moisture (E_M). Output to the model is runoff (Q) which is composed of surface runoff (Q_s) and baseflow (B).

Nine parameters control the functioning of the model with R* being the threshold surface storage which must be satisfied before overland flow (Q_1) occurs. The infiltration (F) is calculated using a three parameter Horton infiltration relationship (Fig. 4-8). The channel storage (S) is assumed to be linear having a storage constant (K_s). Field moisture capacity is defined as M* acting as a threshold on groundwater recharge. The seventh parameter is the groundwater capacity (G*) and the maximum rate of capillary rise (C_{max}) simulates the water loss from the groundwater during very dry periods. The groundwater storage (G) is a linear reservoir having a storage delay time (K_G).

Parameter optimization in hydrologic models has been a topic of wide investigation. Dawdy and O'Donnell used a direct search technique based on Rosenbroek's (1960) method. In all parameter optimization methods, it is easier to optimize some parameters than others, especially if the values as initialized were close to the true value. All of the optimization schemes result from the systematic search techniques made possible with the use of digital computers. In some instances, the optimum values can be obtained only with specific data. For example, the maximum capillary rise (C_{max}) in the Dawdy-O'Donnell model would become apparent only if a record was available containing a long dry spell. Dooge (1973) discusses some optimization problems of hydrologic models in considerable detail.

The non-linear optimization scheme developed by Betson and Green (1967) and later described by De Coursey and Snyder (1969) is an example of an excellent method for routine parameter optimization. Care must be taken to ensure that the values obtained are realistic if physical significance is to be attached to the parameters.

Subsequent developments and improvements in the model were reported by Dawdy, Lichty, and Bergmann (1972). They also reported on testing the model on the Santa Anita Creek Basin in California, a 25 km^2 (9.6 sq. mile) area in the San Gabriel Mountains of southern California, as well as Beetree Creek in North Carolina and Little Beaver Creek in the Ozark Mountains of Missouri. The result of their analysis was very encouraging and their treatment of errors, error sources and error impact on the model and its predictions are important.

Utah State University Model

Utah State University scientists have developed a deterministic, lumped, small watershed, rainstorm-hydrograph model which they have used in both teaching and research applications. The model, which is based on the tank analogy (reservoir) of watershed hydrology, has some similarities to the Dawdy and O'Donnell (1965) work and the digital storm runoff model used by Dawdy, Lichty, and Bergmann (1972). The model contains no provisions for snowmelt or evapotranspiration (which is minimal during a rainstorm), and thus, its use is limited, but as Riley and Hawkins (1976) state: ". . .The model. . .seems to work without the complications of evapotranspiration." Examples of the application of this model to the 217 acre (87.8 ha) Chicken Creek are available in Johnston and Doty (1972) with one set of storms shown in Figure 11-27. It should be emphasized that the same coefficients were used from storm-to-storm with only the initial conditions and inputs varying. Better fits could be made for individual storms.

Table 11-14 contrasts the hydrologic characteristics and watershed specifications which might be encountered in range and forest watersheds. It presents in idealized terms general natural conditions useful in modeling.

Riley and Hawkins (1976) assumed new parameter values for their model (Table 11-15) and compared the resulting responses as shown in Figure 11-28. Needless to say, the conclusions are are only as reliable as the model structure is valid and as trustworthy as the coefficients are, but in general, the range conditions react more to rainstorms than do the forested watersheds as illustrated in Figure 11-28.

Stochastic Methods

In the use of stochastic models, the investigator moves from the concept of

Table 11-14. Generalized and idealized contrasts between natural forest and range conditions.[1]

Item	Range	Forest
Precipitation		
Annual	Low to moderate	Moderate to high.
Rainstorms	Often entire water supply	May include rain and snow.
Intensities	High, short duration	Longer durations, ground intensities modified by cover.
Snow	May be absent	Often a significant role
Soil and Surface Properties		
Surface texture	Fine	Coarser
Depth	Shallow	Deep
Presence of pans	Sometimes	Seldom
Organic matter	Poorly incorporated	Well humified
General development	Poorly developed, accumulation horizons grass and/or shrubs	Well developed, leached horizons; trees; and grass and understory vegetation
Ground cover	Often sparse	Seldom sparse
Intercepting canopies	Single	Often multiple
Runoff and Hydrology		
Infiltration rates	Low	High
Overland flow	Common during rainstorms	Rare
Base flow	Seldom, intermittent	Typical condition
Permanent stream	Absent or intermittent	The usual case
Water quality	Often salty, warm, etc., High natural erosion	Usually good quality Low natural erosion, high if aggravated
Land Use	Consistent annual harvest	Sporadic, occasional harvest at rotation
Initial Conditions		
Soil moisture	Low	Variable
Groundwater	None	None or some
Channel surface area	None	Small, but present

[1]From Riley and Hawkins 1976.

establishing input-output relationships with certainty to one of establishing the level of probability with which various sequences of output may occur in the future. In other words, given a historical sequence of events, what inferences can be derived from their statistical distribution so that the probabilities of future sequences can be assessed?

Amorocho and Hart (1964) indicate that in some of the classical procedures, it is assumed that the past record fully describes all possible combinations of events, and that such a chronology can, therefore, be utilized without transformation as a basis for planning future operations. The weakness in the theoretical foundations of this assumption has been recognized for some time; nevertheless, the procedures (mass curves, etc.) based on the hypothesis are simple, and they are still widely employed. In the last few years, the rapid development of mathematical statistics has opened new avenues of approach

Table 11-15. Parameter and initial conditions (coefficients) values used for forest and range watersheds.[1]

Parameter		Units	Range	Forest[2]	Comments
Symbol	Description				
ACHP	Channel Interception Fraction	inch/inch	0	0.00066	
FMAX	Maximum watershed infiltration rate	inch/1/2hr	4	10	
FMIN	Minimum watershed infiltration rate	inch/1/2hr	.25	10	
SAT	Soil water content at saturation	inch	7	7	
FC	Soil water content at field capacity	inch	5	5	
WP	Soil water content at wilting pt.	inch	0	0	
FQF	Fraction of soil water (>FC) that becomes quick flow	inch/inch	.0020	.0039	
FK	Fraction of soil water (>FC) that contributes to ground water reservoir	inch/inch	.15	.445	
AGW	Ground water reservoir recession coefficient	inch/inch	.00028	.00028	
SMO	Soil water content at beginning of storm event	inch	Variable		Calibrated for individual storm
GWLO	Ground water content at beginning of storm event	inch	0	Storm Specific	Calibrated for individual storm

[1]From Riley and Hawkins 1976.
[2]Parameter (coefficient) values from west branch Chicken Creek calibration.

to the probabilistic study of natural phenomena, and new tools are now available for handling problems of considerable complexity. Using these devices it is possible to generate hypothetical sequences of events which have the same probability characteristics of the past, so that the likelihood of many possible combinations of outputs can be studied in detail.

There now exist many stochastic models of hydrologic processes. Two typical approaches are the Monte Carlo method and the Markov process. The Monte Carlo method uses tables of random numbers to produce artificial sequences having the same stochastic structure as the record. The application of this method is valid for data which are statistically independent; however, in hydrology we must remember that the output of systems containing storage elements depends on processes which are not purely stochastic. The causalistic influences are quite strong, in fact, and are revealed by persistence effects commonly called "carry-over." Therefore, in the selection of the sampling procedures, carry-over effects must be minimized by proper grouping of the data. For example, if 3-year carry-over effects are detectable in the output history, the sampling must be made based on the total output for periods not shorter than 3 years, so that the period totals may approach statistical independence. The Monte Carlo methods are particularly useful for studying non-Markovian processes; that is, processes in which the state of the system at a particular time does not depend on previous states.

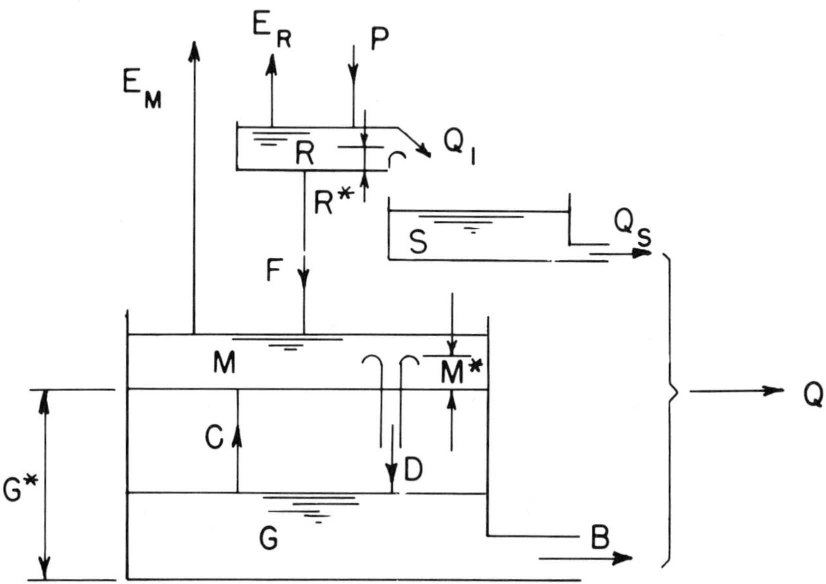

Fig. 11-26. Dawdy-O'Donnell Model (Dawdy and O'Donnel 1965. Symbols are: B = baseflow, C = capillary rise, D = soil moisture lost to groundwater, Em = transpiration from soil moisture, E_R = surface storage evaporation, F = infiltration, G = groundwater storage, G* = groundwater capacity, M = soil moisture storage, M* = field capacity of soil, P = precipitation, Q = runoff, Q_i= channel inflow, Q_5= surface runoff, R = retention storage, R* = threshold surface storage below which channel inflow does not occur, and S = channel storage.

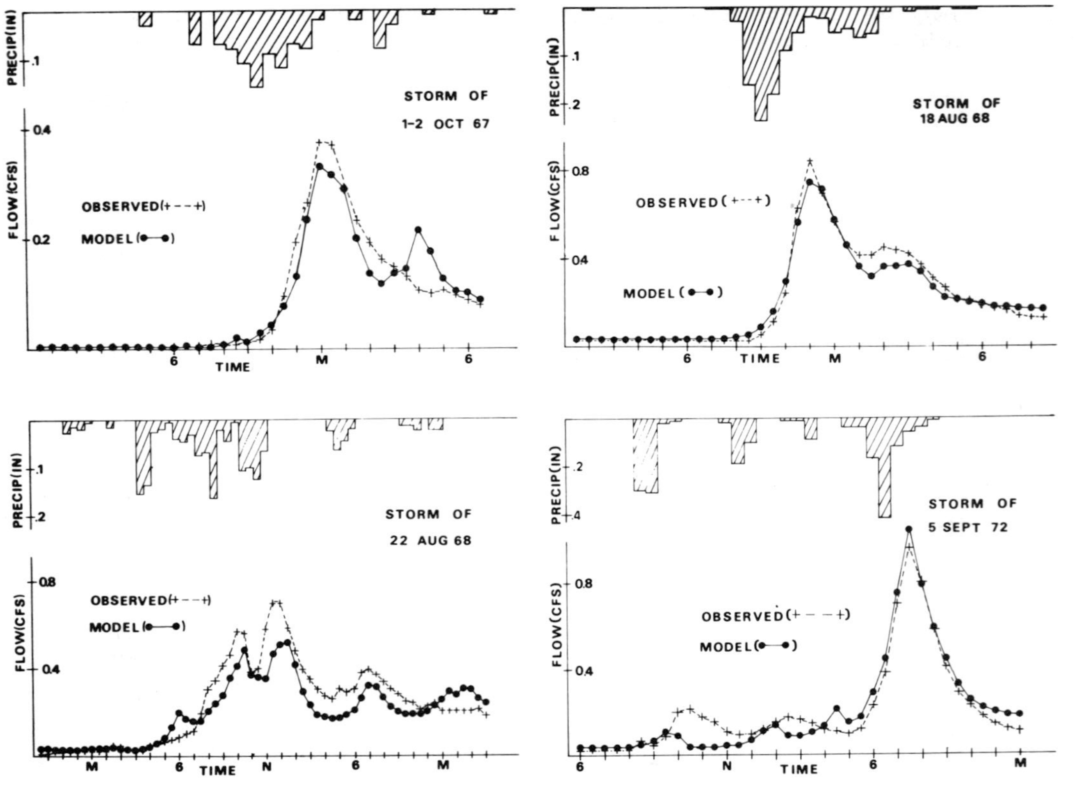

Fig. 11-27. Recorded and model-fit hydrographs for the West Branch of Chicken Creek, Davis County Experimental Watershed, Utah (from Riley and Hawkins 1976). (Multiply ft^3/sec by 0.028 to obtain m^3/sec).

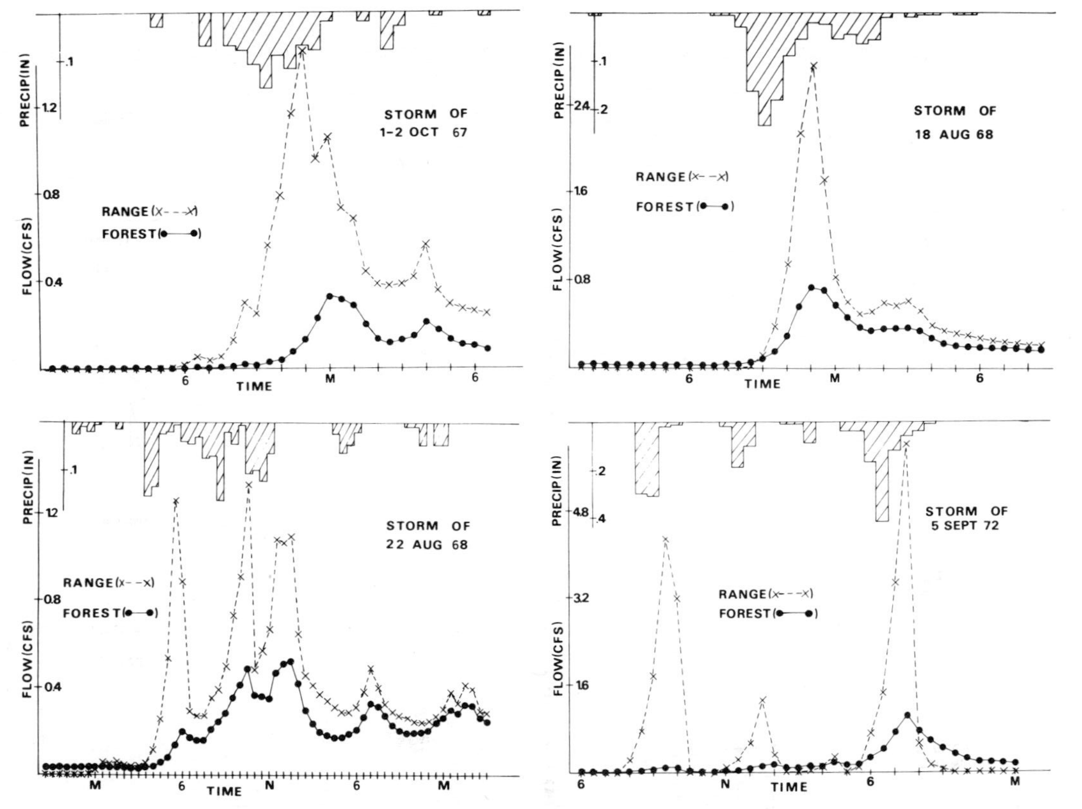

Fig. 11-28. Storm reactions on a forested watershed as represented by the West Branch of Chicken Creek and for a hypotehtical "rangeland" watershed, simulated by adjustment of parameters (from Riley and Hawkins 1976). (Multiply ft³/sec by 0.028 to obtain m³/sec).

The persistance (carry-over) effects, which require special treatment for the application of Monte Carlo techniques, suggest, on the other, that hydrologic processes can be treated by the theory of transition probabilities. Probability studies of streamflow may be undertaken under the assumption that the output sequences represent time series, in which each successive value of the variable depends only on its present value plus a random component. Thus the Markov process can be expressed by the equation:

$$y_{t+1} = py_t + n\,\bar{E} \qquad \textbf{(11-23)}$$

where y_t = runoff at time t,
y_{t+1} = runoff at time t + 1,
p = constant,
n = constant dependent on the distribution of the recorded runoff data, and
$\bar{E}$ = a random variable.

This simple model can be used to generate sequences of outputs by random sampling of the variable $\bar{E}$. When the coefficients are properly chosen, these simulated sequences have the same stochastic properties as the recorded series.

As mentioned, stochastic hydrologic models can be used to generate synthetic hydrologic data which may preserve selected statistical properties of the observed data on the prototype system. Parameters of the distributions associated with the original data are used as stochastic model inputs. The model goodness-of-fit can be judged in terms of the agreement between the statistical parameters of the synthetic data and corresponding parameters of the observed data, including parameters which were not used as part of the input.

Stochastic models are widely used in precipitation modeling because the physical processes connected with rainfall are difficult to quantify even when they are known. In addition, models based on the best known physics of the precipitation may be expensive to operate. For example, Gauntlett and Leslie (1975) reported that weather prediction using a hemispheric, six-level numerical weather prediction model required approximately 100 minutes to complete a 24-hour forecast an IBM 360/65 computer. Such costs would often be impractical for most hydrologic problem solving. Renard and Brakensiek (1976) discussed many stochastic type models used in rangeland aeas to describe precipitation depth and area probabilities as well as the probability of a storm occurrence.

Stochastic models are valuable in runoff studies because they may facilitate extending existing shorter data series. Stochastic models of runoff are also valuable when they are developed so that synthetic data can be generated on basins for which there is no runoff data.

Diskin and Lane (1972) and Lane and Renard (1972) developed a runoff model to describe the runoff encountered on the semiarid regions of Southeastern Arizona where runoff results from thunderstorm rainfall. A flow chart of the model is shown in Figure 11-29.

Table 11-16 shows the probability distributions used to model the inde-

Table 11-16. Probability distributions adopted for runoff variables.

Runoff variable	Symbol used	Theoretical distribution	Parameters
Start of runoff season	S	Normal	Mean, standard deviation
No. of events at outlet per runoff season	N	Poisson	Mean
Begin time of each event	T	Normal	Mean, standard deviation
Logarithm of volume of runoff for each event	L	Normal	Mean, standard deviation
Interval between events	D	Exponential	Mean

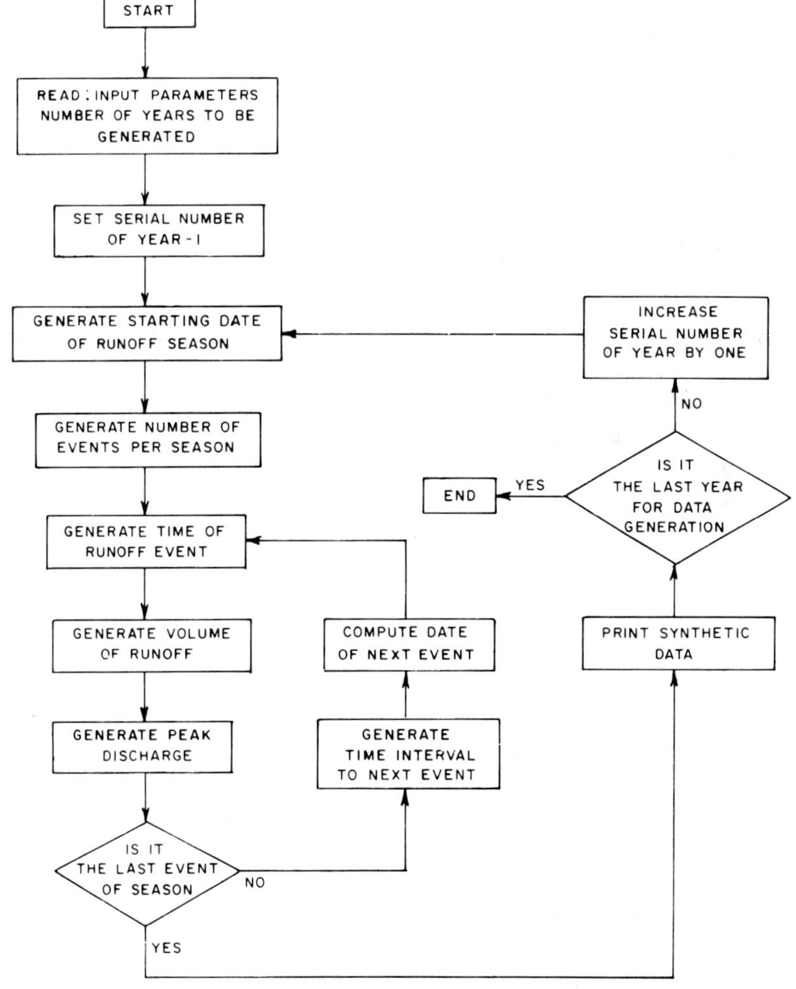

Fig. 11-29. Flow chart of stochastic model for runoff events (from Diskin and Lane 1972).

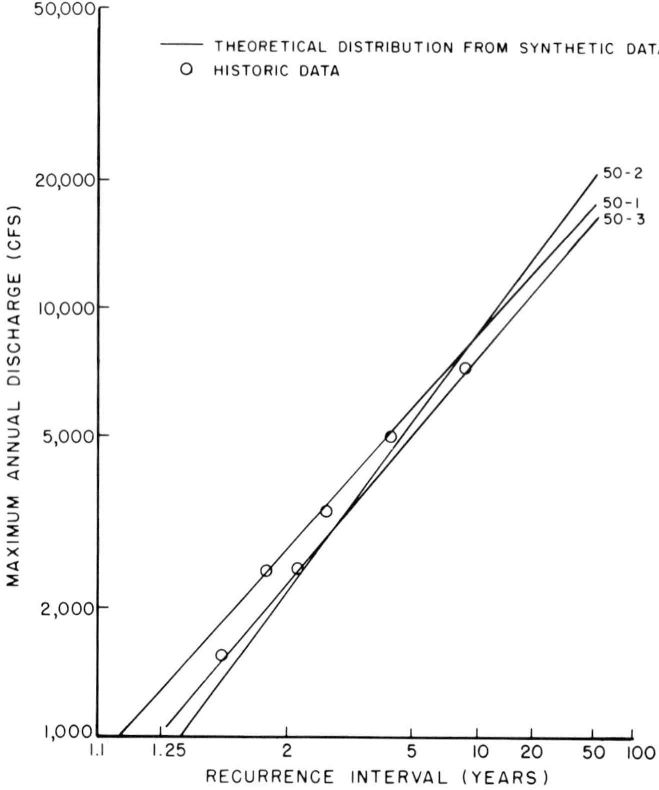

Fig. 11-30. Walnut Gulch Watershed 6, maximum annual discharge, using a log-normal distribution (from Lane and Renard 1972) ($ft^3/s \times 28.32$ = liters/sec). The three lines (50-1, 2, and 3) represent three synthetically generated lines.

pendent runoff events within the season. They then observed that the parameters for each distribution could be related to the drainage area of the watershed which facilitated extending the use of the model to other watersheds in the immediate area. They also showed that the model results could be compared to prototype data for watersheds not used to develop the model and for parameters not used in the model. Lane and Renard (1972) showed how synthetic data sets could be compared with actual data of shorter length as an indication of the sensitivity of the results for three data sets of 50 years (Fig. 11-30).

Modeling Error

A model, whether it is a physical model or a mathematical model, is a representation of something, usually a real world situation. No model is ever an exact image of the real world, but for reasons of cost or knowledge or both, is an approximation to the actual situation. Modeling error is an inevitable result.

Grayman and Eagleson (1969) point out two types of errors that are common: a modeling type error and a parametric estimation error. The modeling type error is the irremovable or least error that can result from choosing a certain (imperfect) type of model. Parametric estimation error is the additional error that results from not choosing the "best" parameters. When the best parameters are chosen, this latter error is equal to zero and the total error is simply the irremovable error.

Grayman and Eagleson (1969) give the following simple example of the two types of errors which involves a linear representation of an non-linear function. In this case the actual function is the parabola $y = x^2/16$. Of course, the actual function would be unknown when the samples are taken. Figures 11-31a, b, c, and d are plots of the best linear model (in the least squares sense) for varying numbers of sample points. As the number of sample points, n, gets very large, the error approaches a limiting value, namely the modeling type error. The additional error is caused by choosing the wrong slope or y-intercept in the linear representation. This type of error has a random component since the error caused by one set of n points may differ from the error caused by another set of n points. In Figure 11-31a and 11-31b are the number of sample points are two, yet the linear relationships and resulting errors are quite different. Figure 11-32 is a plot of the average error as a function of sample size.

Since errors in data are always present, they can affect any of the system investigation methods. However, in synthetic models where the recorded input and output play a basic role on the adjustment of the system functions

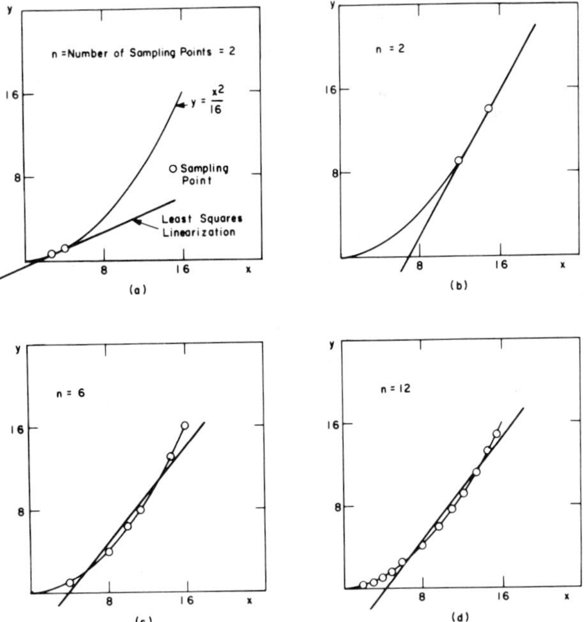

Fig. 11-31. The effect of sample size on the linearization of a prabola (from Grayman and Eagleson 1969).

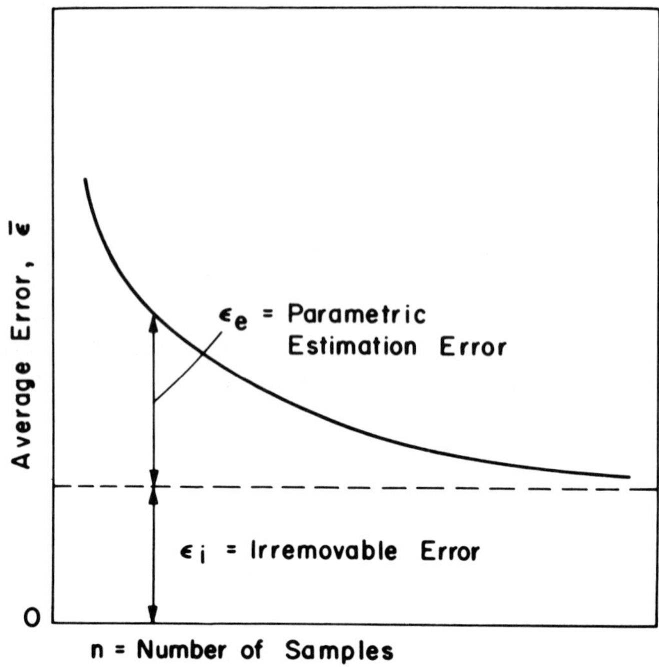

Fig. 11-32. Average error as a function of the number of sampling points (from Grayman and Eagleson 1969).

of the component subsystems, their effects may be exaggerated or compounded out of reasonable proportion. The result is that if the synthetic system appears to give discontinuous matching with respect to recorded outputs, compensating adjustments of one or more of the subsystems may alter these subsystems unrealistically, relatively to their physical function. A false impression of knowledge about the mechanism of the components may result. Most attempts to limit or reject adjustments that intuitively appear unreasonable are biased by subjective considerations that escape quantitative measure, which is unfortunate.

The imperfections of the model structure stem from the fact that considerable simplification must be introduced in the synthetic system for practical reasons. It is necessary, not only because the mathematical and/or analog manipulation of data becomes extremely difficult beyond a certain level of complexity, but primarily because sufficient physical data on the natural systems are very hard to collect and evaluate. Hence, many of the fine but highly sensitive and significant characteristics of the natural systems are lost. Since these characteristics may have strong influence on the output, all modeling activities will be affected.

Finally, the nonuniqueness of the processes of synthesis has a bearing on the reliability of a model; while various synthetic assemblies of components may produce equivalent outputs for the same input within the range of available data, the outputs may diverge strongly outside of this range. Since there is no assurance that any intuitively built model is a faithful image of the natural system, the uncertainties regarding its out-of-range performance must always exist.

Summary

Simulation modeling has made tremendous advances in hydrology in the immediate past and has taken the field of hydrology from an intuitive guesswork stage of a few decades ago to a more exact science. The next problem facing modeling activities are the scarcity of data, poor quality data, and the lack of temporal and spatial resolution. Thus, modeling attempts must still be based partially on the intuition of the modeler. Some modelers are now saying that the modeling capabilities have advanced far beyond our ability to collect prototype data to verify the model's interior points as well as the model as a whole.

Modeling critics accuse it of being subjective, "black box dial twisting", not subject to unique solutions, a collection of nonsense coefficients, and an expensive "game". Computer models are certainly not magic and the user must be aware of expecting too much. The alternatives to using simulation modeling techniques is to apply trial and error methods or exercising sound judgement in making decisions. Riley and Hawkins (1976) stated:

> Modeling is still (to a large degree) the sophisticated application of sound judgment. Shih, Hawkins, and Chambers (1972) refer to computer modeling as "the finest products of the hydrologist's art." Thus, perhaps the best that we can do is to use the basic tools which are at hand, while at the same time recognizing their current limitations and striving for improvement in terms of specific needs as they arise.

Literature Cited

Amorocho, J., and B. Espildora. 1966. Mathematical simulation of the snow melting process. Water Sci. and Eng. Univ. of Calif., Davis. Pap. 3001. 165 p.

Amorocho, J., and W.E. Hart. 1964. A critique of current methods in hydrologic systems investigations. Trans Amer. Geophys. Union 45: 307-321.

Anderson, E.A., and N.H. Crawford. 1964. The synthesis of continuous smowmelt hydrographs on a digital computer. Stanford Univ., Dep. Civil Eng. Tech. Rep. No. 36. 103 p.

Benson, M.A. 1962. Factors influencing the occurrence of floods in humid regions of diverse terrain. U.S. Geol. Surv. Water-Supply Pap. 580-B. 30 p.

Betson, R.P., and R.F. Green. 1967. DIFCOR-4 program to solve nonlinear equations. Tenn. Valley Authority Res. Pap. No. 6.

Brakensiek, D.L. 1967. A simulated watershed flow system for hydrograph prediction: A kinematic application. p. 18-24. *In:* Internat. Hydrol. Symp., Ft. Collins, Colo.,

Chapman, T.G., and F.X. Dunin. 1975. Prediction in catchment hydrology. Australian Acad. Sci. The Griffin Press, Netley, South Australia. 482 p.

Chow, V.T. 1964. Handbook of applied hydrology. McGraw-Hill Book Co., New York. 1418 p.

Crawford, N.H., and R.K. Linsley. 1966. Digital simulation in hydrology: Stanford Watershed Model IV., Standford Univ., Dep. of Eng., Tech. Rep. No. 39. 210 p.

Dawdy, D.R., R.W. Lichty, and J.M. Bergmann. 1972. A rainfall-runoff simulation model for estimation of flood peaks for small drainage basins. U.S. Geol. Surv. Prof. Pap. 506-B. 28 p.

Dawdy, D.R., and T. O'Donnell. 1965. Mathematical models for catchment behavior. Amer. Soc. Civil Eng., J. Hydraul. Div. 91(HY4): 123-139.

DeCoursey, D.G., and W.M. Snyder. 1969. Computer oriented method of optimizing hydrology model parameters. J. of Hydrol. 9(1): 34-56.

Diskin, M.H., and L.J. Lane. 1972. A basinwide stochastic model for ephemeral stream runoff in southeastern Arizona. Hydro. Sci. Bull. 17: 61-76.

Diskin, M.H., and J.R. McCarthy. 1972. An analog computer demonstration of double-peaked instantaneous unit hydrographs. Water Res. Bull. 8: 1144-1156.

Dooge, J.C.I. 1973. Linear theory of hydrologic systems. U.S. Dep. Agr., Agr. Res. Serv., Tech. Bull. No. 1468. 327 p.

Eagleson, P.S. 1972. Dynamics of flood frequency. Water Resources Res. 8: 878-894.

Enderlin, H.C., and E.M. Markowitz. 1962. The classification of soil and vegetative cover types of watersheds according to their influence on synthetic hydrographs. Pap. presented to Second National Meeting of Amer. Geophys. Union at Stanford Univ., Dec. 27-29.

Fogel, M.M., and L. Duckstein. 1970. Prediction of convective storm runoff in semiarid regions, p. 465-478. *In:* Proc. Symp. on the Results of Research on Representative and Experimental Basins. Int. Ass. Sci. Hydrol. Pub. 96.

Frere, M.H., C.A. Onstad, and H.N. Holtan. 1975. ACTMO—An Agricultural Chemical Transport Model. U.S. Dep. Agr., Agr. Res. Serv. H-3. 54 p.

Gauntlett, D.J., and L.M. Leslie. 1975. Numerical methods for predicting precipitation in catchment hydrology, p. 33-45. *In:* Australian Acad. Sci., Prediction in Catchment Hydrology. Chapman and Dunin, Editors, Griffin Press, Netley, South Australia.

Gifford, G.F., R.H. Hawkins, and J.S. Williams. 1975. Hydrologic impacts of livestock grazing on natural resource lands in the San Luis Valley. Utah State Univ. Found., Dec. 370 p.

Gifford, G.F., and R.H. Hawkins. 1978. A preliminary approach towards hydrologic modeling of rangeland grazing management systems, p. 279-283. *In:* Proc. First Internat. Rangeland Congr., Denver, Colo., Aug. 14-18.

Gifford, G.F., and R.H. Hawkins. 1979. Deterministic hydrologic modeling of grazing system impacts on infiltration rates. Water Resources Res. Bull. 15: 924-934.

Gray, D.M. 1961. Synthetic unit hydrographs for small watersheds. Amer. Soc. Civil Eng., J. Hydraul. Div. 87(HY4): 33-54.

Grayman, W.M., and P.S. Eagleson. 1969. Streamflow record length for modeling catchment dynamics. Mass. Inst. Tech. School of Eng., Hydrodynamics Lab. Rep. No. 114. 137 p.

Hawkins, R.H. 1975. The importance of accurate curve numbers in the estimation of storm runoff. Water Resources Bull. 11: 887-891.

Henderson, F.M., and R.A. Wooding. 1964. Overland flow and groundwater flow from a steady rainfall of finite duration. J. Geophys. Res. 69: 1531-1540.

Holtan, H.N., G.J. Stiltner, W.H. Henson, and N.C. Lopez. 1975. USDAHL-74 Revised model of watershed hydrology, U.S. Dep. Agr., Tech. Bull. 1518. 99 p.

James, L. 1970. An evaluation of relationships between streamflow patterns and watershed characteristics through the use of CPSET. A self-calibrating version of the Stanford Watershed Model. Univ. of Kentucky, Water Resources Inst., Res. Rep. No. 36.

Johnston, R.S., and R.D. Doty. 1972. Description and hydrologic analyses of two small watersheds in Utah's Wasatch Mountains. USDA Forest Serv. Res. Paper INT-127. 53 p.

Kibler, D.F., and D.A. Woolhiser. 1970. The kinematic cascade as a hydrologic model. Colo. State Univ., Ft. Collins, Colo., Hydrol. Pap. No. 39., 27 p.

Lane, L.J., and K.G. Renard. 1972. Evaluation of a basinwide stochastic model for ephemeral runoff from semiarid watersheds. Trans Amer. Soc. Agr. Eng. 15: 280-283.

Lane, L.J., D.A. Woolhiser, and V. Yevjevich. 1975. Influence of simplifications in watershed geometry in simulation of surface runoff. Colo. State Univ., Ft. Collins, Colo., Hydrol. Pap. 81. 50 p.

Liou, E.Y. 1970. OPSET: Program for computerized selection of watershed parameter values for the Stanford Watershed Model. Univ. of Kentucky, Water Resources Inst., Res. Rep. No. 34. 299 p.

Nash, J.E. 1958. Determining runoff from rainfall. Inst. of Civil Eng. (Ireland), Proc., 10: 163-184.

Nash, J.E. 1959. Systematic determination of unit hydrograph parameters. J. Geophys. Res. 64(1): 111-115.

Osborn, H.B., and L.J. Lane. 1969. Precipitation-runoff relationships for very small semiarid rangeland watersheds. Water Resources Res. 5: 419-425.

Phillip, J.R. 1957. The theory of infiltration: 4. Sorptivity and algebraic infiltration equations. Soil Sci. 84: 257-264.

Renard, K.G., and D.L. Brakensiek. 1976. Precipitation on intermountain rangeland in the western United States, p. 39-59. *In:* Proc. Fifth U.S./Australia Rangelands Panel, Utah St. Univ., Logan, Utah.

Ricca, V.T. 1972. The Ohio State University version of the Stanford streamflow simulation model, Office of Water Resources, Res., Dep. of the Interior, Part i—Tech. Aspects. May 1972; Part II—User's Manual. Aug. 1972.

Riley, J.P., and R.H. Hawkins. 1976. Hydrologic modeling of rangeland watersheds, p. 123-138. *In:* Proc. Fifth U.S./Australia Rangelands Panel. Utah St., Univ., Logan, Utah.

Rosenbroek, H.H. 1960. An automatic method for finding the greatest or least value of a function. Computer J. 3: 175-184.

Ross, G.A. 1970. The Stanford watershed model: The correlation of parameter values selected by a computerized procedure with measurable physical characteristics of the watershed. Univ. of Kentucky, Water Resources Inst., Res. Rep. No. 35.

Schreiber, H.A., and D.R. Kincaid. 1967. Regression models for predicting onsite runoff from short-duration convective storms. Water Resources Res. 3: 389-395.

Shih, G.B., R.H. Hawkins, and M.D. Chambers. 1972. Computer modeling of a coniferous forest watershed, p. 433-452. *In:* Age of changing priorities for land and water. Amer. Soc. Civil Eng., Irrigation and Drainage Div., Specialty Conf. Proc., Spokane, Wash.

Simanton, J.R., K.G. Renard, and N.G. Sutter. 1973. Procedures for identifying parameters affecting storm runoff volumes in a semiarid environment. U.S. Dep. Agr., Agr. Res. Serv.-W-1. 12 p.

Simons, D.B., R.M. Li, and M.A. Stevens. 1975. Development of models for predicting water and sediment routing and yield from storms on small watersheds. Colo. State Univ., Eng. Res. Center, Rot. CER 75-75 DBS-RML-MAS 24. 130.

Singh, V.P. 1974. A nonlinear kinematic wave model of surface runoff. Ph.D. dissertation, Colo. State Univ., Ft. Collins, Colo., 282 p.

Singh, V.P. 1975. Hybrid formulation of kinematic wave models of watershed runoff. J. Hydrology 27: 33-50.

Smith, R.E. 1976. Simulating erosion dynamics with a deterministic distributed watershed model, p. 166-173. *In:* Proc. Third Interagency Sedimentation Conf., Denver, Colo.

Sobhani, G. 1975. A review of selected small watershed design methods for possible adoption to Iranian conditions. M.S. thesis, Utah State Univ. 120 p.

Striffler, W.D. 1973. User manual of the Colorado State University version of the Kentucky watershed model. Colo. State Univ., Ft. Collins, Colo., Sept.

U.S. Soil Conservation Service. 1966. Methods of flow frequency analysis. Notes of Hydrological Activities, Bull. 13. 42 p. Soil Conserv. Serv., Washington, D.C.

U.S. Soil Conservation Service. 1972. National Engineering Handbook, Sec. 4, Hydrology. Soil Conserv. U.S. Dep. Agr. U.S. Government Printing Office, Washington, D.C.

Wallace, D.E., and L.J. Lane. 1976. Geomorphic thresholds and their influence on surface runoff from small semiarid watersheds. Hydrology and water resources in Arizona and the Southwest. Amer. Water Resources Ass., Arizona Sec.—Arizona Acad. Sci., Hydrol. Sec. Proc. of the April 28-30 Meeting, V. XI.

Wooding, R.A. 1965a. A hydraulic model for the catchment-stream problem: 1. Kenematic wave theory. J. Hydrol. 3: 254-267.

Wooding, R.A. 1965b. A hydraulic model for the catchment-stream problem: 2. Numerical solutions. J. Hydrol. 3: 265-282.

Wooding, R.A. 1966. A hydraulic model for the catchment-stream problem: 3. Comparison with runoff observations. J. Hydrol. 4: 21-37.

Woolhiser, D.A. 1969. Overland flow on a converging surface. Trans. Amer. Soc. Agr. Eng. 12: 460-462.

Woolhiser, D.A. 1973. Hydrologic and watershed modeling—State of the art. Trans Amer. Soc. Agr. Eng. 16: 553-569.

Woolhiser, D.A. 1974. Simulation of unsteady overland flow. Chapter 12, Inst., of Unsteady Flow, Colo. State Univ., Ft. Collins, Colo.

Woolhiser, D.A., and Ligget, J.A. 1967. Unsteady, one-dimensional flow over a plane—The rising hydrograph. Water Resources Res. 3: 753-771.

Wu, I-Pai. 1963. Design hydrographs for small watersheds in Indiana. Amer. Soc. Civil Eng., J. Hydraul. Div. 89(HY6): 35-66.

Subject Index

Other Available Publications of the Society for Range Management

A Glossary of Terms Used in Range Management

Proceedings: First International Rangeland Congress

Rangeland Plant Physiology

Special Management Needs of Alpine Ecosystems

Rangeland Reference Areas

The Jornada Experimental Range

Proceedings of US/Australia Workshop: Arid Shrublands

The above publications may be ordered from the **Society for Range Management, 2760 West Fifth Avenue, Denver, CO 80204.** The Society also publishes a technical publication, *Journal of Range Management,* and the nontechnical *Rangelands.*

The Society for Range Management is an international professional organization with headquarters at Denver, Colorado. Its members include range scientists and technicians, ranchers and range managers, teachers and students of range science and management, personnel of government agencies, businessmen, and individuals interested in the rangeland resource.